A deeply compelling book about how humans came to be in a vast, seemingly unfathomable universe, which also explains how each one of us can find meaning and purpose in our universe. Written by one of the world's leading astrophysicists and humanists, the book deftly guides the general reader through cosmology, biology, history, literature, and social sciences to paint in crystalline prose a picture of the human condition which champions the possibilities for human creativity and free will. The finest attempt to bridge the humanist/literary and scientific cultural divide since Aldous Huxley's *Brave New World*.

Fen Osler Hampson FRSC, Chancellor's Professor & Professor of International Affairs, Carleton University

This is a truly remarkable book. It is remarkable for its breadth, scale, and insight, but it is also remarkable for its logical structure, clarity, and accessibility. It has brought together an immense body of knowledge from many different disciplines and moulded it into a coherent, compelling account of our existence. Its message about the determinants of our existence has profound significance for how we should seek to influence our evolution. There are few things that can be of greater importance.

Colin Mayer, Professor of Management Studies, University of Oxford

This must surely be one of the most magnificently interesting and mind-stretching books I have ever read. It has a truly breathtaking ambition, for it aims to show how everything in our lives hangs together—from the galaxies in our universe to the movement of subatomic particles, from materials to living entities, from individuals to societies. Changes at the micro level of atoms, molecules, and genes cause changes at the macro level of organisms and societies, and in reverse, as when the decision to turn on a light leads to a distinctive flow of electrons. George Ellis's effortless familiarity with the natural and social sciences is nothing short of astounding. By explaining how we have agency in a world governed by laws of nature, this book can change the way we view the world and our place in it.

Dennis Snower, President and Founder of the Global Solutions Initiative

No one else on the planet could have written this book. George Ellis traces how human existence emerges from the fabric of the universe with clarity, depth, and astonishing range—from quantum fields to the human brain, information, and meaning. A truly extraordinary and far-ranging book. A must-read.

Teppo Felin, Anderson Endowed Professor, Utah State University & Associate Scholar, Oxford University

How We Come to Be is a true tour de force, bringing together clearly and concisely much of what science as a whole has to say about our presence and our place in the universe. At the same time, it presents fresh and challenging views on profound issues such as free will, consciousness, morality, and meaning, restoring a humane and compassionate perspective at the heart of the scientific enterprise.

Philip Ball, science writer

Many readers with backgrounds in physics, the other sciences, or in philosophy, will already know about Ellis's enormous scientific achievements in his original field, cosmology, and also about the range and depth of his more recent work in other fields, such as biology and computer science. In that work, key themes have been emergence, downward causation, and the reality of moral and other values. With this splendid book, Ellis places a keystone at the top of this arch of thought. It is full of detail and passion: learned, thought-provoking, and moving.

Jeremy Butterfield, Trinity College, University of Cambridge

This book needs to be used for continual reference, providing a rich resource for those working in education. It demonstrates how we came to be through physical foundations, emergent biological levels, and social and personal dimensions. It gives educators in their professional lives the evidence, language and support needed to take greater agency, which the book argues is key to being human. Examples which aid understanding include upward and downward causation, in addition to symbolic and abstract causation. The book helpfully clarifies the difference between reductionist and integrative approaches in neuroscience which impact on, for example, introducing literacy, and raises questions of whether public policy takes a wrong turn by promoting the former.

Tina Bruce CBE, Honorary Professor, University of Roehampton

Few physicists have been so impressively engaged in philosophical questions in and beyond physics, and in the questions and challenges confronting wider society. In *How We Come to Be*, Ellis explores the human condition with style and verve.

Jim Baggott, Logos Consulting Ltd

How We Come to Be is a good starting point for the ultimate questions of why we are here and what we are for. Professor Ellis brings a magisterial approach, one which recognises the significance of every level of explanation and the bidirectional causal connections between them. Underpinning every page is the understanding that there is a reality which our knowledge seeks to describe accurately—a reality that includes the material but extends beyond it—and that humans have an agency for actions for which they have responsibility. If you want to understand how we emerge from almost everything that leads to our existence, this is the book to read.

Andrew Briggs, Professor Emeritus of Nanomaterials, University of Oxford

This book is an astonishingly comprehensive summary of the structures of the natural world and of human society, emphasizing how higher-level behaviours emerge in any system composed of vast numbers of interacting parts. The book highlights ethical issues, on which the author has been a moral leader throughout his distinguished scientific career.

Joel Primack, professor of physics and astrophysics at the University of California, Santa Cruz

I know of no one more qualified to address the audacious question of meaning in the universe. George Ellis knows the worlds of mathematics and physics, including cosmology, through a lifetime of front-line research. He has interacted deeply in his collaborations with biologists and philosophers. The outcome is a book of immense depth. Naive reductionism has had its day: here is the book for an age that is crying out for this kind of wisdom in making sense of one of the most difficult questions we face.

Denis Noble CBE FRS, Professor, Balliol College

A century of cosmology has led to the conclusion that the early universe was too hot for any complex structure to exist: no atom, no molecule, no planet or star, no cell, no words and thoughts. Nature challenges us to provide scenarios of origin for each of them. But equations are not enough. They need a context to flourish and let the structures of our reality emerge. This book by George Ellis analyses with great care the properties of physical laws and the role of context, and how both are needed to explain our world. The author takes us on a journey from physics to chemistry and biology, brains and computers and the organisation of societies, revealing a world of nested dolls, each which its own irreducible complexity. This tour de force dissolves the walls of disciplinary thinking to offer a holistic vision of almost everything that exits, according to science. This will change the way you look at the world.

Jean-Philippe Uzan, CNRS & Institut d'Astrophysique de Paris

George Ellis is a leading cosmologist. Cosmology is the study of the ordered universe. This book gives new meaning to the term: it really is a theory of everything. A masterpiece!

Mark Solms, professor in neuropsychology, Neuroscience Institute,
University of Cape Town

How We Come to Be

How We Come to Be

Almost Everything That Leads to Our Existence

GEORGE F. R. ELLIS

With Illustrations by
MAURO CARFORA

OXFORD
UNIVERSITY PRESS

OXFORD
UNIVERSITY PRESS

Great Clarendon Street, Oxford, OX2 6DP,
United Kingdom

Oxford University Press is a department of the University of Oxford.
It furthers the University's objective of excellence in research, scholarship,
and education by publishing worldwide. Oxford is a registered trade mark of
Oxford University Press in the UK and in certain other countries.

Published in the United States of America by Oxford University Press
198 Madison Avenue, New York, NY 10016, United States of America

British Library Cataloguing in Publication Data

Data available

Library of Congress Control Number: 2025949870

ISBN 9780198950172

DOI: 10.1093/9780198950189.001.0001

Printed and bound by
CPI Group (UK) Ltd., Croydon, CR0 4YY

The manufacturer's authorized representative in the EU for product safety is
Oxford University Press España S.A. of Parque Empresarial San Fernando de Henares,
Avenida de Castilla, 2 – 28830 Madrid (www.oup.es/en or product.safety@oup.com).
OUP España S.A. also acts as importer into Spain of products made by the manufacturer.

This book is dedicated to my mother Gwen Ellis,
a caring and generous person with a strong social conscience,
and my climbing leader and good friend Kenneth Lambert,
who introduced me to the joys of mountain climbing.
I am very grateful for how each of them were such
positive supportive presences in my life for so many years.

Short Contents

Detailed Contents

List of Figures

6.1   The relation between values, norms, roles and organizations ('situational facilities')   187

6.2   Individuals function in meso-level organizations   189

6.3   The physical hierarchical structure of an aircraft   193

6.4   A copper atom on the nose of a statue of Winston Churchill   197

6.5   The societal quality-of-life feedback system   199

6.6   Components of each level   201

6.7   An overcrowded classroom in Malawi   205

6.8   The inverse feedback system   207

6.9   The feedback between social and individual values, leading to causal closure   207

6.10 Dimensions of welfare and poverty   208

6.11 Three interacting groups and the environment   210

6.12 The scales of social interactions   212

6.13 The spectrum of values   214

6.14 The global heat budget underlying climate change   224

6.15 Evidence of global warming   225

6.16 Global climate change has various effects   226

6.17 Predicted global temperature rise   226

6.18 Water stress predictions   227

6.19 World population growth scenarios   229

6.20 Planetary boundaries   231

6.21 Environment and food: the interacting factors   232

6.22 An oil refinery   240

6.23 Vaccinations save lives   246

6.24 The wealth of the ultra-rich   249

7.1   Trusting and care   270

7.2   Molecular collisions   271

7.3   James Webb Space Telescope image   271

7.4   Anthropic limits on physical constants   273

7.5   Anthropic limits on cosmological dynamics   274

7.6   Pendulum phase space   276

7.7   Mathematical realism: some examples   280

7.8   Mathematical realism: the basic relationship   280

7.9   Moral realism: the basic relationship   282

7.10 The context of informal teaching   288

7.11 The elements of Theory U   289

7.12 Mountain views   291

7.13 Pets and humans   292

List of Tables

Thanks

I thank Mauro Carfora for all the care and ingenuity he put
into the illustrations,
and his thoughtful contributions while he did so.
They have made a great difference to the book.

I thank my wife Carole Bloch for her love, care, and support
during the writing of this book.
She is a solid rock for me at this time of my life.

Preface

Our existence as human beings—conscious, emotional, active, responsive, self-aware—is a most remarkable thing. It depends on a very large number of different aspects of how things are, ranging from the existence of the Earth in a typical galaxy in an expanding and evolving Universe; the way that all biology including the mind/brain emerges out of physics; and the evolution of living beings to be what they are through the process of natural selection, as momentously discovered by Darwin and Wallace. It includes the extraordinary way we each come into being, from simply not existing at all before conception and birth, to becoming breathing living beings; and then developing our capacities and awareness via developmental processes and social interactions of great complexity. This is all based in the underlying molecular biology—storage of information in the double helix of DNA, and reading that information via complex interactions mediated by RNA and proteins in the context of a living cell. All of this is in turn is based in the underlying quantum physics, with interactions between electrons, protons and neutrons described by the Schrödinger equation—the basic equation describing how all things evolve at very small scales.

In Chapters 1 to 5 of this book I give a solid scientific summary of the physics and biology that enables our existence. As the title states, I give a solid view of *Almost Everything That Leads to Our Existence*. While I have aimed to be comprehensive, I have included the word 'Almost' because my wife Carole has reminded me that any such attempt is bound to be incomplete. I must not overstate my case.

If any of the features discussed in these chapters was missing, we simply would not be here. They are all required for us to come into being and live our lives, as walking, talking, thinking and acting humans. Without a Universe we would not be here; the same applies to the Sun and Earth, the stars that provided the elements out of which we are made and the evolutionary and developmental processes whereby human beings emerged as a symbolic species able to produce complex technology that transformed the nature of life on the Earth. In Chapters 6 and 7, I extend the discussion to considering our social and individual existence, enabled by what has already been covered, but now reaching to subtle and complex interactions and features. Society continuously influences us in key ways, at the same time as we are shaping society as individuals and groups. Our own individual lives have many social aspects that are key to our being, superimposed on the extraordinary basic feature mentioned above—our coming into existence, developing through puberty to adulthood, living a social life that may include ourselves having children and then in the end facing the inevitability of our own death. These social interactions cannot in the main be captured in a replicable scientific way, but are still essential aspects of our existence, and I try to sketch their dimensions in these two chapters.

But much more than this: there are a series of abstract possibility spaces underlying what actually exists that are the framework for what is possible. They include

How We Come to Be. George F. R. Ellis, Oxford University Press. © George Ellis (2026).
DOI: 10.1093/9780198950189.003.0001

possibility spaces for physics, biology and social systems. There are even possibility spaces for thoughts, such as thoughts of scientific and engineering possibilities, of cooking and playing sports, of social structures and economies, of good and evil, love and hate. Why they exist and have this nature is a metaphysical issue: it cannot be resolved by science itself, because there is no relevant experiment that can be carried out. And the key issue is this: is there meaning in the Universe? If so, why? That is the topic of Chapter 7.

This book thus sets out a foundational view on the basics of the human condition and our existence. It gives a more global view of the features leading to our existence compared, for example, with the excellent books *The Emergence of Everything: How the World Became Complex* (Morowitz 2002) and *Origins: The Scientific Story of Creation* (Baggott 2015). Everything covered in this book is needed for our coming into being, and then continuing existence as human beings. The picture of the Cosmic Uroboros created by Nancy Abrams and Joel Primack illustrates this.[1] It depicts all the scales of existence from the very smallest (on the top left) to the very largest (on the top right), where the notation 10^{-15}, 10^{15} and so on indicate the relevant physical scales (see the note on magnitudes at the end of the book for an explanation of this notation).

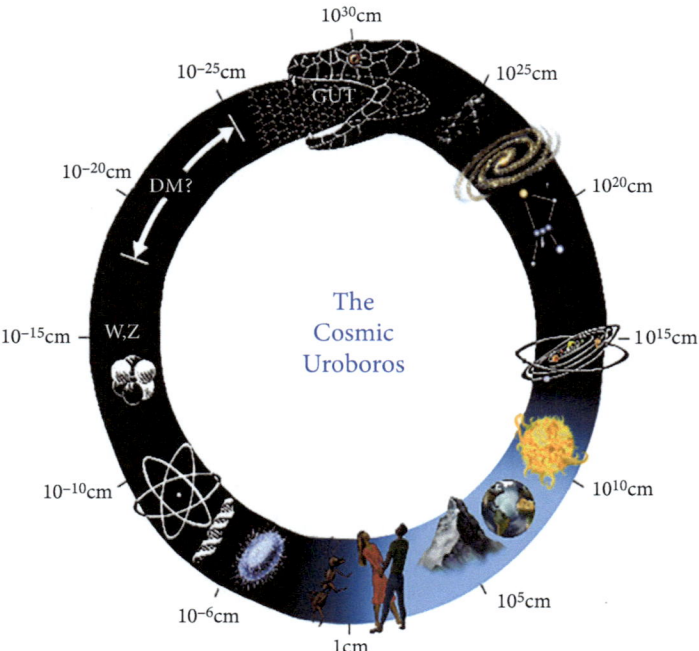

[1] Joel R. Primack and Nancy Ellen Abrams, *The View from the Center of the Universe: Discovering Our Extraordinary Place in the Cosmos* (New York: Riverhead Books, 2006); Nancy Abrams and Joel Primack, *The New Universe and the Human Future: How a Shared Cosmology Could Transform the World* (New Haven: Yale University Press, 2011), reproduced with permission of the authors.

The Uroborus is an ancient symbol depicting a serpent eating its own tail. This modern version by Abrams and Primack depicts the relation between all the scales that affect our existence. The smallest and largest scales close up, because quantum gravity, which occurs at the very smallest scales (shown on the left at the top), plays a key role in the origin of the Universe, and so underlies what happens at the very largest scales (shown on the right at the top). These scales thus interact to create the Universe as a whole, which is the context of our existence. Note that 'GUT' stands for a grand unified theory of physics, not for anything to do with eating! Human beings occur at the bottom because, in terms of the exponential scale used (powers of centimetres), they exist halfway between the smallest and largest physically meaningful scales (10^{-30} cm and 10^{30} cm). Abrams and Primack state 'the Cosmic Uroboros is one of multiple ways our books symbolize the universe as a whole, based on the meaning of Size. Our other images symbolize other aspects of the universe, and collectively they represent reality.'

The core of the book is the phenomenon of emergence: how totally new kinds of behaviour emerge out of simpler components at higher levels of structure. Reductionism has importantly allowed us to understand how many aspects of higher-level behaviour is based in structure and function at lower levels. We are all made of atoms, DNA is the genetic material, flows of electrons enable our brains to work and so on. But this is not the only thing that is happening. Causation does not only reside at the lower levels. Higher emergent levels come into being—cells, physiological systems, organisms and societies—that have causal powers in their own right. Their function and agency emerge from the underlying dynamics, but is of a completely different nature. Reductionism is applicable also in these cases, but it must be applied intelligently. Its laudable aim is, as far as possible, to understand and account for the properties and behaviour observed at every emergent level by reducing these to some set of lower-level governing principles and variables within a model framework. But strong reductionist claims that it is only the lower levels that matter in causal terms are simply wrong. They don't take into account how emergence works in biology and technology and society.

The key phrase 'nothing but' does not even include how emergence takes place in condensed matter physics, soft matter physics or superconductivity, let alone how chemistry and molecular biology emerge from quantum physics. Although reductionism is a key aspect of how we understand things, a serious problem occurs when it is claimed to be all there is. The laws of physics by themselves characterize *possibilities*, rather than determining the specific outcomes that occur in particular contexts, such as existence of a teapot.[2] The growing realization of the limitations of reductionist views include Hartwell et al., who emphasize the importance of emergent function in cell biology,[3] and Noble's 'Principle of biological relativity' emphasizing the functional role of all emergent biological levels.[4] Furthermore Nurse emphasizes the

[2] George Ellis, 'Physics, Complexity and Causality', *Nature*, 435 (2005), 743.

[3] Leland Hartwell, John Hopfield, Stanislas Leibler and Andrew Murray, 'From Molecular to Modular Cell Biology', *Nature* 402, Suppl 6761 (1999), C47–C52.

[4] Denis Noble, 'A Theory of Biological Relativity: No Privileged Level of Causation', *Interface Focus*, 2 (2012), 55–64.

significance of information in biology,[5] and Dasgupta how abstract causation occurs in the case of digital computers, where algorithms and data have causal powers.[6]

In the Prologue that follows, I outline how this emergence occurs. A central theme of this book is that downward causation underlies emergence of higher levels with causal powers. It occurs in biology, in digital computers and in physics and cosmology, where it is a key aspect of both cosmic nucleosynthesis and structure formation. And above all it underlies the agency of each individual person, realizing their vision and passion. This must be one of the most wonderful and awe-inspiring things in the whole vast physical Universe. Each human being has their own integrity that expresses who they are—their individual nature and purpose and character. Each of us is valuable.

I illustrate this here with a poem and a photograph, both trying to capture the extraordinary nature of the emergence of beauty and meaning in the vast physical universe. The meaning is not apparent on either very large scales, where we see galaxies and galaxy clusters, or very small scales, where we see just particles and atoms and molecules. But it is central to what happens at human scales on Earth.

The poem is by Brian Bilston. It is called *Three Postcards.*

Three Postcards
The first came from Weston-Super-Mare
with the Grand Pier—newly built—in view,
shining, stretching out into the distance,
and the sea, an unknowable blue.

Unfamiliar, that neat hand of his, the black fountain pen.
But he was the one: she knew that even then.

The one after that she received two years on:
Tidworth station, as viewed from the Church Hill.
A row of thatched cottages in the foreground,
The barracks beyond, then the fields, silent, still.

She propped it against a vase on their mantelpiece,
A wedding present from her niece.
The last was a busy port scene from Boulogne,
a censor-passed, heaven-sent souvenir.
'Crossing rough—but I made it!' he'd written,
'When it's over perhaps we can all come here'.

She pressed it against her stomach, the baby moved once more.
The telegram had arrived the day before.

I find it deeply moving because it is so understated. The Grand Pier at Weston-super-Mare opened in 1904. To decipher the poem, remember that the First World War started in 1914; British soldiers were then sent from their barracks in Britain to fight

---

[5] Paul Nurse, 'Life, Logic and Information', *Nature*, 454 (2008), 424–6.
[6] Subrata Dasgupta, *Computer Science: A Very Short Introduction.* (Oxford: Oxford University Press, 2016).

in the trenches in France. When soldiers were killed in combat, their families were advised of their death by telegram.

The photograph is of my wife Carole Bloch.[7] I will not embarrass her by extolling all her achievements in terms of what she has done in terms of promoting children's literacy at a national level in South Africa, but just comment that the photograph is an example of how the inner beauty of a deeply caring nature can be expressed in a human face and be visible externally.

I thank many people, some who have passed away, who have helped me develop this world view: my research supervisor Dennis Sciama[†], who wrote in depth about *The Unity of the Universe* (Sciama 2012); Stephen Hawking[†] and Brandon Carter at the Department of Applied Mathematics and Theoretical Physics (DAMTP) at Cambridge University, who both wrote about the Anthropic Principle in cosmology; members of the Vatican Observatory/Centre for Theology and Natural Sciences (CTNS) ten-year project on theology and science: Nancey Murphy, Bill Stoeger[†], George Coyne[†], Bob Russell and Philip Clayton, from whom I learned a great deal about many aspects of science and its relation to emergence and philosophy; Reza Tavakol, Jean-Philippe Uzan and Barbara Drossel, from whom I learnt much about emergence in physics itself; Denis Noble, Paul Davies, Sara Walker and Philip Ball, from whom I learnt about biology and the emergence of biology from physics; David

---

[7] *Source*: Yvonne Schmedemann @ THE NEW INSTITUTE.

Kibble, Judy Toronchuk, Mark Solms and Jaak Panksepp[†], from whom I learned much about how the brain works; Teppo Felin, Stu Kauffmann and Kevin Mitchell for wide-ranging discussions; Dennis Snower, Colin Mayer, Paul Collier and Fen Hampson for key discussions on social and economic issues; Anthony Zee and David Tong for discussions on quantum field theory; and Carole Bloch from whom I learnt essentials of early literacy.

I am very conscious of the danger that someone like myself, who started off in physics and developed expertise there and then moved on to explore other fields, could write about them in a very naive fashion. I have tried to avoid that pitfall. It is not possible for anyone nowadays to become an expert in the details of topics in all scientific areas other than one's own: there is far too much information needed. It is, however, practicable, once one has developed detailed knowledge and research expertise in some specific field, to gradually expand one's horizon to understanding the general nature and methods of many other major areas of science, gaining a broad overview of how they work at a reasonable depth. As an example, I believe Philip Ball has achieved this in his great book *How Life Works* (2023).

It is through the many interactions with the people just mentioned that I have tried to develop a broad understanding across the many fields touched on in this book. In some cases, I have written about these issues at a professional level, but always with a co-author who is an established expert in that specific area. For instance I have written with David Dewar on low-income housing policy,[8] with Judith Torunchuk on the neuroscience of the primary emotional systems[9] and with Mark Solms on what is innate in the brain.[10]

Ultimately, this book stands on what is presented here. I have confidence in it, but it may be mistaken in some places. If so, I am willing to learn, and when errors are pointed out to me, I will correct it in my own continually updated version of the text. Living is after all a life-long process of learning.

I thank Margaret Ellis, Dennis Snower, four referees and particularly Carole Bloch for very helpful comments on earlier versions of this text. Members of the Human Condition project at the New Institute, Hamburg, commented usefully on the first draft of this text, which was written as a contribution to that project in 2022 while I was a Fellow at the New Institute. I particularly thank Mauro Carfora (Pavia) for many of the diagrams illustrating this book, and associated discussions.

In the following, I start with a **Prologue**, setting the stage for the rest of the book by discussing **emergence, its principles and operation**.

Then in Part I: Physical Foundations, I look at the cosmological and astronomical context that is the framework for our existence (Chapter 1) and the underlying physics and chemistry (Chapter 2). Those who find this heavy going can just omit **Part I** and move straight to **Part II**, but bearing in mind the key link between physics

---

[8] David Dewar and George Ellis, *Low Income Housing Policy in South Africa with Particular Reference to the Western Cape* (Cape Town: University of Cape Town Urban Problems Research Unit, 1979).
[9] George Ellis and Judith Toronchuk, 'Affective and Immune System Influences', in N. Newton and R. D. Ellis, eds, *Consciousness & Emotion* (Amsterdam: John Betjemans, 2005). Judith Toronchuk and George Ellis, 'Affective Neuronal Selection: The Nature of the Primordial Emotion Systems', *Frontiers in Psychology*, 3 (2013), 27021.
[10] George Ellis and Mark Solms, *Beyond Evolutionary Psychology* (Cambridge, UK: Cambridge University Press, 2018).

and chemistry is represented by the Periodic Table of the Elements (see Figure 2.22). The different natures of these elements, as characterized there, underlies biology, technology and all daily life.

In Part II: Emergent Biological Levels, I give solidity to the understanding of emergence in biology by looking in turn at each of the emergent biological levels (Figure P.1 in the Prologue) and what they each do (Chapter 3). Then I turn to, How did we get here? Natural selection and developmental processes (Chapter 4); and How our brain shapes our actions: perception, rationality, emotions, and values (Chapter 5.)

In Part III: Social and Personal Dimensions I look at aspects of society (culture, values, narratives, technology) and the environment (Chapter 6), presenting a spectrum of values underlying our actions, and discuss the crises facing humanity at the present time. I close the book with a discussion of whether there is meaning in the Universe (Chapter 7), which is a personal exploration of the many issues that face us today individually and communally as we live our lives. Chapters 6 and 7 represent my personal view on society and individual life, supported by much evidence, but not based in solid science as are the previous chapters. These chapters describe significant features we share as humans, presented from my own, particular historical and cultural context. A wide diversity of narratives describes similar features from other viewpoints.

A key point I discuss in the last chapter is what should be regarded as data regarding the nature of the Universe? In my view, this includes—as well as data from laboratories and particle-physics colliders and telescopes—history and the arts and humanities, and indeed the entire scope of daily individual and communal human experience, with all its joys and sorrows, beauty and ugliness, achievements and failures, love and hate. They should all be taken into account, because they all exist in, and are enabled by, the particular nature of our Universe. Certainly our evolutionary history plays a key role in our existence and all these features, but then why is that evolutionary story possible? The nature of the Universe could have prevented anything like this occurring. Biology and evolution depend on the specific nature of physics and cosmology.

The possibility of the existence of consciousness, moral reflection, actions for good and bad are written into the foundations of the Universe—in possibility spaces of various kinds that may be claimed to have existed before space and time began. Nothing can happen unless it is possible that it happens. The laws of physics themselves, crucial as they are, represent only a small part of this vast terrain.

The book concludes with a note on **magnitudes**, a **glossary**, an **appendix** about the need for a Basic Income Grant and an **appendix** reproducing an article I wrote with my wife Carole Bloch about the neuroscience of learning to read. This makes clear why a reductive viewpoint can have negative outcomes when it shapes public policies.

References enabling you to follow topics to a deeper level, should you wish to do so, are spread throughout the text, and with a set of selected references given at the end of each chapter. However, it is not essential to consult any of these articles or books in order to comprehend the story of our nature that I present here. I have tried to make it a stand-alone presentation. I hope you enjoy it.

Cape Town, 2025 August 8

Prologue
Emergence, its Principles and Operation

Life is enormously complex and diverse. How does it work? How can life arise out of the underlying physics? The fundamental point is that all living beings, including mice and elephants and ourselves, are built of lower-level elements in a hierarchical way (Figure P.1). This is a profound principle underlying all truly complex systems. The same is true of complex artefacts we have created such as digital computers and aircraft, cities and economies, languages and writing.

Emergence takes place. Biological emergence involves higher-level principles and functions that do not occur at lower levels. As stated by Philip Ball[1],

> Emergence refers to the appearance of overall behaviour in a complex system of many parts that can't be predicted or understood by focusing just on what those parts themselves are like. That such phenomena occur is beyond doubt ... we have to acknowledge that there are higher-level entities and influences that must be recognised as every bit as real and fundamental as their constituent parts. These are the causes of what transpires: and they are not just the sum of lots of microcauses.

Or in brief:

The whole is more than the sum of its parts.

Emergent entities have functions. The purpose of eyes is to see, the purpose of the heart is to pump blood, and so on. Studying a living being at the physiological level will reliably establish what these functions are, and how they emerge from the relevant structure at that level. These emergent properties arise out of lower-level structures and functions, but they do not occur at the lower levels, which have quite different dynamics. At the molecular level, DNA stores genetic information, while proteins act as enzymes[2] and structural elements and light detectors and play a key role in gene regulation. At the cellular level, cells inter alia do things like conveying information (neurons) and transporting oxygen in blood (using haemoglobin) and making muscles work. Above all, they replicate themselves—which is the basis of growth. Studying living beings at the atomic or molecular levels will not reveal higher-level principles such as how a cell works, physiological principles such as homeostasis, or physiological functions such as breathing or walking or thinking.

There is a large philosophical literature on whether the impossibility of determining higher-level functions and outcomes from the lower levels per se is a matter of principle ('strong emergence') or only practicality ('weak emergence'), and whether it

---

[1] Philip Ball, *How Life Works* (Chicago: University of Chicago Press, 2023), 214.
[2] Catalysts that speed up reactions a huge amount: without them life would be impossible.

How We Come to Be. George F. R. Ellis, Oxford University Press. © George Ellis (2026).
DOI: 10.1093/9780198950189.003.0002

Figure P.1 The hierarchy of biological emergence. Atoms (left) and molecules (next) make up cells such as neurons (third from left at the bottom). Cells are the first level where all the functions of life emerge, with neurons making up networks and brains. Organisms, such as the elephant in the middle, are the first emergent level of holistic entities with agency. They in turn are imbedded in populations, ecosystems, and the biosphere (top).

Figure by Mauro Carfora.

Figure P.2 Galileo and Aristotle at the Leaning Tower of Pisa. The force of gravity is universal, as Galileo demonstrates. The pigeon is no exception, but additionally the pigeon is an extremely complex system where emergent phenomena take place. As demonstrated by Aristotle, a complex chain of interconnections driven by the underlying physics emerge at the biological level and enable the agency that allows the pigeon to counteract the grip of gravity.

Figure by Mauro Carfora.

is ontological (it is a real feature out there) or epistemological (it's only about what we can know). I will not enter that debate here, because the outcome does not undermine what is stated above. Neither physics nor molecular biology can predict the theory of evolution by natural selection (inter alia because they don't include the concepts of being alive or of reproducing). They also can't predict the existence of brains able to sustain minds experiencing feeling pain and seeing the colour red, let alone engaging in philosophical thoughts or creating iPhones.

Agency occurs. In a little-known event,[3] Galileo and Aristotle once met at the top of the Leaning Tower of Pisa. Galileo dropped simultaneously a heavy metal ball and a light one, and they both reached the ground at the same time. Galileo said to Aristotle, 'The force of gravity is universal. Nothing can evade it! Drop anything and it will fall just as these two metal balls did'. Aristotle replied, 'Watch this!', and dropped a surprised pigeon from the top of the tower. The pigeon looked around, opened its wings, and flew away to land on a rooftop nearby. The moral of the story is that it has agency and can (within limits) decide what it will do. The different outcome occurred because the pigeon did not want to fall (**Figure P.2**). How is this agency compatible with the laws of physics? That is a major part of what this book is about.

---

[3] I am indebted to Philip Ball for the idea.

This Prologue briefly summarizes the principles of emergence that underlie the rest of the book. Three themes will be covered here: (a) how structure and function are related in complex systems; (b) how they get to be the way they are; and (c) symbolic causation and the causal power of abstract entities.

P.1 Structure and Function

The first foundational principle is this: all truly complex systems, including ourselves, are **adaptive modular hierarchical structures,**[4] for good functional and evolutionary reasons.

Each word is important. The same principle underlies biology,[5] technology,[6] and societies. It holds for elephants and birds and human beings, digital computers and aircraft, cities and large corporations. Furthermore, this is also true for abstract structures such as algorithms and computer programs and language, written or spoken. I'll focus on biology and digital computers in this Prologue, but the same principles apply in those other contexts. In biology, a hierarchy of modules is created (**Figure P.1**), described in detail in textbooks on physiology. We are made of electrons and protons, which make atoms, which make molecules, which make cells, and so on, as depicted in **Figure P.1** and summarized in **Figure P.3 (left)**. The case of digital computers is similar,[7] summarized on the right of **Figure P.3**.

Physically, chemical elements are put together in different ways in these different contexts. Symmetry breaking is key to emergence of higher levels out of lower levels.[8] Symmetries that occur at lower levels do not occur at higher levels. This takes place at **Level L3** by forming crystals in the case of digital computers, and molecules in the case of biology. These are then assembled to form emergent structures including physical and interaction networks. Multiple scales occur, and a key feature is that the numbers of components at the lower levels is huge. We are made of about 3.7×10^{13} cells, each cell containing 42 million proteins, totalling 10^{14} atoms per cell and 10^{27} atoms per person.[9]

None of this is obvious. We only understood the cellular nature of life and the atomic nature of matter relatively recently, because the cells and particles are so small and not visible to the naked eye. Thus vast number of particles underlie our existence. That is not surprising: if we were made of much smaller numbers of components (think LEGO), really complex behaviour would not be possible. Furthermore, this

---

[4] Herbert A. Simon, *The Sciences of the Artificial* (Cambridge, MA: MIT Press, 2019); Grady Booch et al., *Object-Oriented Analysis and Design with Applications* (Boston: Addison Wesley, 2007), Chapters 1 to 3.

[5] George Ellis, *How can Physics Underlie the Mind. Top-Down Causation in the Human Context* (New York: Springer, 2016).

[6] W. Brian Arthur, *The Nature of Technology: What it is and How it Evolves* (New York: Simon and Schuster, 2009).

[7] Andrew S. Tanenbaum, *Structured Computer Organisation* (Englewood Cliffs, NJ: Prentice-Hall, 2006); G. Ellis and B. Drossel, 'How Downwards Causation Occurs in Digital Computers', *Foundations of Physics*, 49 (2019), 1253–77.

[8] Philip W. Anderson, 'More is Different: Broken Symmetry and the Nature of the Hierarchical Structure of Science'. *Science*, 177 (1972), 393–6.

[9] These huge numbers are written out in full in the note, 'Orders of Magnitudes', at the end of this book.

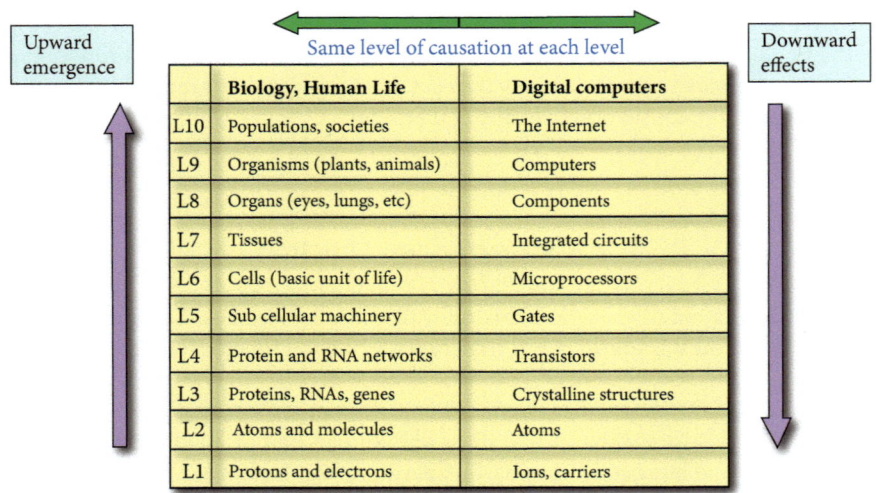

	Biology, Human Life	Digital computers
L10	Populations, societies	The Internet
L9	Organisms (plants, animals)	Computers
L8	Organs (eyes, lungs, etc)	Components
L7	Tissues	Integrated circuits
L6	Cells (basic unit of life)	Microprocessors
L5	Sub cellular machinery	Gates
L4	Protein and RNA networks	Transistors
L3	Proteins, RNAs, genes	Crystalline structures
L2	Atoms and molecules	Atoms
L1	Protons and electrons	Ions, carriers

Figure P.3 Multilevel causation. The existential hierarchy of emergent levels in biology and digital computers, with same level functioning occurring at every level. Chaining up of functional processes and down of constraints and adaptive processes creates a system with multilevel causation underlying emergent complexity. Causal closure only occurs when all the interacting levels are taken into account.

vast number of molecules in biology enables the crucial randomness discussed later. Our multilevel nature is indicated in **Figure P.4.**

Why modularity? In essence the principle is this:

> In order to obtain genuinely complex behaviour, one must split a complex task up into simpler tasks, design units to handle the simpler tasks and then knit the results together so that the complex task is accomplished. This will require multiple levels, each contributing to the overall task in essential ways.

Each level **LN** will be composed of modules at the next-lower level **LN-1.** Submodules of modules do even simpler tasks, down to the physical level **L1** where the needed underlying work is carried out by physical forces. Thus a modular hierarchy of emergent levels is needed as in **Figure P.3** and **Figure P.4**, with each level demonstrating different kinds of behaviour, as shown there for both digital computers and life.

This modular logic leads to **multiple realizability**. Every physical level **LN** in **Figure P.3** can be realized in different ways at the next-lower level **LN-1**. This effect chains down to the lower levels: any higher-level state can be implemented by a huge number of states at the molecular and particle levels. Which specific lower-level state realizes a higher-level state is immaterial as far as higher-level causation is concerned. The system as a whole only cares that the module does what it is needs to do, not how it does it, although implementations are preferred if they are faster or use less resources. It follows that the higher-level dynamics is not equivalent to lower-level dynamics in any simple way. In the case of digital computers, higher-level structures can be based in different computer chips with different instruction sets. In the case

Figure P.4 Our modular nature. Vast numbers of each lower level element underly our physical existence, and hence enable our biological functioning at each emergent level.
Figure by Mauro Carfora.

of biology, vast numbers of different gene regulatory networks can produce the same biological outcome, which is vital to evolutionary processes.[10] Because of this multiple realizability with huge numbers of lower-level elements (cells, molecules, atoms, particles) involved, the function and agency so patent at higher levels is indiscernible at these lower levels. That does not mean it does not exist.

The key role of stochasticity. There is a huge amount of randomness (stochasticity) at the molecular level in biology,[11] and evolution has taken advantage of it. This

---

[10] Andreas Wagner, *Arrival of the Fittest: Solving Evolution's Greatest Puzzle* (London: Penguin, 2014).
[11] Peter M. Hoffmann, *Life's Ratchet: How Molecular Machines Extract Order from Chaos* (New York: Basic Books, 2012); Ball, *How Life Works*, 214, 256–9.

enables life to react flexibly, taking advantage of the multiple realizability of higher levels, with stochastic dynamics at the molecular level being harvested by higher-level emergent dynamics for their purposes. An example is the way the protein dynein extracts energy from the molecular storm in order to transport cargoes in a cell.[12] In biology, noise at the molecular level is a key resource. This introduces a freedom in interlevel interactions that underlies the possibility of agency emerging at higher levels (see §5.6). Randomness can be a useful resource in computer programs too, using Monte Carlo simulations to determine probable outcomes, and in using ensemble-based models to estimate weather probabilities.[13]

Causal emergence via multilevel causation. Every emergent level is causally effective, as indicated by the horizontal arrow in **Figure P.3**. The whole would not work if every emergent level was not there and functioning properly. Each level plays a key role in determining physical outcomes via effective laws expressed in variables appropriate to that level: blood pressure, blood temperature, heart rate and so on. The same is true in the cases of technology: digital computers, aircraft, even cities. The same level functioning is enabled by multilevel causation: the combination of upward emergence and downward causation, as indicated by the up and down arrows in **Figure P.3**.

Upward causation takes place by the kind of functional and structural emergences that are the core of reductionist explanations. They are characterized by the coarse graining of particles at the lower levels, as in the kinetic theory of gases; existence of crystals or molecules made out of atoms; and at higher levels by mechanical forces, temperature differences, fluid flows, electric currents, electric and magnetic fields, combinations of chemicals and so on. In the case of biology, it occurs by combining molecules into cells, cells into tissues, tissues into organs and organs into organisms. In the case of digital computers analogous physical emergence occurs, with transistors combined into integrated circuits, these then combined with peripheral devices into computers, these combined into local networks and the Internet. Additionally, the logical combination of lower-level effects occurs in the associated logical functioning. Combinations of operations at the gate level produce operating systems, languages and applications programs at higher emergent levels.

Downward causation takes place by two key physical processes and two geometrical effects. The two physical effects are time-dependent constraints, and creation, modification or deletion of lower-level elements. The two geometrical effects are the existence of networks, and topological effects.

Time-dependent constraints are exerted by emergent higher levels on lower levels.[14] This shapes lower-level outcomes by setting their context. Electron flows cause electric currents when they flow along a wire. The wire is a higher-level structure that determines where the current flows, because its insulating sheath is a constraint determining where electrons go. **Figure P.5** illustrates an electric circuit with time-dependent constraints acting as a form of downward causation.

If wires link a battery to a light bulb and back via a circuit with a switch, then turning the switch on and off at the macro level turns the light on and off: it allows currents to

---

[12] Richard B. Vallee and Peter Höök, 'A Magnificent Machine' *Nature*, 421 (2003), 701–2.
[13] Tim N. Palmer, *The Primacy of Doubt* (Oxford: Oxford University Press, 2022), 57–63, 84–108.
[14] Alicia Juarrero, *Dynamics in Action: Intentional Behaviour as a Complex System* (Cambridge. MA: MIT Press, 1999).

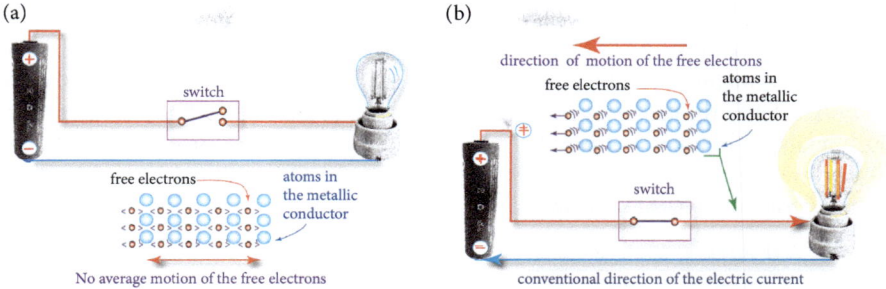

Figure P.5 Macro-level effects causing micro-level effects that in turn cause macro-level results. (left) With the switch open, circuit is open, light is off, no electrons flow on average. (right) With the switch closed, the circuit is closed. The light is on because electrons are flowing on average—which constitutes a current.

Figure by Mauro Carfora.

flow at the macro level or not, according to the state of the switch. Closing the switch at the macro level causes electrons to move on average at the micro level. Then as the electrons flow through the light bulb they generate heat in its filament due to macro-level resistance. The time-dependent macro constraint downwardly controls whether electron flows take place at the micro level and so whether the light bulb is glowing at the macro level. This is also an example of mental causation from your mind to the electron level.

A biological example is the functioning of voltage-gated ion channels in neurons. They control the flow of ions in and out of neuronal axons in a brain as they change shape,[15] and so enable the action potential spike chains in the neural networks that are the basis of thought.

Creation, modification or deletion of lower-level elements can take place to meet higher-level needs. This determines what entities lower-level laws can act on. Thus the action of metabolic networks and gene regulatory networks determines what molecules will exist and interact in a cell; this determines how an embryo grows into an adult.[16] The arctic cod survives in water that chills below 0° Celsius, where ice crystals should destroy their cells, but the cod produces anti-freeze proteins that lower the freezing temperature of its body fluids, and so prevents this destruction taking place.[17]

A more homely example is cooking and baking. Here carefully thought-out human action, developed by a long process of trial and error, modifies the molecules that make up our food so that the outcome is tasty. An example of deletion of lower-level elements is *apoptosis*: programmed cell death. This plays a key role in developmental processes, for example separating out the digits in your hand.

Existence of networks. These can only be characterized at emergent levels, and play a key role in physical outcomes. The key physical network is the human brain—an immensely complex interconnected set of neurons, enabling us to think and act. Its

---

[15] George Ellis and Jonathan Kopel, 'The Dynamical Emergence of Biology from Physics: Branching Causation via Miomolecules', *Frontiers in Physiology*, 9 (2019), 1966.

[16] Sean B. Carroll, *Endless Forms Most Beautiful: The New Science of Evo Devo and the Making of the Animal Kingdom* (New York: Norton, 2005).

[17] Wagner, *Arrival of the Fittest*, 107.

connectivity determines who we are—and we determine that connectivity by how we think and act, because the brain is plastic: it changes over time as we learn things, take action and remember what happened. Functional networks include gene regulatory networks, determining what genes get read at what position in the body at what time, and metabolic networks, determining how molecules needed for cells to work get created as needed out of other molecules.[18] The connectivity of these networks plays a central role in physical outcomes.

Topological effects. Many systems have properties based in topology—that is, the way objects are shaped when we disregard their size, and examine how they are connected.[19] Do they have no holes through them, like a tennis ball or football, or one hole, like donut or teacup? This is what topology classifies. Simple examples are electric circuits that turn devices on (the circuit is closed so current runs round it) or off (the circuit is open so no current flows), as discussed earlier. The circuit topology is different in the two cases. But topology can be much more complex than that. Communication networks and computer networks can be characterized by their topology, which is very complex if there are a large number of nodes. A domestic example is knitting a one-dimensional woollen yarn into a two-dimensional fabric, that is then formed into a three-dimensional garment with nontrivial topology because of arm and neck openings. This three-dimensional structure determines the possible future relative positions of the particles that make it up. The topology changes if you button or unbutton a cardigan—and all the particles it is made of follow as you do so.

Causal closure only takes place when all the levels linked in this way by upwards and downwards causation are taken into account, for they all affect the outcome. Some key causal effects will be omitted if this is not done: the time change of some significant variables will be indeterminate. Outcomes are determined by what matter is present where, together with boundary conditions and constraints at all relevant scales. A persuasive example is Denis Noble's analysis of heart function, taking into account the dynamics ranging from genes to cells to the whole organ.[20] Consider then how cell dynamics at level **L6** is crucially affected by heart dynamics at level **L8**: if the heart ceases to function at level **L8**, then all the cells in the body (level **L6**) will die within minutes.

Consequently, contrary to what many claim, no underlying physics level is causally closed by itself. They represent possibilities of what might occur, rather than determining what will occur in specific contexts at particular times and places. For example, Maxwell's equations of electromagnetism do not by themselves determine any specific outcomes at all, such as light from the Sun powering ecosystems on Earth, or an eye seeing a picture.

Branching dynamics occurs at every emergent level in both the biological and digital computer cases. Outcomes at any level depend on conditions either at that level, or at a higher level. We have already seen examples of this. This is a key feature enabling complex behaviour to arise out of underlying physics that does not, by itself, show such contextually determined branching behaviour. It is enabled at the lower levels in the biological case by change of conformation of proteins such as voltage-gated ion

---

[18] A key example is the citric acid cycle in cells, which is a closed loop associated with many interactions.
[19] Joseph Howlett, 'The Two Mathematical Perspective on Shape', *Quanta Magazine*, 2 September 2024.
[20] Denis Noble, 'Modelling the Heart—From Genes to Cells to the Whole Organ', *Science*, 295 (2002), 1678–82.

channels, and in the case of digital computers by the switching effects of transistors and gates. At the highest levels in biology, agency arises: choices of actions are made consciously in order to attain optimal outcomes, as I discuss below (§5.6). This agency has arisen through evolutionary processes.

P.2 How Did They Get to be the Way They are?

All truly complex systems get to be the way they are firstly by an evolutionary process of trial and error, optimizing the way a class of objects is structured and functions, and secondly by a developmental or manufacturing process, constructing members of this class so that they have this nature. Furthermore, this process of construction is itself also optimized by an evolutionary process of trial and error. Thus two timescales are relevant—the times for evolutionary processes on the one hand, and developmental processes on the other. They must both be adaptive in that the whole organism exists in an ecological and social context which may change in crucial ways over time: the hierarchy as a whole must adapt to new contexts in order to continue to function and survive.

Evolutionary processes over long timescales lead to the genetic and cellular inheritance that each generation of living beings passes on to the next. Random variation of genes leads to organisms with small variations, and those comparatively better suited to the environment (they produce more offspring) tend to be preferentially selected as the next generation. Thus genes to be passed on to the next generation are selected in this way, via survival of relatively better-adapted organisms.

An example is the selection of genes for polar bears, as opposed to brown bears (**Figure P.6**). The result will be different in a Canadian forest (brown or black bears) and the Arctic (white bears) because this enhances hunting prospects. The polar bear has white fur because that makes it easier for it to creep up on its prey than if it were brown, which would be good in the Canadian forest.

Thus a combination of the nature of the environment (snow and ice) together with physiological needs (food to eat in this frozen context) chains down, through evolutionary processes, to determine genes at the molecular level that will produce proteins for white fur.

Overall, this is a process of adaptation to the environment; thus it is downward causation,[21] whereby organisms that are more viable in a specific environmental context because of their physiology and their mental and social abilities tend to be selected. Advantageous proteins and genes are selected, but it is not just a process at the molecular level. It also involves cellular structure, tissues, physiology; indeed all levels must evolve harmoniously together.

However, the process is not always adaptive: some drift takes place, with some traits being there not because they are particularly useful, but just by chance. Furthermore not all variation is random: sexual selection takes place, where male and female choose mates via complex rituals that influence evolutionary outcomes. A key

---

[21] Donald Campbell, '"Downward Causation" in Hierarchically Organised Biological Systems', in F. J. Ayala and T. Dobzhansky, eds, *Studies in the Philosophy of Biology: Reduction and Related Problems* (London: Palgrave, 1974), 179–86.

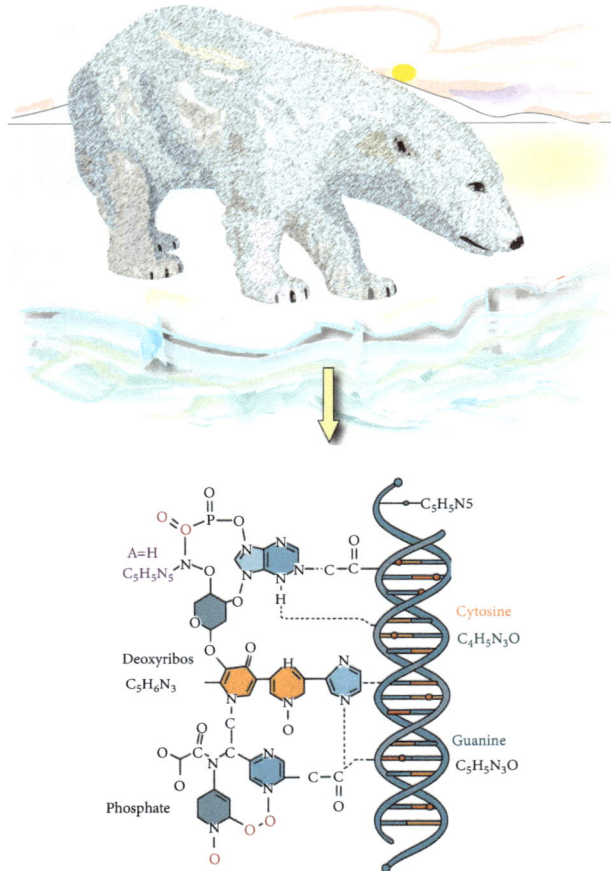

Figure P.6 Evolutionary determination of genes and proteins as a form of downward causation (above) Polar landscape is an ecological context where being a white bear is better for hunting than being a black or brown bear. (below) A segment of DNA coding for proteins that will produce white fur rather than brown fur.

Figure by Mauro Carfora.

point is that it is not just the genetic material that gets passed on to the next generation. The whole structure of the cell, including its DNA but also the cell wall, nucleus and all the organelles gets duplicated at each cell division. There is much more information there than is stored in the DNA: for example the ability for cell division to take place is passed on to the daughter cell.

Developmental processes on shorter timescales take place that create individual organisms out of a fertilized egg. The core of the process is firstly cell division, and secondly cell specialization. Each cell gets adapted to a specific type (bone, blood, muscle, neurons and so on) as needed to perform specific functions at particular places in the developing body. These arise through developmental processes

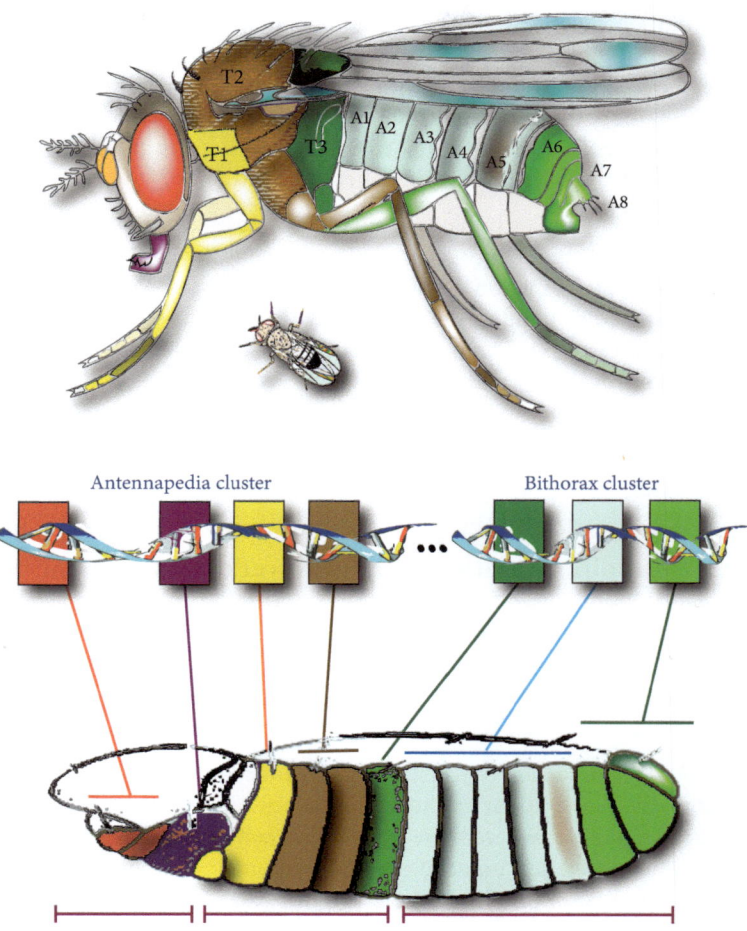

Figure P.7 Developmental determination of cell types as a form of downward causation. Development of the segments of the fruit fly *Drosophila melanogaster* larva (below) and adult (above) from the organism's DNA (middle). Positional signals result in different genes being read at different positions in the embryo, resulting in different proteins being expressed there. This leads to different cell types, and hence different organs, developing at the corresponding positions in the adult. Position along the anterior–posterior axis—a higher-level variable—establishes the functional differences that develop along this axis.

Figure by Mauro Carfora.

leading to needed proteins and cell types at that location, using the inherited genetic information (**Figure P.7**).

EVO–DEVO. These two processes interact with each other: the relative success of developmental processes increases the likelihood of reproductive success, and hence of evolutionary selection. The developmental processes themselves therefore become a target of evolutionary selection.[22] Furthermore, there is a two-way interaction between organisms and the environment. Organisms adapt to ecological niches (opportunities) that exist in the environment, but they also create such niches because they have agency, enabling them to make nests and shelters (think birds and ants and people) and adapt rivers and lakes (think beavers) and so on.

Technology. In the case of technology, similar processes of evolutionary development occur.[23] In this case there are processes of adaptive evolution both of the technologies (such as structure and function of passenger aircraft, of digital computer hardware and software and so on) and of their manufacture (such as manufacture of aircraft and of integrated circuits). The manufacturing processes themselves are subject to evolutionary improvement, and the machinery used for manufacture is itself also manufactured and subject to evolutionary improvement. Similar remarks apply to mechanical and civil and chemical engineering. They are all products of the human mind. Mental causation is certainly possible—it clearly occurs.

P.3 Symbolic Causation and the Power of Abstract Entities

In the case of human beings, quite different kinds of causation arise at levels **L9** (individuals) and **L10** (societies) through our agency, and the particular nature of our brains. Because we are a symbolic species,[24] both symbolism and abstract entities have causal powers in a social context.

Language. As to symbolism, our first basic symbolic system is language, which has made a huge difference to our evolutionary success. It has enabled cooperation to underlie social structures whereby we can achieve communal goals. Symbolic causation occurs, whereby we communicate social functions and opportunities. Thus store signs indicate what is for sale in the store (**Figure P.8a**) as do advertising handouts, local newspapers and so on; maps indicate how to get from here to there; cooking recipes tell us how to bake cakes; radio broadcasts tell us news such as road closures; and so on.

Language enables abstract entities to have causal powers; for example, the rules of chess (stated in some language) determine the movements allowed of chess pieces on a chessboard (**Figure P.8b**). These rules are abstract because they are not the same as any single individual's brain state; the rules of chess do not cease to be when any specific individual dies. They are written in chess rule books, are realized in the brains of all people who play chess and in chess-playing computer programs, are described in videos on teaching chess and so on, and can be conveyed in any language. They are multiply realizable in many different formats, but are not the same as any one of those realizations: they are essentially the equivalence class of all such realizations. To be sure they derive their causal powers through being activated in the brain states

[22] Gerd Müller, 'Evo–Devo: Extending the Evolutionary Synthesis', *Nature Reviews Genetics*, 8 (2007), 943–9.
[23] Arthur, *The Nature of Technology*.
[24] Terence Deacon, *The Symbolic Species: The Co-evolution of Language and the Brain* (New York: Norton, 1998).

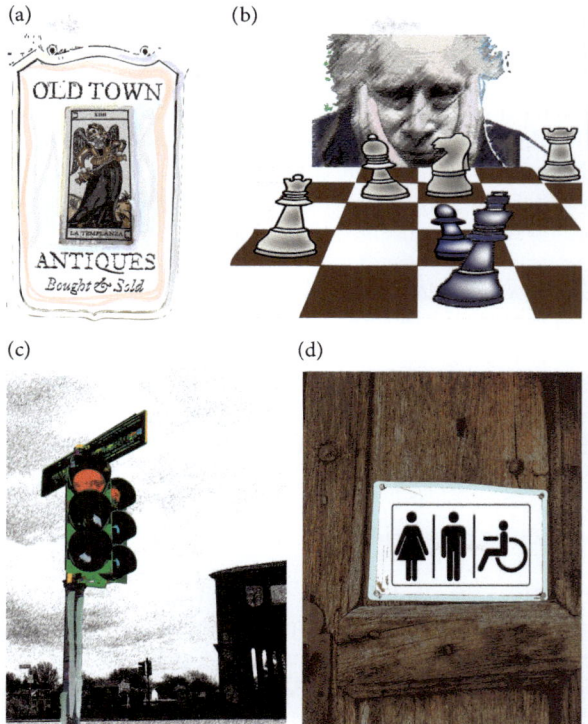

Figure P.8 Symbolic and abstract causation. (a) Store signs indicate what is for sale in the store. (b) Chess rules determine the movements allowed of pieces on a chessboard. (c) Traffic lights control the movement of cars and pedestrians via symbols that change over time. (d) Restroom signs indicate who they are for, influencing where people go.
Figures by Mauro Carfora.

of individuals, but they only attain that causal power through a social agreement that they are indeed the rules of chess.

Other examples of abstract causation are the effects of laws such as when one can and cannot cross an intersection controlled by a traffic light (**Figure P.8c**), and signs indicating which group of people a restroom is intended for (**Figure P.8d**). Through social agreement, these symbols have causal powers.

Mathematics. Our second basic symbolic system is mathematics. The laws of physics are accurately expressed in mathematical terms—a deep feature of nature which we do not understand, but is very well confirmed. This has enabled development of the science and technology that has enabled us to dominate the Earth.

Abstract causation. Digital computers represent a whole new realm of abstract causation. Turing's great discovery was essentially that arbitrary data, representing text, images, music, videos, equations, together with arbitrary programs to manipulate them, can all be expressed as bits—sequences of 0's and 1's. Because of this discovery, algorithms, computer programs, and data—all abstract entities—have causal powers.

They alter physical outcomes in a real-world social context. Digital computers are, at their heart, symbol processing systems. As expressed by Dasgupta,[25]

> Some computational artefacts are entirely abstract: they not only process symbol structures, they themselves are symbol structures and are devoid of any physicality, though they may be made visible by physical media such as marks on paper or computer screens. Physico-chemical laws do not apply to them. They neither occupy physical space nor do they consume physical time.

However, by the processes detailed in this book these abstract entities have causal powers via what Dasgupta refers to as 'liminal structures' where the abstract becomes physical and can result in images on a screen, marks on paper, three-dimensional printing of artefacts, automatic landing of an aircraft and so on. Thus it is not true that only physical entities have causal powers, as some would have us believe. Algorithms, and the thoughts that create them, have them too.

I discuss all of this in depth in what follows. Please note the following:

None of this modifies or overrides the lower-level physics in any way.

Some of my critics have implied that this is what I am claiming, but this is not correct. In material terms, animals and humans and digital computers and aircraft are all solutions of the standard Schrödinger equation of quantum theory. The lower-level physics implied in **Figures P.1**, **P.3** and **P.4** is absolutely standard. But this does not mean that materialism is correct: abstract entities such as thoughts and the rules of chess and algorithms have causal powers—including determining how much money you can withdraw electronically from your bank account. The underlying physics enables this. It has no idea that this is what it is doing.

The essential point is that **higher-level emergent features alter the context in which physics functions,** and this is what determines specific outcomes in particular contexts. Emergence is contextual. Indeed the underlying physical laws per se do not determine any specific outcomes whatever. They only do so in a context of specific boundary conditions and constraints[26] with specific entities to act on, for example electrons and protons in a digital computer loaded with a particular applications program and data. This multilevel nature of emergence is essential to our existence. It is the centre of what I explore in depth in this book. As emphasized by Philip Anderson, studying this is just as fundamental as studying high-energy physics or cosmology.[27]

Finally, a key point is this: in a recent book,[28] Denis Noble writes about a debate he once had with Charles Taylor. He and others had proposed that a molecular-level explanation in biology could be complete. That would necessarily exclude any higher-level explanation from adding anything of value. They were not needed to describe the situation.

---

[25] Subrata Dasgupta, *Computer Science: A Very Short Introduction* (Oxford: Oxford University Press, 2016), 23.

[26] Juarrero, *Dynamics in Action*.

[27] Andrew Zangwill, *A Mind over Matter: Philip Anderson and the Physics of the Very Many* (Oxford: Oxford University Press, 2021).

[28] Denis Noble, *The Pacemaker Channels of the Heart From Reductionism to Systems Biology* (Singapore: World Scientific, 2026).

Both Tony Kenny and Denis initially thought this was a knock-down argument. They were intrigued by Taylor's reply. He granted the case they made, but replied with a very insightful comeback. He agreed that might be true in any particular case, but what would be true if one considered *a whole series of cases*? Might it not then be true that the higher level shows order while the lower level does not? In that case, how could the lower level explain that higher-level order in multiple cases, since the order clearly did not exist at the lower level? In fact higher-level relations and explanations are key to describing and understanding biology in generic cases as well as specific cases. They do not reduce in any simple way to lower-level causes.

That is the view that I put in this book. Emergent levels show causal structures and relations that apply generally, not just to some specific set of lower-level states. They enable us to understand causation at emergent levels.

Further Reading

I have made a practice for many decades of keeping a golden thread of references that illuminate the way things work on the basis of a deep understanding. I will provide such a list at the end of each chapter.

Here are some related to the Prologue.

Studying emergence is as fundamental as studying basic physics or cosmology:

Andrew Zangwill, *A Mind over Matter: Philip Anderson and the Physics of the Very Many* (Oxford: Oxford University Press, 2021).

Stephen Blundell, *Superconductivity: A Very Short Introduction* (Oxford: Oxford University Press, 2009).

Basic principles of emergence

Paul Humphreys, *Emergence: A Philosophical Account* (Oxford: Oxford University Press, 2016).

George Ellis, *How Can Physics underlie the Mind: Top Down Causation in a Human Context* (Cham: Springer, 2016).

Emergence in biology

Neil A. Campbell and Jane B. Reece, *Biology* (San Francisco: Benjamin Cummings, 2005).

Sean B. Carroll, *Endless Forms Most Beautiful: The New Science of Evo Devo and the Making of the Animal Kingdom* (New York: Norton, 2005).

Denis Noble, *The Music of Life: Biology beyond Genes* (Oxford: Oxford University Press, 2008).

Denis Noble, *Dance to the Tune of Life: Biological Relativity* (Cambridge, UK: Cambridge University Press, 2016).

Philip Ball, *How Life Works. A User's Guide to the New Biology* (Chicago: University of Chicago Press, 2023)

Importance of randomness in biological emergence

Paul M. Hoffmann, *Life's Ratchet: How Molecular Machines Extract Order from Chaos* (New York: Basic Books, 2012).

Emergence in computer science

Subrata Dasgupta, *Computer Science: A Very Short Introduction* (Oxford: Oxford University Press, 2016)

Philosophical underpinnings

Clare Mac Cumhaill and Rachael Wiseman, *Metaphysical Animals: How Four Women Brought Philosophy Back to Life* (New York: Anchor, 2023)

PART I
PHYSICAL FOUNDATIONS

1

The Cosmological and Astronomical Context

We only exist because the Universe exists, and is of such a nature as to allow us to come into being.

Neither of those statements is inevitable. The Universe need not have existed, and given its existence, it need not have been of such a nature as to allow our existence as intelligent beings. The Sun and Earth came into being because of the existence and evolution over time of the physical Universe. We would not be here if that were not the case.

The nature of the Solar System, our Galaxy (the Milky Way, **Figure 1.1**) and the Universe is far from obvious to us as we look up at the sky. The discovery of this astronomical context and its vast scale was a major step in our understanding of our position in nature. We discovered that the Earth revolves round the Sun, which is very much larger than the Earth, rather than the Sun revolving around the Earth. Much more than this, we found out we were not at the centre of the Universe, as had previously been taken for granted. We were demoted from the central place in creation that we had assumed we occupied. Given what we could see with the naked eye, that assumption was not unreasonable. But it was wrong. Telescopes and careful measurements altered our perspective.

We live on a small planet whirling round an average star in an average galaxy of vastly greater than human dimensions, itself just one of billions of galaxies in the vastly greater expanding Universe, with its hierarchical structure (**Figure 1.2**). In terms of physical size, we are negligible and can only affect what happens in a tiny corner of the physical Universe. Discovering this was the Copernican revolution. How do we discover this? By measuring sizes, distances and relative velocities. This demonstrated not only the vast size of the Universe, but its evolution with time and its vast age.

The sections in this chapter are as follows: §1.1, 'Basic Astronomical Measurements: Distance and Velocities'; §1.2, 'The Cosmological Context of the Expanding and Evolving Universe'; §1.3, 'The Universe Allows Our Existence Because of Its Special Nature'; §1.4, 'We Might Live in a Multiverse, But Maybe Not'; §1.5, 'No, We Do Not Live in a Simulation'; §1.6: 'Philosophical Issues Regarding Cosmology: The Limits of Science'; and §1.7, 'The Sun, Earth and Moon are our Local Astronomical Context'.

Overall, this is the large-scale context which makes our existence possible.

1.1 Basic Astronomical Measurements: Distance and Velocities

Our understanding of the Universe depends on (a) being able to measure the sizes and distances of objects we see in the sky, and (b) being able to determine how they are moving.

How We Come to Be. George F. R. Ellis, Oxford University Press. © George Ellis (2026).
DOI: 10.1093/9780198950189.003.0003

Figure 1.1 The Milky Way as seen on a dark night. This is our own Galaxy, seen edge on from our position out near its periphery. Its structure is similar to that of Andromeda (Figure 1.8).

Source: ESO/B. Tafreshi (twanight.org)

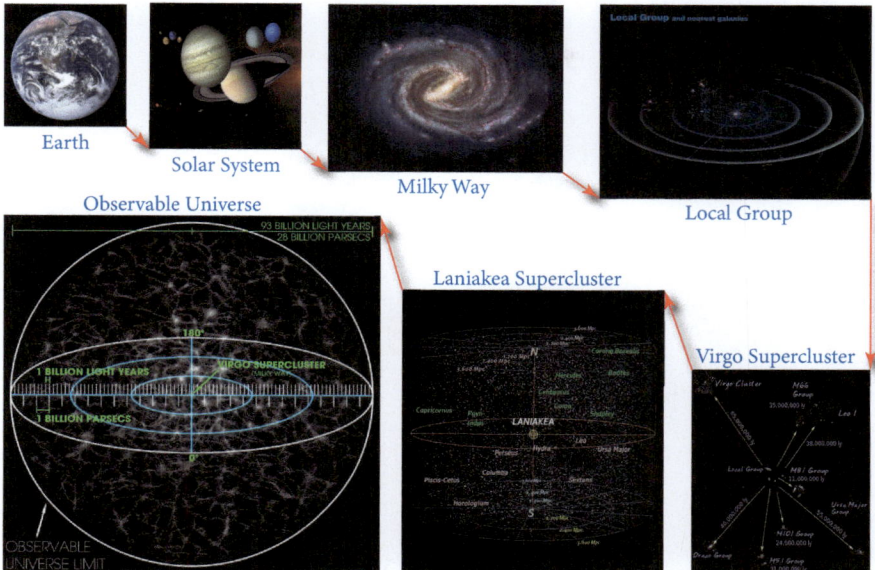

Earth

Solar System

Observable Universe

Milky Way

Local Group

Laniakea Supercluster

Virgo Supercluster

Figure 1.2 The hierarchical structure of the Universe, from the scale of the Earth up.

Adapted from https://simple.wikipedia.org/wiki/Template:LocationOfEarth

1.1.1 Measuring Sizes and Distances

It is not obvious that the Sun is vastly larger than the Earth. It is 1,391,000 km across, about 109 times the diameter of Earth. It looks small because it is 147 million km away from us. By an astronomical coincidence, it appears almost exactly as large in the sky as the much smaller and closer Moon, which is why solar eclipses can happen as they do (the Moon precisely covers the shining disc of the Sun in the sky). How do we determine these scales? Eratosthenes (276–194 BC), a librarian in Alexandria in Egypt, observed the shadows of gnomons (vertical rods) at two different places on Earth (**Figure 1.3**). He thereby calculated the circumference of the Earth to be 42,000 km, and so its radius to be about 6,684 km—which is remarkably accurate.

Aristarchus of Samos (310–230 BC) used the geometry and timing of lunar eclipses to determine the size of the Sun and the Moon. The distance between the Earth and Sun (called an astronomical unit, or **AU**) can be determined by timing transits of Venus across the surface of the Sun from two different places on Earth. Nowadays radar signals can be used to accurately determine various planetary distances in the Solar System, setting the scale for the whole, including our distance from the Sun.

Stellar parallax occurs when one measures the apparent position in the sky of a nearby star relative to a distant star over the course of a year. A star with a parallax angle of 1 arc second is by definition at a distance of one *parsec from us*, which is 3.26 light-years (**Figure 1.4**). Parallax can be used to determine the distance from us

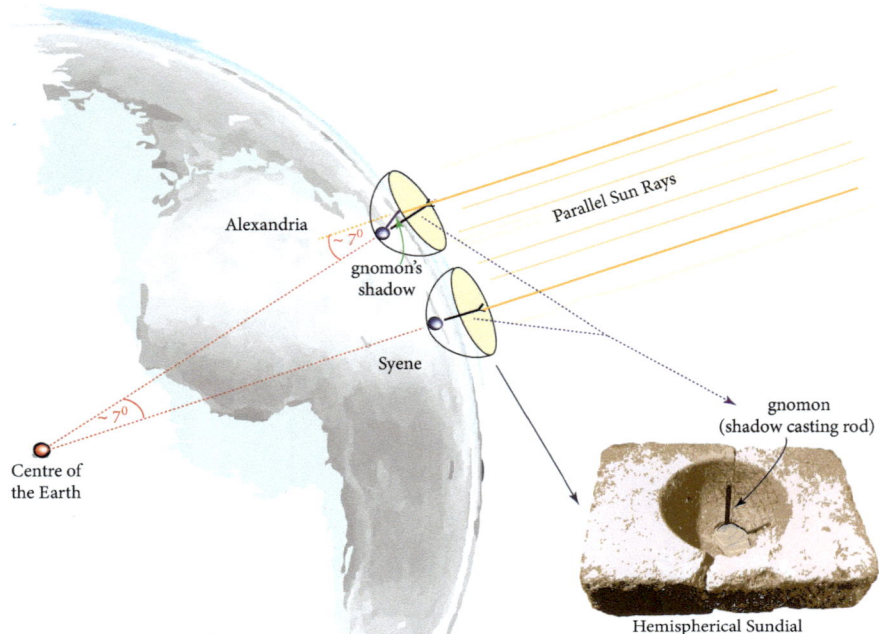

Figure 1.3 Eratosthenes' measure of the circumference of the Earth exploited the fact that at noon on the day of the summer solstice there is no shadow cast by the vertical gnomon of a hemispherical sundial located at the city of Syene in upper Egypt. The gnomon of a similar sundial located at Alexandria would cast, under the same conditions, a spherical angular shadow of 7.2°. Eratosthenes assumed that Alexandria lies due north of Syene (up to a few degrees they lie along the same meridian), and at a distance of 5000 stadia. Since 7.2° is 1/50 of 360°, the geometrical setup shown in the figure determines that the distance of 5000 stadia is 1/50 of the circumference of Earth. This gives an estimate of 250,000 stadia for the circumference (~42,000 km, depending on the assumed length of a stadium).

Figure by Mauro Carfora.

to the centre of the Galaxy, because we know the size of the Sun's orbit. This method only works up to that distance, because parallax is simply too small to be useful for more distant objects.

To determine greater distances we need **standard candles**: objects whose intrinsic brightness is known. Thus measuring their apparent brightness tells us how far away they are because of the inverse square law for radiation: the flux of radiation Sr through a portion of spherical surface received from a source is inversely proportional to the square of its distance r (**Figure 1.5**).

Stars in general have a very wide range of luminosities and so are not good as standard candles, but in 1908 Henrietta Leavitt at Harvard College Observatory showed that a class of variable stars called Cepheids could be used as good standard candles, because their luminosity could be determined from their pulsation period. They were then used by Edwin Hubble to determine the distance to the Andromeda Galaxy, and

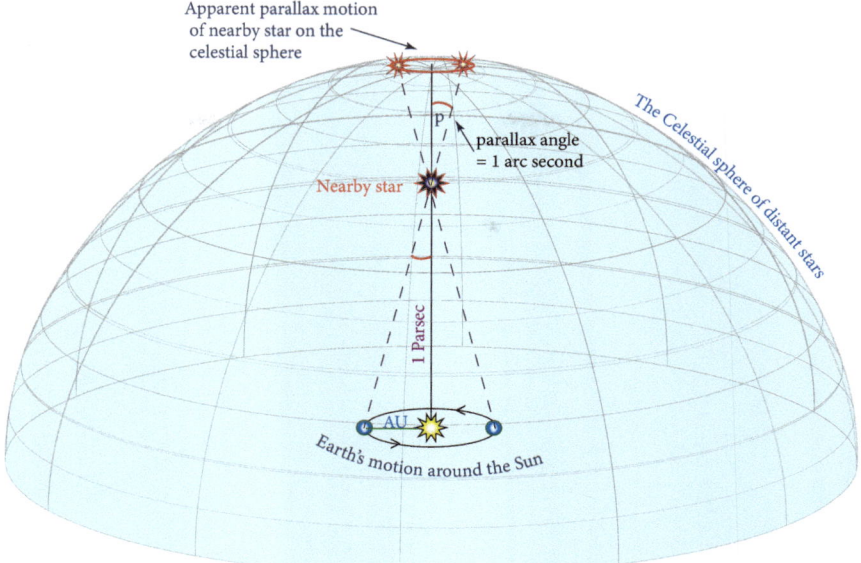

Figure 1.4 Parallax of a star. The parsec (pc) is equivalent to about 30.857×10^{12} km, or about 206,000 AUs, and is itself defined as the distance at which one astronomical unit subtends an angle of one arcsecond.

Figure by Mauro Carfora.

to some more distant ones. Measured brightness of a star or galaxy as a whole is called its magnitude m (**Figure 1.6**): a strange measure (for historical reasons), where the larger the number m is, the fainter and more difficult it is to detect the object.

Galaxies, as well as quasars and radio sources, are not very good standard candles, but some exploding stars (type Ia supernovae) are because their luminosity is tightly related to their decay rate. In many ways, the story of astronomy since Hubble has been the search for ever better standard candles.

1.1.2 Measuring Relative Velocities

As shown by Isaac Newton, a prism breaks up light into different wavelengths, seen as different colours. Astronomical telescopes use diffraction gratings and mirrors to do this very precisely. The light received from a galaxy is made up of the sum of the light from all the stars that make it up, which have many bright emission lines, but with dark absorption lines resulting from light passing through cooler gas as the radiation leaves the galaxy. Emission lines include those from hydrogen, helium, oxygen, nitrogen, carbon, magnesium and neon, while absorption lines include those from hydrogen, potassium, magnesium and sodium.

Figure 1.7(a) shows absorption lines in a galaxy spectrum with intensity plotted as a function of wavelength λ (above) and as dark lines seen against a basically white light spectrum (below). Now the key point is that if the galaxy is moving away from

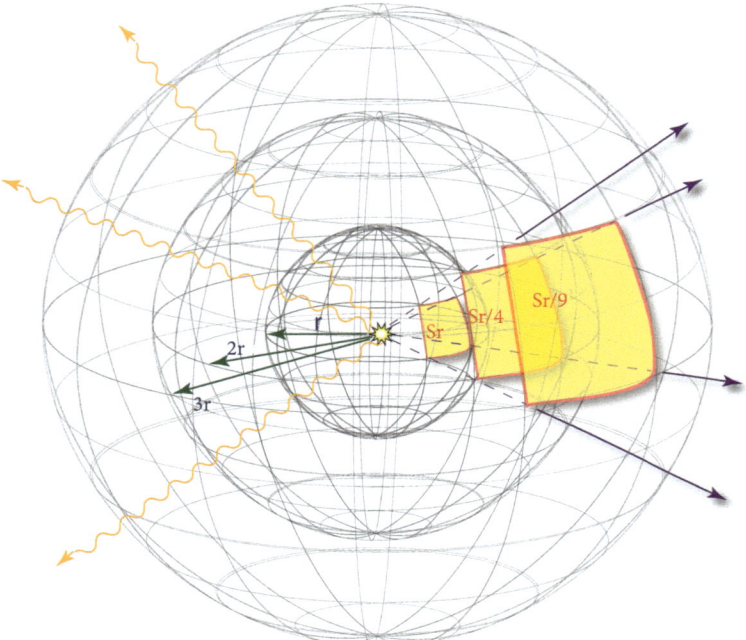

Figure 1.5 Inverse square law for radiation.

Figure by Mauro Carfora.

us, the light will be received at a different frequency from that at which it was emitted. This is essentially the Doppler Effect, which occurs in sound: if a fire engine comes toward you with its horn blaring, the pitch of the noise you hear will drop as it passes you. Similarly the wavelength of light will be different in the rest frame of the galaxy emitting it (as measured in a laboratory on a planet in the galaxy) and as received by us, as shown in **Figure 1.7(b)**.

The galaxy redshift **z** is a measure of the resulting change in wavelength:

$$1 + z = \lambda_{\text{obs}}/\lambda_{\text{emit}},$$

where λ_{emit} is the emitted wavelength and λ_{obs} the observed wavelength. Redshift z determines the relative speed of motion, depending on the context. For motion radially away at small speeds it determines the rate of recession **v** by the relation

$$v/c = z.$$

As recalled earlier, Cepheids are stars that brighten and dim periodically.[1] By measuring the period of 25 Cepheids in the Magellanic Clouds, Henrietta Swan Leavitt[2]

---

[1] See the Wikipedia article Cepheid Variable.
[2] See the Wikipedia article Henrietta Swan Leavitt.

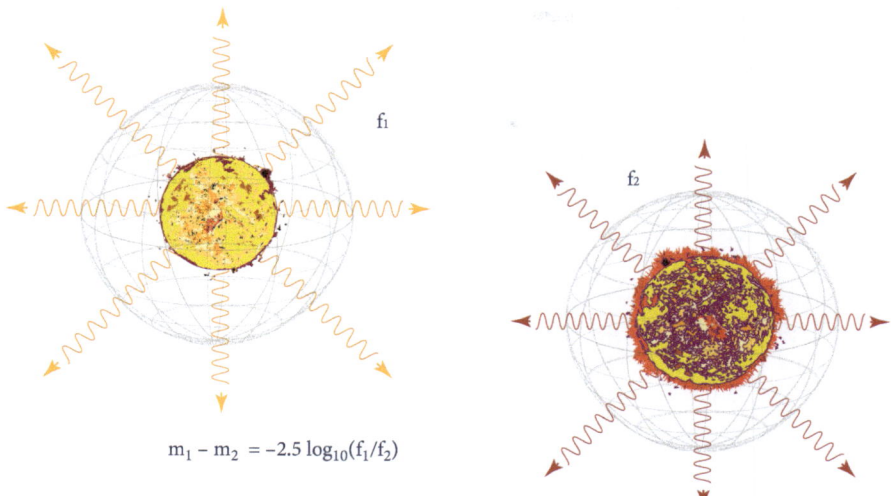

$$m_1 - m_2 = -2.5 \log_{10}(f_1/f_2)$$

Figure 1.6 Magnitude. The magnitude m is a way of writing the ratio between observed energy fluxes (luminosities) of different astrophysical sources. If two stars have fluxes f_1 and f_2, then the magnitude difference between the stars is defined according to the formula in the figure. The magnitude m (sometimes labelled μ) measures the luminosity of a star or of galaxy (for more details, see the Wikipedia article 'Apparent Magnitude').

Figure by Mauro Carfora.

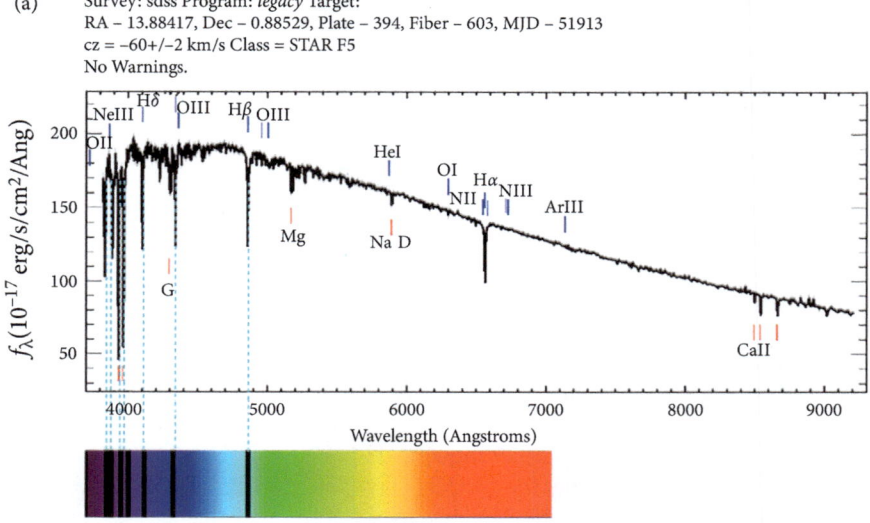

(a) Survey: sdss Program: *legacy* Target:
RA – 13.88417, Dec – 0.88529, Plate – 394, Fiber – 603, MJD – 51913
cz = –60+/–2 km/s Class = STAR F5
No Warnings.

Figure 1.7(a) Intensity of light received from a galaxy as a function of wavelength. Absorption lines from matter are visible.

(b)

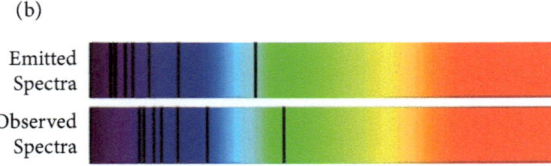

Emitted
Spectra

Observed
Spectra

Figure 1.7(b) Displacement of absorption lines due to relative motion (the Doppler Effect) determines redshift.

(1908) determined that the brighter the Cepheid, the longer its period. This discovery allows Cepheid stars to be used as cosmic yardsticks, paving the way to the result that marked the birth of modern cosmology: Hubble's Law. By exploiting the observations of Vesto Slipher[3] and Milton Humason, Edwin Hubble measured the redshift and distance of nearby galaxies, determining distance by using Cepheid variable stars as standard candles. Hubble's Law (1929) is that for nearby galaxies, redshift is proportional to distance:

$$z = H_0 D / c.$$

Combining these two equations implies an expanding Universe:

$$v = H_0 D.$$

The point of these preliminary remarks is to emphasize that astronomy and cosmology are empirical sciences, based in careful observations of various kinds. They show us the way things actually are.

1.2 The Cosmological Context of the Expanding and Evolving Universe

It is hard to overstate the change that has taken place in our understanding of the relation between the Earth, Sun and stars over the past 500 years. A first central event was the publishing by Copernicus in 1543 of *De Revolutionibus*, presenting his heliocentric theory with the Earth moving round the Sun rather than the Sun round the Earth. We were no longer the centre of things. Second was the use of Cepheid variables by Edwin Hubble in 1924 to determine the distance of the Andromeda Galaxy, proving it was not a cloud of gas in our Galaxy but rather another stellar system of the same size and nature as our own Milky Way. This was followed by Georges Lemaître in 1927 and Hubble in 1929 showing that galaxy redshifts vary linearly with distance—the first evidence that the Universe is changing with time.

We determine the nature of the Universe by a variety of exquisitely sensitive telescopes operating at many different wavelengths (radio, infrared, optical, ultraviolet,

---

[3] See C. O'Raifeartaigh, 'The Contribution of V.M. Slipher to the Discovery of the Expanding Universe', in M. J. Way and D. Hunter, eds. *Origins of the Expanding Universe: 1912–1932*, ASP Conference Series, vol. 471 (2013), 49–63.

X-ray, gamma ray). Some are on Earth, but many are in satellites orbiting the Earth, so as to get above the interfering effects of the atmosphere. These include the Hubble Space Telescope, James Webb Space Telescope and Euclid Telescope imaging distant galaxies and quasars, as well as COBE, WMAP and Planck, measuring the cosmic microwave background radiation.

Three fundamental features of the Universe are (a) its vast size and age, (b) we are not at a special place in the Universe and (c) it is not static but evolving, this dynamical evolution having led to the existence of galaxies, stars and planets such as Earth.

1.2.1 The Vast Size and Age of the Sun, Galaxies and the Universe

When we look at the night sky, it is not obvious that our own Galaxy (which we see on a dark night as the Milky Way, see **Figure 1.1**) is a huge spiral disk made up of billions of stars, with our own Sun just an average star near its outer edge. It is very similar to the Andromeda Galaxy (**Figure 1.8**), each of these galaxies containing about 10^{11} stars.

The Universe is of huge size, containing billions of galaxies like the Milky Way and the Andromeda Galaxy. Each is made of billions of stars together with gas, many with gigantic black holes at their centre. They occur in galaxy clusters (**Figure 1.9**), containing various kinds of galaxies (barred, elliptical, spiral, irregular). Galaxy clusters are grouped in superclusters and larger sheets, walls and filaments separated by vast voids, forming a foam-like structure. On even larger scales, the same kind of structures occur everywhere that we can observe: these structures are not special to our location. The scale of the visible part of the Universe is about 45.7 billion

Figure 1.8 The Andromeda Galaxy and its two dwarf satellite galaxies. It has a diameter of about 152,000 light-years, and is about 2.5 million light-years from Earth. Thus we see it as it was 2.5 million years ago (a light-year is the distance light travels in a year). Our own Galaxy (Figure 1.1) is similar to Andromeda, with us living towards the outer edge of the disc, which we see edge on.
Source: ESO (R. Gendler).

Figure 1.9 A typical cluster of galaxies. The Fornax Galaxy Cluster, one of the closest of such groupings beyond our local group of galaxies. At the lower-right is the barred-spiral galaxy NGC 1365. Each image is a galaxy, except for those that seem to have crosses emerging from them, which are stars with the cross being an observational artefact resulting from the structure of the telescope interacting with incoming light. We see them as they were billions of years ago, because of the time it takes for light to travel to us from them.

Source: ESO (A. Grado and L. Limatola).

light-years—vastly bigger than a galaxy, and containing something like 10^{12} galaxies (about 10^{23} stars). This is not the whole of the Universe. It undoubtedly extends beyond the furthest matter that we can see (visible as the last scattering surface; see **Figure 1.12(a)**).

Associated with these vast scales are the huge ages of astronomical objects. The Universe is about 13.7 billion years old. Galaxies are younger: most are between 10 billion and 13.6 billion years old. The Sun is about 4.5 billion years old, and the Earth about the same age. These are enormously larger than the timescales of daily life, and even of the existence of civilization on Earth. Simple life forms (microbes) existed about 3.7 billion years ago, but it took a long time for complex life to emerge from this beginning. The fossil record and DNA dating suggest people like us (*Homo sapiens*) evolved around 300,000 years ago, while complex technology and cultures evolved much more recently: from 50,000–65,000 years ago before the present day. Due to its complexity, it would take just as long for intelligent life to emerge on planets in other galaxies. For most of its existence, the Universe has been devoid of intelligent life.

1.2.2 We are Not at a Special Place in the Universe

We can determine the distribution of matter in the visible part of the Universe by galaxy, radio source and quasar surveys. It appears that on a sufficiently large scale (larger than that of its foamlike structure) the Universe is almost the same everywhere, and almost the same in every direction we look in the sky. We do not see any indications of the existence of a preferred centre of the Universe. This is confirmed by the very near isotropy of the cosmic background radiation (CBR) we receive from the early Universe (see **Figure 1.12(a)**). A mathematical theorem states that if this near-isotropy holds for all observers, the Universe is (almost) spatially homogeneous.

Various recent observations have upheld this implication: it is probable that we do indeed live in a typical place in such a universe. Thus the Universe can be well represented on large scales by a very simple model: a Robertson–Walker Geometry that is spatially homogeneous (it is the same everywhere at a given time) and spatially isotropic (it looks the same in every direction about us). This formalizes the idea that we are not at the centre of the Universe. Indeed no planet or star or galaxy is, as it has no centre.

1.2.3 The Evolution of the Universe

It was a huge step forward to understand that the Universe was changing with time. Everyone—including Albert Einstein—had taken for granted that it was static. This simply had to be so. But then new theories by Alexander Friedmann and Georges Lemaître showed it might be dynamic, and astronomical evidence confirmed this was indeed the case. The cosmos is changing with time—a completely new vision of the Universe. The Universe is expanding, with galaxies receding from each other. Its scale at each time **t** is given by a function $a(t)$ that has increased continuously with time. We can measure apparent magnitudes of observed sources, an indication of their distance (**Figure 1.5** and **Figure 1.6**), and the redshift of their spectral lines (**Figure 1.7(b)**). Plotting one against the other gives the magnitude–redshift diagram (**Figure 1.10**), in this case for type Ia supernovae in galaxies.

The dynamics of the Universe is governed by gravitation as described by Einstein's general theory of relativity (§2.1). It has evolved through various stages during its evolution from a Hot Big Bang radiation-dominated early era, through decoupling of matter and radiation when CBR was emitted, to the formation of stars and galaxies via gravitational instability.[4] The main epochs in the history of the Universe according to the standard theory are indicated in a famous NASA diagram (**Figure 1.11**). In highly simplified form, they are as summarized in **Table 1.1**.

The Hot Big Bang Era and before. There was probably some kind of start to the Universe, but this is not certain, as we do not know what the theory of quantum gravity was that would have held at that time. At about 10^{-37} seconds after the start of

---

[4] Patrick Peter and Jean-Philippe Uzan, *Primordial Cosmology* (Oxford: Oxford University Press, 2013); Scott Dodelson and Fabian Schmidt, *Modern Cosmology* (London: Academic Press, 2020).

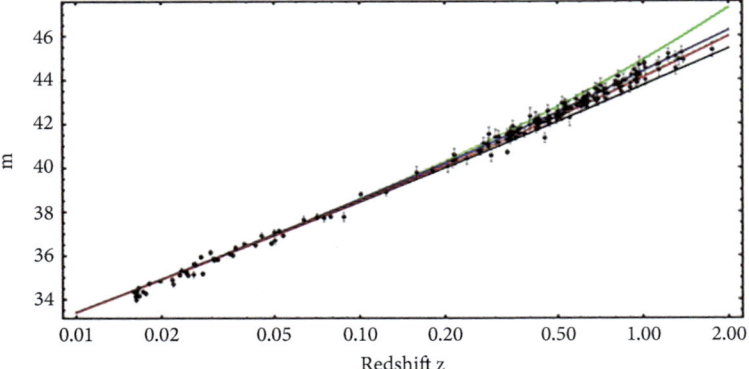

Figure 1.10 The magnitude–redshift relation that gives evidence for both the expansion of the Universe and its speeding up at recent times. Magnitude m on the vertical axis represents distance away from us, determined from observed luminosity; redshift z on the horizontal axis represents relative motion away from us. Fluctuations about the mean curve are shown below the curve. The upward bend on the right indicates accelerating expansion at recent times.

Adapted from Saul Perlmutter, 'Supernovae, Dark Energy, and the Accelerating Universe', *Physics Today*, 56 (2003), 53–60.

the expansion of the universe, a period of incredibly fast exponential growth know as inflation occurred,[5] ending at between 10^{-33} and 10^{-32} seconds after its size had increased by something like a factor of between 10^{30} and 10^{80}. Fluctuations occurred at the end of inflation due to the effects of quantum fields, which later formed the seeds for structure formation. Inflation cooled the Universe to a very low temperature, but was followed by a period of reheating, leading to a quark–gluon plasma with a small excess of quarks over antiquarks arising through a process that is still not fully known. At 10^{-6} seconds, quarks and gluons combined to form baryons, leading to the existence of ordinary matter: protons and neutrons.

Between 20 seconds and 20 minutes after the start, at about 10^9K, nucleosynthesis took place, with neutrons decaying to protons, followed by a chain of nuclear reactions producing deuterium, tritium, helium-3, helium-4 and some lithium. No higher elements were produced then. At this time matter (ions and electrons) and radiation were in equilibrium because they strongly scattered off each other, so the mean-free path of radiation was negligible and the Universe was opaque. Baryon acoustic oscillations—in essence, very large-scale sound waves—then occurred because of the small density inhomogeneities. As the temperature dropped through 4000K at about 370,000 years, electrons and ions combined to form atoms. Matter–radiation scattering ended, and the Universe became transparent.

---

[5] Swagat Mishra, 'Cosmic Inflation' (2024). arXiv preprint arXiv:2403.10606.

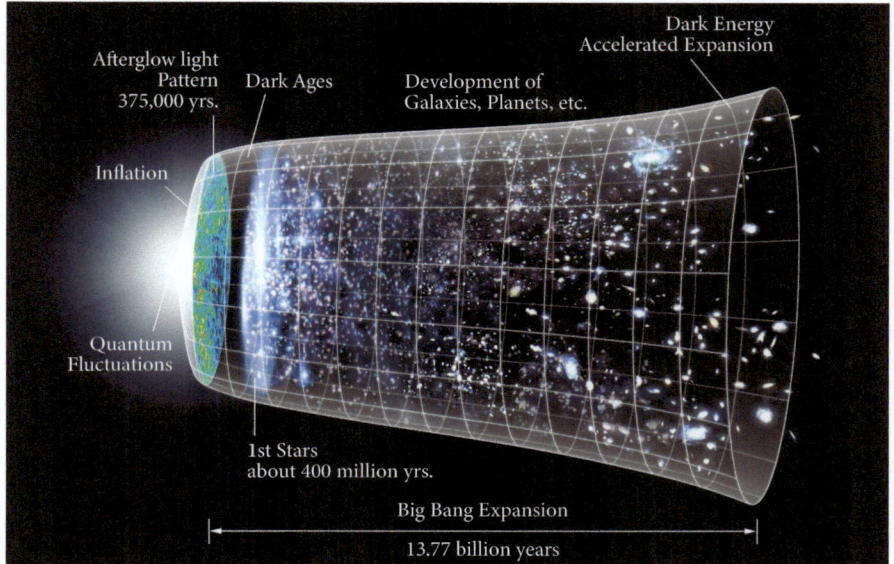

Figure 1.11 The expansion of the universe. In this diagram, time passes from left to right, so at any given time, the Universe is represented by a disk-shaped 'slice' of the diagram

Source: Wikimedia Commons, https://en.wikipedia.org/wiki/Expansion_of_the_universe#/media/ File:CMB_Timeline300_no_WMAP.jpg.

The epochs after. CBR was then emitted, which today is observed in the microwave region, from the *last scattering surface* (LSS) at a temperature T of 4000 K. The temperature T then decayed inversely with the scale of the Universe as expansion took place. Due to the fluctuations that occurred at the end of inflation, the gravitational potential fluctuates slightly on the LSS, providing the seeds for structure formation later. Gravitational attraction then caused inhomogeneities to grow in a wavelength-dependent way, with non-baryonic cold dark matter (CDM) and a cosmological constant Λ shaping the outcome.[6] This is called the ΛCDM theory. 'Dark ages' occurred after decoupling until between 200 and 500 million years, as there were not yet any stars and the black body radiation was cooling. Hierarchical structure formation then took place. Gravitationally bound stars and star clusters formed first. These then merged to create galaxies with black holes at their centres, and sometimes created quasars.

Massive first-generation stars burned their hydrogen and helium rapidly, with heavy elements such as carbon, nitrogen and oxygen forming in their interiors by stellar nucleosynthesis. At the end of their lives, these stars exploded and spread those elements through space, providing the clouds from which second-generation stars surrounded by planets could form from accretion discs, with the elements needed for life available on these planets. This is the origin of the Earth on which we live.

---

[6] P. James E. Peebles, *Cosmology's Century: An Inside History of our Modern Understanding of the Universe* (Princeton, NJ: Princeton University Press, 2020).

Table 1.1 Events in the history of the universe (depicted in Figure 1.11)

Time	Event	Outcome
$t = 0$	Start to Universe	Spacetime and matter exist
$t = 10^{-37}$ to 10^{-32} secs	Inflation ending in quark–gluon plasma	Universe very smooth but with small density fluctuations
$t = 10^{-6}$ secs	Quarks and gluons combine	Existence of protons and neutrons as well as electrons
$t = 20$ secs to 20 mins	Nucleosynthesis	Light elements: D, T, ^{3}He, ^{4}He, ^{7}Li
$t = 20$ mins to 370k years	Baryon acoustic oscillations	Larger scale density fluctuations
$t = 370{,}000$ years	Decoupling	Black body background radiation
370k years $< t < 4{\times}10^6$ years	Dark ages	Inhomogeneity growing
370k years $< t < 10^9$ years	Structure formation	Stars, galaxies, larger scale structures
$10^6 \le t \le 360 \times 10^6$ years	Elements (C, N, O, etc) form inside stars, spread in space	Planets with elements of life come into being
$t > 7.8 \times 10^9$ years	Acceleration epoch	Structure formation ends

Life eventually arose on Earth and presumably other planets, the whole being an irreversible process.

Recently, the cosmological constant, or some other form of 'dark energy', caused the expansion of the Universe to start speeding up. This late time acceleration is evidenced in the right-hand side of **Figure 1.10**. The age of the Universe at the present time is estimated to be between 13.7 and 13.8 billion years. A key point is that we would have a major problem if our estimates of stellar ages were larger than this. With current estimates of the rate of expansion of the Universe (the Hubble constant), this is not the case.

Evidence for the standard theory comes from many observations. They include redshift–magnitude diagrams for distant sources, such as **Figure 1.10**, and number counts of discrete sources, such as the Sloan Digital Sky Survey.[7] This makes it possible to derive the spatial power spectrum of the relevant discrete sources (galaxies, radio sources, quasars) to compare with structure formation scenarios, using baryon acoustic oscillations as standard rulers. Measurement of primordial element abundances restrict the density of baryons at the time of nucleosynthesis in the early Universe, and hence gives estimates of baryon densities today. The apparent variation

---

[7] See the Sloan Digital Sky Survey (SDSS) website: https://www.sdss.org/.

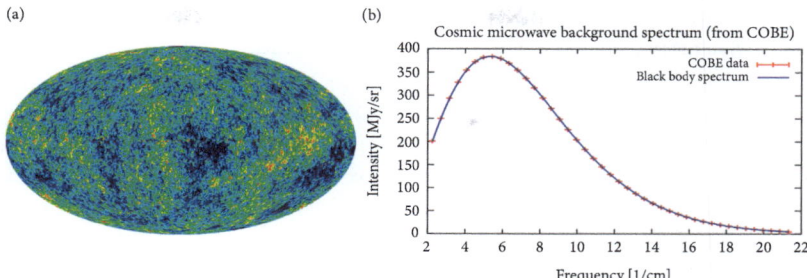

Figure 1.12 Cosmic background radiation (CBR). (a) Temperature variation over the sky, with the dipole and our galaxy removed. This is an image of the last scattering surface (LSS) as it was when light was emitted from it 13.7 billion years ago. It is incredibly homogeneous: the fluctuations seen are at a level of 1 part in 500,000. (b) The radiation spectrum is a perfect black body spectrum, such as we can measure in a laboratory on Earth. This proves that physics there at these early times was the same as it is on Earth today.

of the CBR temperature over the LSS[8] (**Figure 1.12(a)**) and CBR polarization data can be analysed in terms of angular scales in the sky. A key observation is the CBR energy spectrum (**Figure 1.12(b)**). This is a perfect black body spectrum, as predicted by quantum physics together with statistical physics, proving physics was the same back then in the very early Universe as it is today. Strong and weak gravitational lensing observations due to bending of light by massive objects give data on the distribution of matter. An exciting new frontier is the detection of gravitational radiation emitted from binary black hole mergers and black hole–neutron star mergers. The technology involved in making these extremely delicate observations via interferometers with a 40-km baseline is truly extraordinary. Inter alia it will help determine the limits on the Hubble constant as well as on black hole masses.

All this carefully collected and analysed evidence is the basis of the statements above about the nature and history of the Universe, summarized in **Figure 1.11** and **Table 1.1**.

Limits on observations of its nature. However, there are various limits on what we can determine about the Universe because of its scale and dynamic nature.

Limits in time occur because the earliest stages are invisible to us: the very early Universe was opaque to radiation due to Thomson scattering (light bouncing off electrons) as mentioned earlier. We cannot 'see' to times before the LSS by photons of any wavelength. It may yet prove possible to observe the cosmic neutrino background which would push the limits of observation back to earlier times, and in principle gravitational waves allow observations to even earlier times. But that is the ultimate limit.

---

[8] There is a dipole in this anisotropy which indicates our galaxy is moving at 350 km/sec relative to the Universe. This dipole has been removed in Figure 1.12(a).

Correspondingly, there are also **limits in space.** Our ability to see out to the furthest reaches of the Universe is limited by **visual horizons**, because light travels at the speed of light since the decoupling of matter and radiation, when the Universe became transparent. The corresponding distance is 43 billion light years—greater than the age of the Universe in years, because the Universe was expanding faster in the past. We can't see further by any electromagnetic radiation whatsoever. Similar horizons apply to observations by neutrinos and gravitational waves. Particle horizons limit what matter can have causally affected us since the beginning of time, and so limit what we can deduce about such matter from any data at all.

The Universe may or may not be spatially infinite. It is impossible to prove that it is spatially infinite if this is indeed the case, for we have no data for regions beyond our visual horizon, and have had no interactions with matter beyond our particle horizon. The Universe might, for example, close up spatially on scales much larger than the visual horizon, or end on a time-like singularity located where we can't observe it. The spatial topology could be such as to make the Universe finite: spatial sections could be like a torus or a Möbius strip, for example.

We would have shown it was finite if we proved that its spatial curvature is positive, but we don't know at present whether that is the case. If it is indeed spatially finite, we could possibly live in a **small universe**, where the Universe closes up on itself on a scale smaller than the visual horizon. Then we would see the same galaxies—including our own—in different directions in the sky at different stages in their history. Neither outcome—finite or infinite—affects our physical existence. If it did, the answer would be knowable. The part of the Universe that significantly affects us physically is much smaller than our particle or visual horizons.[9] It is a comoving volume of about two light-years around us; almost all the matter on Earth has come from this region. The radiation that significantly affects us all comes from the Sun, not from further out. In cosmic terms, we lead a very local existence.

The physics horizon. A further issue is that limited knowledge of physics restricts our ability to probe the dynamics of the very early Universe. This is the case for events before nucleosynthesis, for we don't have colliders with high enough energy to probe the relevant physics in such extreme circumstances. In particular, we do not know the details of how matter/antimatter asymmetry arose in the very early Universe. We have an outline of a theory, but no solid tie-in to established physics.

This asymmetry is crucial to our existence: if it did not exist, matter and antimatter would have annihilated each other as the Universe evolved producing radiation, with no ordinary matter left over to enable the Sun, Earth and ourselves to exist. We also can't test theories that try to unify the four forces with gravity such as string theory/M-theory because the energies we'd need to do so are far higher than we can attain in colliders such as the Large Hadron Collider (the LHC) at Geneva. We can't even reach the energies needed to test the physics of inflation.

Unresolved issues and tensions. Our current cosmological standard model is supported by a huge amount of data, as just discussed. Underlying basics are agreed, but

---

[9] George Ellis and William Stoeger, 'The Evolution of our Local Cosmic Domain: Effective Causal Limits', *Monthly Notices of the Royal Astronomical Society*, 398 (2009), 1527–36.

there are both unresolved issues and some actual tensions.[10] A thoughtful review is given by Jim Peebles,[11] who states,

> Empirical tests give excellent reason to expect that a more advanced physical cosmology will look much like the theoretical ΛCDM universe, because many well-checked tests show that the ΛCDM universe looks much our universe. But the great progress in cosmology and the other physical sciences has left anomalies, some of which have been troubling for a long time.

The issues are summarized in **Table 1.2**.

Unresolved issues relate to key aspects of the standard cosmological model. The big three are that the nature of dark matter, dark energy and the inflaton (the field that caused inflation) are all unknown. It is agreed dark matter has to exist in order that structure formation and galactic dynamics can work as shown by observations, and it is also required by CBR observations, but is quite unlike the ordinary matter around us because it does not interact with ordinary matter or with light: it is therefore non-baryonic (i.e. unlike ordinary matter). It might be weakly interacting massive particles (WIMPS), axions or primordial black holes. Many experiments are being carried out to explore the various particle alternatives, without success so far. The nature of dark energy is also unknown. It could just be a cosmological constant, but it could be a dynamic field that changes with time. Then the issue is 'What is its equation of state?' Observational analyses are trying to establish which is the case, and explore the possible equation of state of dynamical dark energy.

It could be, however, that instead of dark energy or dark matter being needed, the problem is that on these scales we need an alternative theory of gravity to general relativity on cosmological scales. Many options, such as scalar-tensor theories, modified Newtonian dynamics (MOND) and massive gravity, are being actively explored.

Table 1.2 Unresolved issues, actual tensions, and underlying basics in cosmology

Unresolved issues	Actual tensions	Underlying basics
Dark matter?	Hubble tension	Physics same everywhere
Dark energy?	Dark matter halos	RW geometry
The inflaton?	Too early structure?	Theory of gravity
End of inflation → classical?	Dipole tension?	Equation of state
Matter/antimatter asymmetry?		Multilevel causation
Spatial curvature, topology?		Time passes
Start to the Universe?		Future not yet determined

[10] E. Abdalla et al., 'Cosmology Intertwined: A Review of the Particle Physics, Astrophysics, and Cosmology Associated with the Cosmological Tensions and Anomalies', *Journal of High Energy Astrophysics*, 34 (2022), 49–211.

[11] P. J. E. Peebles, 'Anomalies in Physical Cosmology' (2022), arXiv:2208.05018.

While most cosmologists agree that inflation took place in the early Universe, there is no agreement as to what the dynamical field that caused it is; indeed over 150 different inflationary theories have been proposed. Much less commented on, but a key issue, is how did the quantum fluctuations that occurred during inflation, and are supposed to have provided the seeds for structure formation later on, become classical? Actually there were no such fluctuations: rather there was a potential for such fluctuations.[12] The dark secret of inflationary cosmology is that while wave function collapse to produce classical fluctuations must have occurred via some dynamical process—without this, classical structures would not exist—we don't know what that collapse process was.[13] Additionally, we don't know whether the Universe had a beginning because we don't know what the theory of quantum gravity is.

Unresolved tensions. There are some cases where there seems to be an actual conflict between observations and theory. There is a Hubble parameter tension, where estimates of the Hubble constant based on relatively local objects, particularly type SN1a supernovae, disagree significantly with those arising from far more distance measurements, namely those from CBR observations. One possible resolution is that we don't live in a Roberston–Walker geometry—a basic assumption underlying standard cosmology. Maybe the Universe is inhomogeneous on very large scales. Also recent observations from the amazing James Webb Space Telescope have identified massive galaxies that formed less than 700 Myr after the Big Bang. There does not seem to have been enough time for them to have formed within the standard cosmological model. The same may apply to the emergence of super-massive black holes early in the Universe. Both these observations significantly challenge standard cosmology. Are we using the wrong theory of gravity in the cosmological context? The wrong equation of state? The answer will become clear as we gather more data. These are key challenges for cosmologists, but will not overall change the grand picture of how the evolving and expanding Universe is the context for our existence.

The overall understanding presented above will surely remain, even if details differ. Some foundational issues underlie all this.

Physics is the same everywhere in the Universe, as far as we can tell. This is an important underpinning of our standard analysis of cosmology: we assume physics has not changed since the start of the Universe, and is the same in distant places. Important support for this idea comes from the extraordinarily accurate CBR spectrum (**Figure 1.12(b)**). This shows that quantum physics and statistical physics were the same 13.7 billion years ago as they are at the present time. Some claims have been made that there is a variation of fine structure constant, basic to some physics interactions, but that claim is not commonly agreed.

Emergence and multilevel causation. The astronomical hierarchy of emergence is shown in **Figure 1.2**, from the scale of the Earth up. There are underlying physics levels below, down to the atomic and particle levels (Chapter 2). Emergence takes place by physical forces underlying existence and dynamics of higher-level entities

---

[12] Andrew Steane, *Liberating Science: The Early Universe, Evolution and the Public Voice of Science* (Oxford: Oxford University Press, 2023).

[13] R. L. Lechuga and Daniel Sudarsky, 'Eternal Inflation and Collapse Theories', *Journal of Cosmology and Astroparticle Physics* (2024), 038.

such as planets, stars, galaxies and so on. These obey effective laws at each level that arise out of the combined effects of the lower-level laws and structures. Examples are the orbital motion of planets around the Sun, of stars in the Galaxy and the rotational motion of the Galaxy.

Does downward causation occur in cosmology and astrophysics, as it does in biology? (See the Preface.) Indeed it does. An example is nucleosynthesis in the early Universe. The rate of nuclear reactions at the nuclear physics level is determined by the rate of change with time of the temperature, $T(t)$, which is a cosmological-level variable, in turn determined by the rate of expansion of the Universe, determined by the overall matter density. Another is the creation of structure in the Universe (**Figure 1.2** and **Figure 1.11**), which depends again on the rate of expansion of the Universe, which in turn depends on its average density (both cosmological-level variables). This dependency enables us to use measurements of primordial nucleosynthesis and of large-scale structures to determine cosmological parameters, in particular dark matter and baryon densities.

Time passes. Contra claims by some physicists and philosophers,[14] time passes because we live in an evolving block universe,[15] which is a spacetime with a future boundary constantly extending (the right-hand side of **Figure 1.11** will lie further to the right a million years from now). The spacetime grows with time, and is ever getting older. Nowadays the Universe has an age of 13.7 billion years. This would not be the case in a block universe, which is a spacetime that has no concept of the present time and conceptually already extends an infinite time in the future. But the real Universe *never* becomes infinitely old, because infinity is not just a very big number: it is bigger than any number that can possibly exist. An infinite age is never attained no matter how long you wait. It's always in the future.

One of the most fundamental features of daily experience is the direction of time. You can't remember what has yet to happen, but can remember the past; you can change the future, but cannot alter the past. If you break an egg or shatter a glass, you can't put them together again. Milk diffuses in your coffee. The puzzle is that the underlying physics is time symmetric. If you change the time t to time $t' = -t$, the basic equations of physics are invariant. So where does the direction of time come from? The answer is that the emergence of the direction of time in daily life is a result of the cosmological context. The globally determined **direction of time** results from the expansion of the Universe. It determines local **arrows of time**: thermodynamic, electrodynamic, sound, chemical, biological and mental.[16]

Emergence of life, and then intelligent life, occurred on Earth and presumably on other suitable planets in the expanding Universe. The crucial implication of the fable regarding agency discussed in the Prologue is that **the future of the Universe on human scales is not determined until it happens**. It is not possible that the specific detailed content of this book is somehow written into some kind of initial data (perhaps some form of hidden variables for quantum theory) at the start of the Universe.

---

[14] See the Essentia Foundation discussion: https://www.youtube.com/watch?v=2Bnktj5_o6M.
[15] George Ellis, 'Physical Time and Human Time', *Foundations of Physics*, 54 (2024), 1.
[16] George Ellis and Barbara Drossel, 'Emergence of Time', *Foundations of Physics*, 50 (2020), 161–90.

Firstly, the issue of coding arises: there is no plausible way that the initial data needed for this to happen are coded in the fluctuations we see on the LSS (**Figure 1.12(a)**). Secondly, there is no plausible mechanism of any kind whereby such data, if it indeed existed, could get transferred from there into our brains. The developmental processes whereby our brains arise completely preclude this, as explained in my book with Mark Solms.[17] The contents of that book were not determined until the authors discussed it and wrote it. I return to this issue in Chapter 5.

And thirdly, if we ignore all these overwhelming problems, and such data did indeed exist on the LSS, the crucial issue that then arises is this: *how did this data get there*? What kind of demi-urge wrote that meaningful data—plans for iPads and aircraft, musical scores and Shakespeare's poetry, the rules of chess and football, the occurrence of the Second World War and of Brexit—into the LSS? Nothing in standard cosmological theory of structure formation[18] can lead to such outcomes. Theories such as 'superdeterminism' are, in fact, highly implausible versions of Intelligent Design whereby the random fluctuations on the LSS somehow get overwritten by some kind of intelligence.

The only possible way that all we see around us on Earth today can come into existence is that agency arises via evolutionary processes,[19] leading to outcomes not directly determined by the initial data of the Universe. Such outcomes are allowed by, but not uniquely determined by, that data. Discussion of the processes whereby this agency occurs is a central feature of this book.

1.3   The Universe Allows Our Existence Because of Its Special Nature

This need not have been the case: our existence would not have been possible if some of the dimensionless constants of nature, such as the fine structure constant and the ratio between the electromagnetic and strong forces, were different than they are. Thus in a scientific sense, the Universe is fine tuned in such a way that life can come into existence. This is the **anthropic principle**.[20] As stated by Stephen Hawking:[21]

> The laws of science, as we know them at present, contain many fundamental numbers, like the size of the electric charge of the electron and the ratio of the masses of the proton and the electron. ... The remarkable fact is that the values of these numbers seem to have been very finely adjusted to make possible the development of life.

Special values of physical constants are needed for stars like our Sun to exist, atoms to exist and habitable planets to exist. Furthermore spacetime must have four dimensions for stable structures to arise.

---

[17] George Ellis and Mark Solms, *Beyond Evolutionary Psychology* (Cambridge, UK: Cambridge University Press, 2017).

[18] Peebles, *Cosmology's Century*.

[19] Kevin J. Mitchel, *Free Agents: How Evolution Gave us Free Will* (Princeton, NJ: Princeton University Press, 2023).

[20] Yuri Y. Balashov, 'Resource LetterAP-1: The Anthropic Principle' (1991). Martin J. Rees, *Just Six Numbers: The Deep Forces That Shape the Universe* (New York: Basic Books, 2008).

[21] Stephen W. Hawking, *A Brief History of Time* (New York: Bantam Books, 1988), 7, 125.

A second set of such fine tunings arises from special cosmological conditions which allow structure formation in the Universe to lead to the existence of habitable planets. If the cosmological constant Λ were too large this would not be possible, as no structures would form. If the amplitude of primordial fluctuations is too large, then only black holes form, so no stars or planets would exist; if it is too small then no structures at all form as the Universe expands and dilutes matter.

A comprehensive overview of such fine tunings is given in a recent edited book.[22] Many claim the solution to the anthropic issue is that we live in a multiverse (§1.4). I return to this issue in §7.1.

1.4 We Might Live in a Multiverse, But Maybe Not

A multiverse—a vast number of expanding universe domains like ours but with different physical properties—will result from some inflationary universes. The relevance to us on Earth is that it might solve the anthropic problem by making a biofriendly universe highly probable. However, the existence of a multiverse is not directly observationally provable, because the supposed other evolving domains lie beyond the visual horizon. Its existence is not directly testable by observations or experiments.

What about indirect evidence? The underlying property of self-replication is claimed to be a generic property of many inflationary universes; indeed, existence of a multiverse is claimed to be a necessary outcome of the Starobinsky models, which are the best fit to the data.[23]

However, claims that a multiverse exists are controversial. The proposal is certainly not part of standard cosmological analysis, as summarized by the Planck team:[24] the idea is simply not mentioned there. It contains no attempt to determine parameters describing a multiverse. Supposing a multiverse exists, to solve the anthropic problem underlying our existence one also needs a mechanism for populating the various bubbles with different effective physics in each bubble. Such mechanisms have been proposed, specifically the KKLT theory based on an extension of string theory (an untested add-on to an unverified proposal for quantum gravity) but this is not an established scientific result. It is a theory of a possible mechanism of a controversial nature. But most importantly, the claim that quantum fluctuations bubble in and out of existence in the early Universe, providing the seeds for both structure formation and for other expanding universe domains, is simply wrong. Certainly the **potential** for such fluctuations to exist and create classical perturbations that leads to galaxies must be right. It is part of the standard theory,[25] even though the key step of showing how the potential quantum fluctuations become classical is missing. Furthermore, only local physical effects, unrelated to a multiverse, determine what happens locally

[22] David Sloan, Rafael Batista, Michael Hicks, and Roger Davies, *Fine Tuning in the Physical Universe* (Cambridge, UK: Cambridge University Press, 2020).

[23] Will Kinney, *An Infinity of Worlds: Cosmic Inflation and the Beginning of the Universe* (Cambridge, MA: MIT Press, 2023).

[24] N. Aghanim et al., 'Planck 2018 Results—VI. Cosmological Parameters', *Astronomy and Astrophysics*, 641 (2020), A6.

[25] Patrick Peter and Jean-Philippe Uzan. *Primordial cosmology*. Oxford University Press, 2009.

in a cosmological context.[26] Work has to be done to show that potential observations can verify a multiverse exists.

I find this explanation of fine tuning unconvincing, but I record here that the distinguished cosmologist Martin Rees strongly believes that the multiverse must exist because it is needed to solve the anthropic problem. This was also the position of the Nobel Prize-winning physicist Steven Weinberg, who showed how an analysis of the anthropic issue might be taken as predicting a small positive cosmological constant—as is now known to be the case. Thus one can make the claim that even if solid evidence is not available, one can support the conclusion by inference to best explanation. But then the issue is 'Who is going to define "best"?' Does a multiverse satisfy Occam's Razor because a single entity (the multiverse) is invoked to solve the anthropic issue? Or is this not the case, because the alleged multiverse contains billions of universe bubbles, perhaps even an infinite number, in contradiction to Occam's razor ('entities should not be multiplied unnecessarily')?

The issue is thus controversial, but with strong claims being made that the multiverse must exist because it is a solid physically based way to solve the fine-tuning issue. But if both necessary conditions are indeed true—a multiverse exists containing many expanding universe domains, and some have suitable physics for life to exist—this just displaces the explanatory problem one level up: **Why does the multiverse itself allow life to exist? What explains the fine tuning of the multiverse?** Any multiverse theory is based on some proposal for fundamental physics at the multiverse level, which in turn have constants with specific values that either do or do not result in a multiverse with bubble domains that are biofriendly. Exactly the same issue recurs at one higher explanatory level. The philosophical issue remains.

One other justification that has been given for a multiverse is based in the Everett (many-worlds) interpretation of quantum mechanics. From this viewpoint, it is claimed that a multiverse is inevitable, even that there are billions of copies of you out there somewhere in the multiverse (this is the many-minds interpretation, which relies solely on quantum mechanical concepts and makes no attempt whatever to consider how the brain or mind actually works). However, in my view these proposals are simply wrong. They assume without justification that the entire universe and everything in it can be described by a single quantum wavefunction. There are very good reasons to believe this cannot be the case.[27]

There is no way that a single linearly evolving wavefunction can account for the emergence of the complex nonlinear systems we see around us. Rather the true situation must be that there are local wavefunctions everywhere, defined in local domains which taken together cover the whole Universe, but there is no single universal wavefunction for all that exists. Those supporting an Everett-type interpretation have to justify their belief in the existence of this single universal wavefunction—and they don't do this. They just take it for granted, no matter how implausible it is. I claim it cannot exist.

I return to this issue in Chapter 7, where I discuss the anthropic issue (§7.1).

[26] George Ellis and William Stoeger, 'The Evolution of our Local Cosmic Domain: Effective Causal Limits', *Monthly Notices of the Royal Astronomical Society*, 398 (2009), 1527–36.

[27] George Ellis, 'Quantum Physics, Digital Computers, and Life from a Holistic Perspective', *Foundations of Physics*, 54 (2024), 56.

1.5 No, We Do Not Live in a Simulation

It is extraordinary that the idea we might be living in a simulation has gained major traction in certain techno circles, presumably inspired by the film *The Matrix*. This idea is physically and biologically absurd. One would need all the emergent biological levels **L1** to **L10** discussed in the Prologue to be simulated, down to the last proton and electron, in every galaxy in the Universe. This requires simulation of the position and dynamic state of about 10^{80} protons in the visible Universe, together with many more electrons and at least 4×10^{84} photons. This is completely unfeasible in engineering terms.

What size is this computer? How much memory does it use for storing data for 10^{85} or so particles and what's the access time? Where does it get its energy supply from? Even simulating a single individual requires simulating the position and dynamics of 10^{27} atoms, subject to both quantum uncertainty and major stochasticity at the molecular level. It is unfeasible in computational terms: how large is the program to handle all this detail? Who wrote it? How can it possibly not contain errors? Why does it not crash every fraction of a microsecond?

Above all, it is an inane proposal which does not solve any physical, biological or philosophical problems. Where does this computer exist, in what universe? How did *that* universe come into existence, and what are its properties? How did the computer come into being? And who is responsible for it?

It's basically a nerd's Intelligent Design argument: fun for late night at the pub, but not a serious cosmological or biological or philosophical proposal. You can safely ignore it.

1.6 Philosophical Issues Regarding Cosmology: The Limits of Science

There is one and only one Universe, and that makes the study of philosophy a unique subject amongst all the sciences. We can't rerun the Universe, or compare it with similar objects, as in all other sciences. We can't look at it from outside. While we can do a great many observations of the visible part of the Universe (§1.1), and experiments on its local components at all levels, we can't do any experiments on the Universe as a whole per se. We can't rerun it again to determine laws of the Universe, or effects of different initial conditions. We can, of course, run many simulations of its dynamics, and indeed do so, but that is not the same. A model isn't the territory. And we can't compare our Universe with any other universe—for no other universe is visible to us. This underlies unique issues that cosmology faces as a science.[28] We can't determine laws for the creation of universes, or for the possible nature of universes as a class.

Science per se cannot explain why the Universe exists, or has the nature it has. These are philosophical issues, not scientific, because we can't do any experiments on why the Universe exists. The underlying issue is that physics itself only came into being when space and time came into existence. There was no physics before the Universe existed: indeed there was no 'before' then!

---

[28] George Ellis, 'On the Philosophy of Cosmology', *Studies in History and Philosophy of Science Part B: Studies in History and Philosophy of Modern Physics*, 46(A) (2014): 5–23.

Claims that the Universe 'came into being out of nothing' are not scientific claims. Specifically, a 'vacuum' is not nothing: it exists in spacetime.[29] We can make models of how and why the Universe came into being, but they are not, and cannot be, scientifically tested theories. They relate to contexts where no scientific laws, or anything else, existed. Even the word 'exists' comes into contention.

1.7 The Sun, Earth and Moon are our Local Astronomical Context

The Sun is a typical star: a thermonuclear reactor transforming hydrogen to helium and other elements up to iron, and giving off solar radiation at a temperature of 5000°C. It is losing mass due to its emission of radiation, and has a finite lifetime: eventually, it will run out of nuclear fuel and end its life. It is vastly larger than the Earth, which orbits it at a distance of 93 million miles with a period of one year. This orbit is determined by the gravitational force, as shown by Isaac Newton, which also keeps our feet on the ground and prevents our atmosphere from floating off into space.

    The Earth (Figure 1.13) is one of a number of planets orbiting the Sun, with two planets (Mercury and Venus) on orbits closer to the Sun and mainly larger planets (Mars, Jupiter, Saturn, Uranus, Neptune and perhaps Pluto[30]) on orbits further from the Sun, as well as a belt of asteroids mainly between the orbits of Jupiter and Mars. The Earth is almost spherical, with a mean radius of 6371 km, but a bulge at the equator due to its rotation. This radius determines the force of gravity G at the surface of the Earth via Newton's gravitational law (Eq. (2.4)). The Earth in turn is orbited by its satellite the Moon, which is much smaller than the Earth. Many of the other planets also have moons circulating them. Earth spins on an axis that is tilted relative to its orbit round the Sun, the spin resulting in the cycle of day and night **(Figure 1.14(a))**. The tilt results in the annual cycle of Earth seasons: summer, autumn, winter, spring **(Figure 1.14(b))**, which are a key feature of our lives and shape much activity.

    The Moon is a satellite of the Earth that circles the Earth with a period of one month at a distance of about 385,000 km (1.28 light-seconds), causing tides that play a key role in marine and aquatic life, and hence in the evolutionary emergence of life on land.

1.7.1 The Earth and Moon: Our Home in the Universe

The Earth provides the main essentials for human existence. It provides our existential needs of air, water and the soils and seasons which are the basis of food production, as well as the resources for modern technological society: energy sources, minerals and renewable resources, enabling production of fertilizers, concrete, steel and plastics that underlie society.[31]

---

[29] David Z. Albert, 'A Universe from Nothing', *New York Times* book review, 23 March 2012.
[30] Famously, its status as a planet has been disputed.
[31] V. Smil, *How the World Really Works: A Scientists Guide to the Past, Present, and Future* (London: Penguin Random House UK, (2022).)

Figure 1.13 The Earth seen from space. Most of its surface is covered by water. The shape of the continents is historically determined, and related to continental drift.

It has land masses (continents) and seas (**Figure 1.13**) surrounded by a very thin atmosphere made mainly of oxygen and nitrogen. Major circulation of the seas is a key driver of weather patterns. This is the geographical basis for our existence, which plays a major role in historical events by constraining interactions between groups (through the configuration of mountains, seas and lakes, rivers, deserts and so on) and governing access to resources.

These range from fresh water, vital for life, plants and trees and animals in ecosystems and on farms, to sand, gravel, rocks and minerals which we use for construction (e.g. silicates, oxides, lime) and technology (e.g. copper), salt, which is vital for nutrition and production of chemicals, oil (used for energy and plastics) and gas (used for energy). Also scarce minerals which we use in technology (antimony, phosphorus, arsenic, boron, gallium, indium) or as a basis for money (gold and silver, valued for their scarcity) or jewellery (diamonds and gems, valued for their appearance), some accessible by open cast mining and others only by deep mine shafts. Economies are built on the basis of access to such resources.

The Earth's interior has many layers and a high-temperature core of molten rock, kept hot by thermonuclear decay that powers continental drift of tectonic plates, still occurring at a very slow rate today. Occasionally this hot material escapes to the surface in volcanic eruptions, disrupting life locally (as in the case of Vesuvius), and some of which (like Krakatoa) have had global effects on weather.

The Earth has various biogeochemical cycles, powered by heat from the Sun, that recycle the elements of life: the water cycle, carbon cycle (key to regulating global temperature cycles), oxygen cycle, nitrogen cycle, calcium cycle and so on. The oxygen, nitrogen and carbon cycles are key to making Earth capable of sustaining life, forming a central element of the biosphere.

(a)

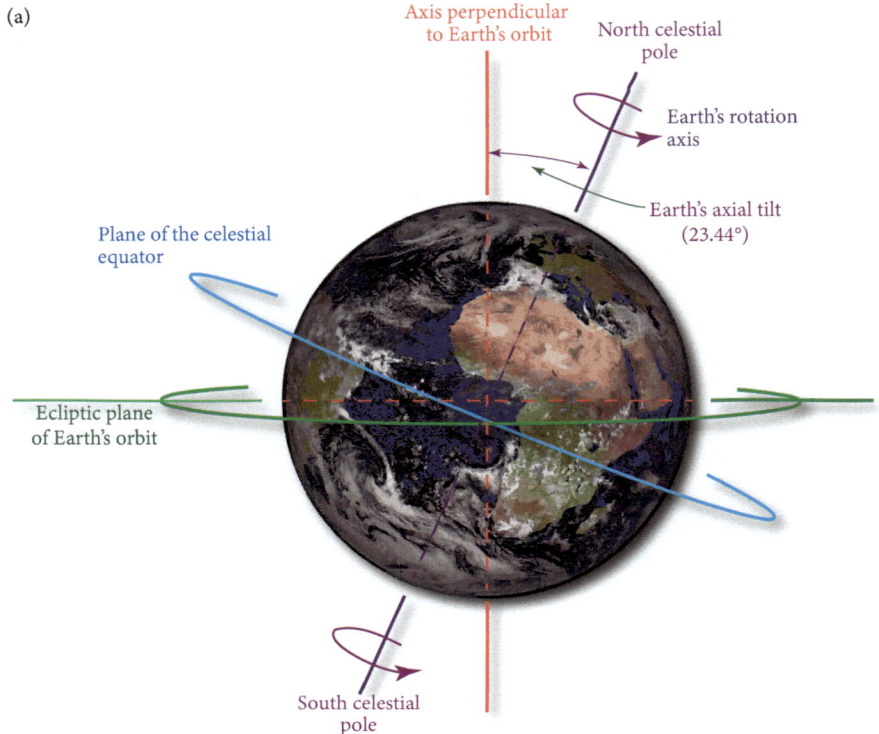

Axis perpendicular
to Earth's orbit

North celestial
pole

Earth's rotation
axis

Earth's axial tilt
(23.44°)

Plane of the celestial
equator

Ecliptic plane
of Earth's orbit

South celestial
pole

Figure 1.14(a) Rotation of the Earth causes the daily day and night.

The Sun provides the further key essential for our existence: solar radiation, which provides all the energy that enables plants to grow and ecosystems to function. Thermodynamically considered, the Earth receives high-grade solar radiation from the Sun and emits low-grade infrared radiation to the sky. Some of this radiation is reflected back to the Earth—the greenhouse effect, without which we could not live because the temperature on the Earth's surface would be too low. However, anthropogenic gas emissions have strengthened this effect so that the amount of reflected radiation is too high, leading to global warming and consequent climate change resulting in floods, droughts, fires and ecological changes.

The Sun–Earth–Moon system is the basis of our physical existence: it is our local astronomical context (**Figure 1.16**). Seen from a great distance, the Earth is a 'pale blue dot', as seen in a famous photo taken from the *Voyager* spacecraft.[32] Carl Sagan wrote in book *The Pale Blue Dot*,[33]

From this distant vantage point, the Earth might not seem of any particular interest. But for us, it's different. Consider again that dot. That's here. That's home. That's us. On it everyone you love, everyone you know, everyone you ever heard of, every human being who ever was, lived out their lives. The aggregate of our joy

[32] The Pale Blue Dot Revisited (Jet Propulsion Laboratory PIA23645).
[33] Carl Sagan, *The Pale Blue Dot* (New York: Random House, 1994).

(b)

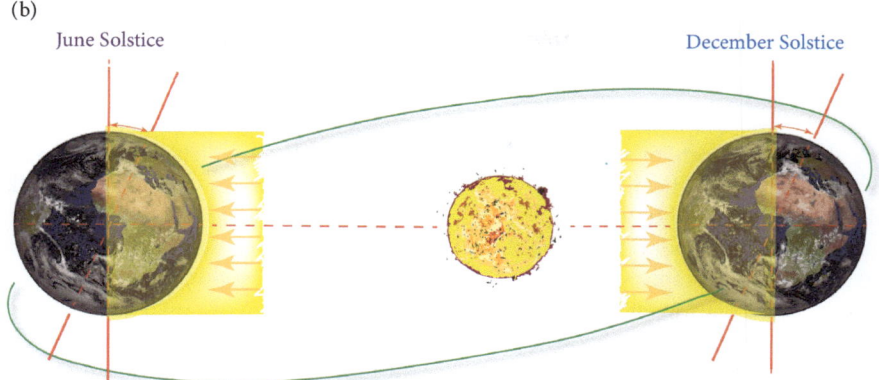

Figure 1.14(b) The cause of the seasons. The tilt of the Earth's axis of rotation relative to its orbit around the Sun results in the annual cycle of summer—autumn—winter—spring. Summer in the northern hemisphere is shown on the left (June solstice), and in the southern hemisphere on the right (December solstice).

Figure by Mauro Carfora.

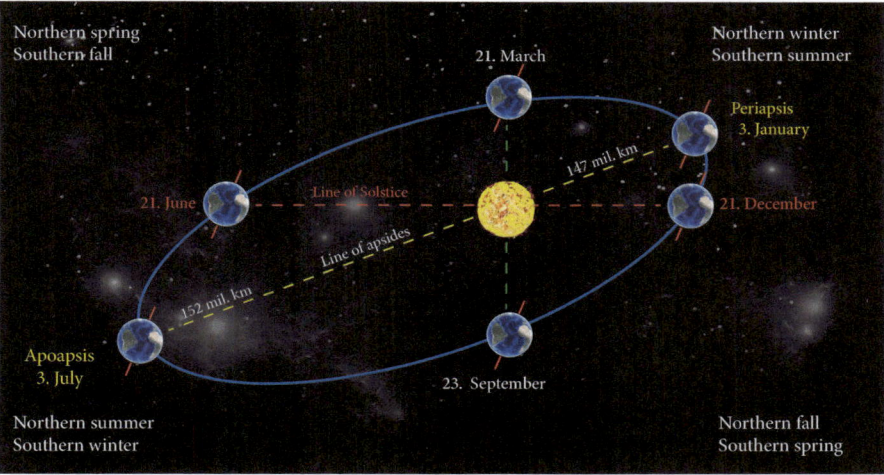

Figure 1.15 Solstices. A winter solstice happens when one of Earth's poles has its maximum tilt away from the Sun and a summer solstice when one of Earth's poles has its maximum tilt toward the Sun. This corresponds to the longest or shortest day or night, respectively.

Figure by Mauro Carfora.

and suffering, thousands of confident religions, ideologies, and economic doctrines, every hunter and forager, every hero and coward, every creator and destroyer of civilization, every king and peasant, every young couple in love, every mother and father, hopeful child, inventor and explorer, every teacher of morals, every corrupt politician, every 'superstar', every 'supreme leader', every saint and sinner in the history of our species lived there—on a mote of dust suspended in a sunbeam.

The Solar System, including the other planets, came into being by astrophysical processes in the expanding Universe; all life on Earth, including humans, then came into being via abiogenesis followed by natural selection.[34] That was possible because the Universe is biofriendly.

No we can't migrate to another planet. We cannot get to very distant planets round other stars because the distances are simply far too large (you can safely discount talks of reaching them by traversable wormholes in spacetime: these are a fiction which cannot exist in physical reality[35]). It would take hundreds of thousands of years to get there. Astonishingly, we now have the technology whereby a few hundred people, maybe even a few thousand, could migrate to other planets in our own Solar System; for example, we could set up a colony on Mars, but conditions are very harsh there. Not many would choose that fate. In any case the numbers of people who could escape this way would make no difference at all to our situation on Earth, with its population of billions: life on Earth would face all the same issues as before irrespective of such migrations.

What might be useful is the ability to mine rare minerals on the Moon, other planets or asteroids: this certainly could be feasible. Apart from this, the existence of the other planets and their moons has no practical significance for our existence: it just provides a basis for imagination and science fiction. But they are a remnant of the processes that led to the formation of the Solar System, including the planet Earth. In that sense they are close relatives. It is the Earth, Sun and Moon that are important for us (**Figures 1.15 and 1.16**).

Figure 1.16 The Earth seen from the Moon: with the Sun, our local astronomical context. The day/night boundary in the Earth's image separates the regions above where it is day because the Sun is shining in the sky above, and the regions below where it is night because the Sun is not visible from there.

Source: Earthrise by astronaut William Anders.

---

[34] Harold Morowitz, *The Emergence of Everything: How the World Became Complex* (Oxford: Oxford University Press, 2002); Jim Baggott, *Origins: The Scientific Story of Creation* (Oxford: Oxford University Press, 2015).

[35] G. Ellis and D. Garfinkle, 'The Synge G-Method: Cosmology, Wormholes, Firewalls, Geometry'. *Classical and Quantum Gravity*, 41 (2024), 077002.

Further Reading

The are many excellent technical books about cosmology. Two I particularly like for broad overviews and physical understanding are these:

Patrick Peter and Jean-Philippe Uzan, *Primordial Cosmology* (Oxford: Oxford University Press, 2013).
Scott Dodelson and Fabian Schmidt, *Modern Cosmology* (New York: Academic Press, 2020).

Here are two good books that are less technical. First,

Steven Weinberg, *The First Three Minutes: A Modern View of the Origin of the Universe*, second edition (New York: Basic Books, 1993).

This focuses on the Hot Big Bang epoch in the early Universe. It contains his infamous statement about the meaninglessness of the Universe towards the end.

An excellent overall book on cosmology, both presenting the physics and considering its broader philosophical side is this:

Edward Harrison, *Cosmology: The Science of the Universe*, second edition (Cambridge, UK: Cambridge University Press, 2001).

For a thought provoking book on the philosophy of cosmology this one is good:

Khalil Chamchan, Joseph Silk, John D Barrow, and Simon Saunders, eds, *The Philosophy of Cosmology* (Cambridge, MA: Cambridge University Press, 2017).

The issue of quantum fluctuations and inflation referred to above is discussed in an excellent book:

Andrew Steane, *Liberating Science: The Early Universe, Evolution and the Public Voice of Science* (Oxford: Oxford University Press, 2023).

Finally a useful discussion of how views of cosmology have changed over time is here:

Joel R. Primack and Nancy Ellen Abrams. 'Cosmic Questions—an Introduction', Annals New York Academy of Sciences, 950 (1999).

2

The Underlying Physics and Chemistry

What is physics? Physics 2024 Nobel Prize winner John Hopfield writes,[1]

> To me—growing up with a father and mother who were both physicists—physics was not subject matter. The atom, the troposphere, the nucleus, a piece of glass, the washing machine, my bicycle, the phonograph, a magnet—these were all incidentally the subject matter. The central idea was that the world is understandable, that you should be able to take anything apart, understand the relationships between its constituents, do experiments, and on that basis be able to develop a quantitative understanding of its behaviour. Physics was a point of view that the world around us is, with effort, ingenuity and adequate resources, understandable in a predictive and reasonably quantitative fashion. Being a physicist is a dedication to the quest for this kind of understanding.

That is the topic of this chapter.

All material entities are made up of atoms—entities so small that we cannot see them with the naked eye. They interact with each other through various forces which determine how they move. Atoms come in a variety of kinds—the chemical elements such as hydrogen, nitrogen, carbon, oxygen, phosphorus and so on. They can be combined to form gases or liquids with no definite shape, as well as crystals and molecules with a specific range of shapes. Atoms in turn are made of smaller particles: protons and neutrons bound in a nucleus with electrons in orbits around them. The nature and interaction of these particles, atoms and molecules is described by physics and chemistry. They are characterized at levels **L1** and **L2** in the hierarchy of existence presented in the Prologue. This is the physical basis of all we see around us.

Forces between atoms result in material things such as rocks, mountains, paper, silk, glass, metals and all living things: grasses, flowers, trees, amoebas, insects, fishes, birds and mammals—including you and me. The great discovery we have made in the past century is this:

> There is no special kind of stuff underlying life. The same kind of matter, described by the periodic table of the elements (Figure 2.22), underlies both living and non-living things. It's just arranged differently in these two cases—with life based on the properties of carbon.

Physics studies how particles and atoms interact to create what we see around us. It has many branches. The three volumes of the famous *Feynman Lectures in*

[1] John J. Hopfield, 'Whatever Happened to Solid State Physics?' *Annual Review of Condensed Matter Physics*, 5 (2014), 1–13.

How We Come to Be. George F. R. Ellis, Oxford University Press. © George Ellis (2026).
DOI: 10.1093/9780198950189.003.0004

Physics[2] deal with mechanics, radiation and light, heat, electromagnetism and quantum mechanics. More applied topics include astrophysics, condensed matter physics, materials science, acoustics, biophysics and the way physics underlies chemistry.

It is a key feature that physics is a quantitative subject, its laws expressed via mathematical equations. As remarked by Galileo[3] in his book *The Assayer*, 'mathematics is the language of nature' (**Figure 2.1**). The behaviour of falling objects, waves, light, heat and so on can be described accurately by suitable equations.[4] To tackle physics fully, therefore, you need a toolkit of a variety of mathematical subjects. However, I will largely avoid equations in what follows.

The nature of physics and chemistry is determined by careful experiments that demonstrate how physical and chemical phenomena occur at a given observational scale and not just someone's opinions about their nature. A key point is that as we move through different energy scales, the corresponding laws of physics and chemistry reflect the experimental reality at that scale to the accuracy our experimental methods allow. Our opinions cannot influence these laws in any way.

The discovery of this impersonal, mathematically describable structure of atoms and forces underlying all of physical existence was a key event in our growing understanding of our physical nature. Human desires and wishes make no difference to this foundation of our existence. However, understanding this underlying basis of

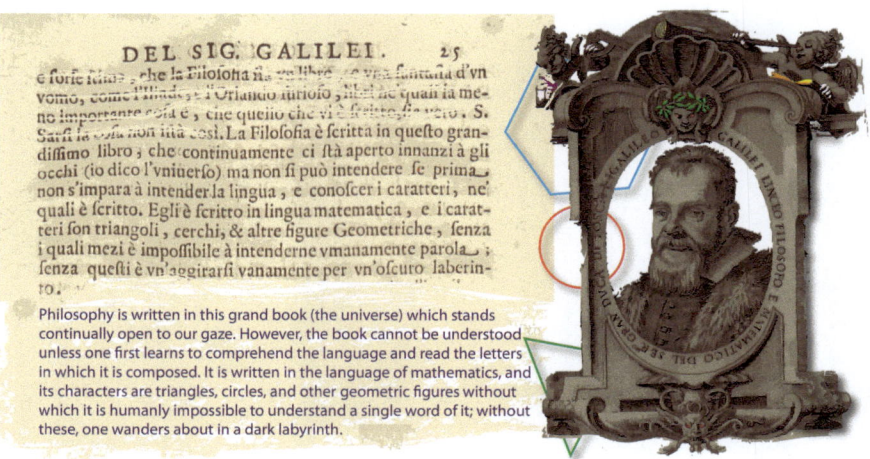

Philosophy is written in this grand book (the universe) which stands continually open to our gaze. However, the book cannot be understood unless one first learns to comprehend the language and read the letters in which it is composed. It is written in the language of mathematics, and its characters are triangles, circles, and other geometric figures without which it is humanly impossible to understand a single word of it; without these, one wanders about in a dark labyrinth.

Figure 2.1 Galileo's comment on the role of mathematics. This is how it appears in the original edition of *Il Saggiatore* (*The Assayer*), published in 1623.

Figure by Mauro Carfora.

[2] Richard P. Feynman, Robert Leighton and Mathew Sands, *Feynman Lectures in Physics*, vol. I: *Mainly Mechanics, Radiation, and Heat*; vol. II: *Mainly Electromagnetism and Matter*; vol. III: *Quantum Mechanics* (1963). All are available online at https://www.feynmanlectures.caltech.edu/.

[3] Galileo's foundational contributions to science are discussed in the Wikipedia article Galileo Galilei.

[4] This role of mathematics is discussed in Eugene P. Wigner's 'The Unreasonable Effectiveness of Mathematics in the Natural Sciences', *Communications on Pure and Applied Mathematics*, 13 (1960), 1–14, and in Richard Feynman's book *The Character of Physical Law* (Cambridge, MA: MIT Press, 1967).

causation allows us to reliably control what happens in many contexts, through our behaviour and careful engineering design. In this way it is the foundation of much of everyday life—indeed, this understanding has transformed our existence.

We can reliably control what happens in everyday physics since we do not need to know the nature of physical laws at every possible scale. Consider, for instance, the physics of skating (**Figure 2.2**), where gravity, the laws of Newtonian dynamics and the rigidity properties of the skate are called into play in a sophisticated way. Yet the ability required for the skater to master these forces does not include needing to being aware of the spatial dependence of the gravitational field, the rotation of the Earth, the tidal gravitational forces exerted by the Moon and the electromagnetic forces responsible for the chemical bonds holding the material of the skateboard together. This information is irrelevant for characterizing the skater's acrobatic movements because the energetic scale involved in skating is very small compared to the scales involved in the spatial dependence of Earth's gravitational field and its planetary motion.

Scale segregation mechanisms of this type provide the underlying basis of causation in everyday physics and engineering. Reliable effective laws emerge at each scale, and one does not have to know how this works, or what the laws at other levels are.

Figure 2.2 Some scales involved in the physics of skating. These include the macro levels of the skater and the skateboard and the scale of the molecular structure of the skateboard. The scale of the Earth is involved because it determines the gravitational field felt by the skater. The scale of the buildings behind is irrelevant.

Figure by Mauro Carfora.

In this chapter, I discuss, in turn, particles, spacetime, forces (§2.1); quantum theory and quantum field theory (§2.2); bulk matter and thermodynamics (§2.3); atoms and the periodic table of the elements (§2.4); molecules, water, crystals, macromolecules (§2.5); the material basis of daily life (§2.6); and outcomes in engineering and society (§2.7).

2.1 Particles, Spacetime, Forces

Physics is based in the relation between particles, forces and fields. Forces make particles move; fields generate forces. The particles that concern us in everyday life, in order of increasing size, are electrons; protons and neutrons; atoms; and molecules. The forces that concern us in everyday life are the gravitational force and the electromagnetic force. They dominate what happens dynamically around us. The associated fields determining these forces fill space, but are invisible; however, we can experimentally show they are there because of their effects. Electromagnetic fields that you cannot see convey messages to your cell phone. The gravitational field is the reason that the gravitational force can act at a distance.

Particles. Everything physical is made of atoms, including you and me. This is not obvious. The air in a room seems continuous, as does glass or a metal. But solids and liquids and gases are made of individual atoms. In a gas, atoms are combined to form billions of molecules buzzing around in space; in a metal, atoms are vibrating in a lattice. There are a vast number of atoms in every object you look at—including ourselves. They are so small that you cannot see them, but their properties are very well understood. In classical terms, atoms are mostly empty space. They can be represented as being made of a nucleus consisting of positively charged protons and uncharged neutrons, and a cloud of much lighter and smaller electrons orbiting the nucleus (**Figure 2.3(a)**). However, they really have a much more diffuse nature (**Figure 2.3(b)**). Individual atoms are too small to image (they are much smaller than the scale of the electromagnetic waves that constitute light (**Figure 2.20**)) but arrays of atoms can be imaged (**Figure 2.3(c)**).

Particles have mass, position, momentum and electric charge (positive, negative or zero). A particle's motion is determined by the forces acting on it, as described by Newton's three laws of motion, illustrated in **Figure 2.4**. The first law is Galileo's principle of inertia, stating that in the absence of external forces, a body moves in a straight line at a constant speed. Galileo describes this basic observation in his *Dialogue Concerning the Two Chief World Systems* (1632) as a famous thought experiment involving the status of rest or uniform motion of a ship.

Mathematically, the second and third laws are

$$\textit{Force} = \textit{mass} \times \textit{acceleration} \Leftrightarrow \textit{acceleration} = \textit{Force}/\textit{mass}; \quad (2.1)$$

If two bodies interact, the forces exerted by each on the other

have the same magnitude but have opposite direction. $\quad (2.2)$

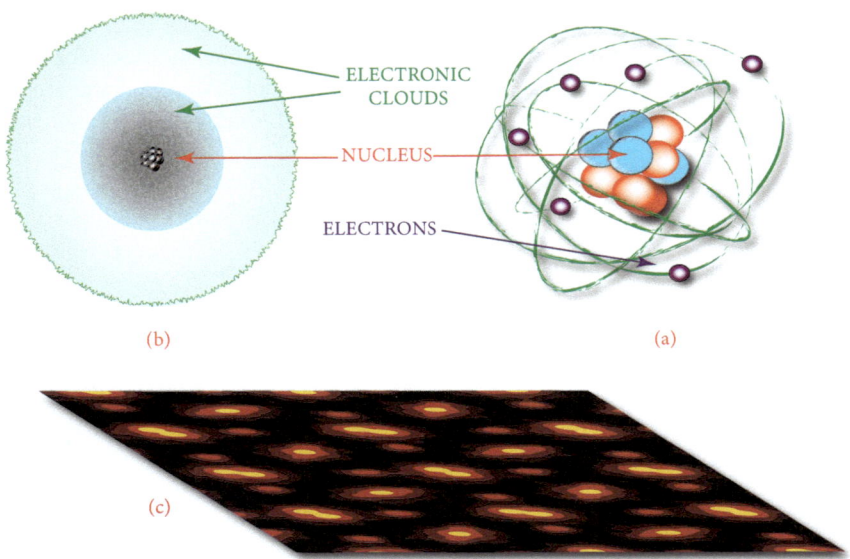

ELECTRONIC CLOUDS

NUCLEUS

ELECTRONS

(b)

(a)

(c)

Figure 2.3 Representing atoms. We typically depict an atom (here, the carbon atom) as a miniature solar system with the atomic nucleus at the centre and electrons orbiting around (a). As familiar as this picture can be, it is pretty inaccurate. A better rendering is provided by (b), where the electrons are not described as orbiting but rather as localized, in the sort of fuzzy way dictated by quantum mechanics, on electronic clouds. Yet, inaccuracies are still present in (b), since the representation of the nucleus is not to scale; since the atom is mostly empty space, the nucleus should be a tiny dot. Finally, (c) is an accurate image of the array of praseodymium, scandium and oxygen atoms in a praseodymium-orthoscandate crystal. [5]

Figure by Mauro Carfora.

Thus, if a car crashes into a wall, both the car and the wall are damaged. The second law implies the first, which is the special cases when there is no force acting (so the acceleration is zero).

According to the law of inertia, the laws of Newtonian mechanics obey the Galileian relativity principle, according to which time is an absolute quantity independent of the status of motion of the observer, and there is no absolute standard of rest (even if Newton himself believed that absolute space existed). Newton's laws of motion are compelling and predictive, but they also contain the seeds that lead to the modern interpretation of the law of inertia in special and general relativity.

Conservation of energy, momentum and matter. Summing up local interactions of all the particles in a physical body and using Eqs. (2.1) and (2.2), it follows that if no forces act on the body, its energy and momentum will be conserved (**Figure 2.5**). Furthermore, matter is conserved in everyday life: the mass of an isolated body does not change.

[5] Zen Chen et al. 'Electron Ptychography Achieves Atomic-Resolution Limits Set by Lattice Vibrations', *Science*, 372 (2021), 826–31. DOI: 10.1126/science.abg2533

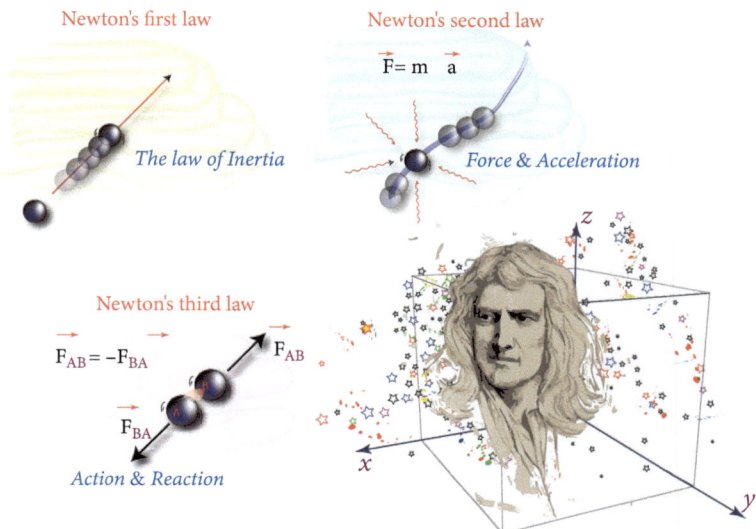

Figure 2.4 Newton's laws of motion. The law of inertia characterizes inertial observers as those who feel no effects of motion. Newton's second and third laws characterize the laws of motion of material bodies as described by the inertial observers.

Figure by Mauro Carfora.

However, bodies do interact, and energy can be interchanged between various forms (gravitational, kinetic, electric, thermal, chemical). **Figure 2.6** is a rough model of what happens in everyday life when two identical cars (A and B), each of mass m traveling at the same constant speed, crash into each other, generating a metallic car wreck (D) at rest at the location of the crash. The mass of the wreck M is the sum of the masses m_A and m_B of the crashing cars, i.e. $M = 2m$, expressing the conservation of matter.

Since the wreck (D) remains at rest after the crash, the kinetic energy of the two cars A and B is transferred to D not as kinetic energy, but rather is transformed into heat, the most degraded form of energy. Its total energy before the crash is just provided by the total kinetic energy of the cars. After the crash, this total kinetic energy is interchanged into an equal amount of thermal energy, which is lost as it is radiated away into the sky as heat.

Spacetime. The first indication that the foundations of classical physics are not what they seem was a key discovery by Albert Einstein.[6] The laws of electromagnetism governed by Maxwell equations do not comply with Galilean relativity. For instance, the usual composition of velocities, so familiar in our everyday experience, does not hold for light signals (**Figure 2.7**). The velocity c of a light signal emitted by an inertial observer in motion is always c for any other inertial observer.

---

[6] Einstein's truly remarkable contributions to science are discussed in the Wikipedia article Albert Einstein.

Figure 2.5 Conservation of angular momentum. A well-known example of a conservation law in everyday physics is provided by figure skating jumps. While moving along a straight trajectory, the skater starts slowly spinning around her vertical body axis, with her arms extended, then jumps, gently pulling her arms in and entering a faster spinning mode. The conservation of angular momentum is the physical law responsible for this pleasant effect of figure skating jumping. Since the ice skate's blades experience low dynamical friction on the surface of the ice rink, momentum conservation is also at work here, and when the skater jumps off-ice, entering in the faster spinning mode, she covers a considerable horizontal trajectory exploiting the (horizontal) momentum she had at the instant of the jump.

Figure by Mauro Carfora.

In addressing this asymmetry between Newtonian physics and electromagnetism, Einstein realized that a relativity principle associated with the law of inertia cannot hold just for Newtonian mechanics; it must be valid for all physical phenomena. This extension of Galilean relativity led to a profound revision of the very nature of space and time as understood in classical physics. To make sense of the invariance of the speed of light, space (three-dimensional) and time (one-dimensional) are not immutable separate entities, as seems to be the case. They are part of a single four-dimensional entity, spacetime,[7] and the way that different observers measure spatial distances and time differences depends on their motion relative to each other. In particular what seems instantaneous to you is not the same as what seems to be instantaneous to me.

This geometry of spacetime leads to the often debated twins paradox (**Figure 2.8**). One twin stays at home doing theoretical exoplanetary research. Meanwhile, the other twin embarks on space travel to reach a candidate exoplanet, make an observational

---

[7] See George Ellis and Ruth Williams, *Flat and Curved Spacetimes* (Oxford: Oxford University Press, 2000).

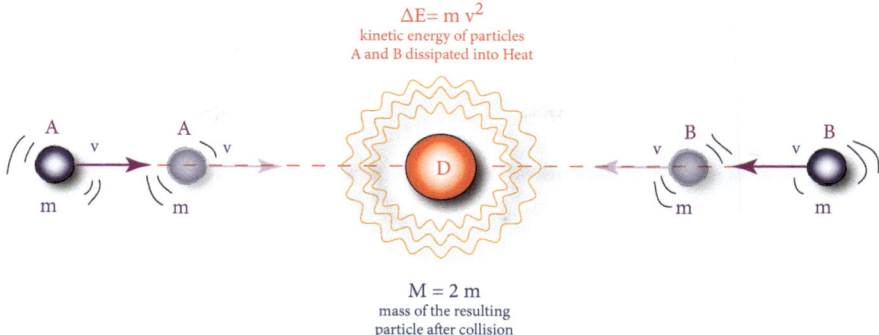

$$\Delta E = m\, v^2$$

kinetic energy of particles
A and B dissipated into Heat

M = 2 m
mass of the resulting
particle after collision

Figure 2.6 Particle collisions. The anelastic point collision of two particles A and B of equal masses, $m(A) = m = m(B)$, moving toward each other at constant speed v, illustrates both the matter conservation law and the way that energy can be interchanged between various forms.

Figure by Mauro Carfora.

Inertial observer in uniform rectilinear motion at a speed U emitting a light signal whose speed in vacuum is c

Inertial observer at rest receiving the light signal. He still measures c for the speed of the light signal and not c+U as expected according to everyday physics

Figure 2.7 Invariance of the speed of light. A moving observer measures exactly the same speed of motion of light **c** as a stationary one, when moving at speed U relative to each other. However, the observer at rest expects the moving observer to measure the light speed c' relative to him to be $c' = c + U$.

Figure by Mauro Carfora.

flyby around the planet and then come back to Earth to report her findings. The identical twins move on separate paths in spacetime.

The one staying at home moves along a straight spacetime path (if we consider the Earth in inertial motion, a good approximation in the situation described), while the other moves rapidly away on a rocket, describing a curved spacetime path. First, she accelerates to reach the exoplanet, makes an exploration flyby and then uses the rocket engine to decelerate and return to Earth to meet her twin and report the gathered data. Not surprisingly upon their meeting, the twins are now of different ages: the one who travelled far away will be younger. The distinction between the two lies in the acceleration caused by the rocket engine that made the traveling twin travel from Earth to the exoplanet and back. The stay-at-home twin never experienced such an acceleration, thus never altering her inertial status.

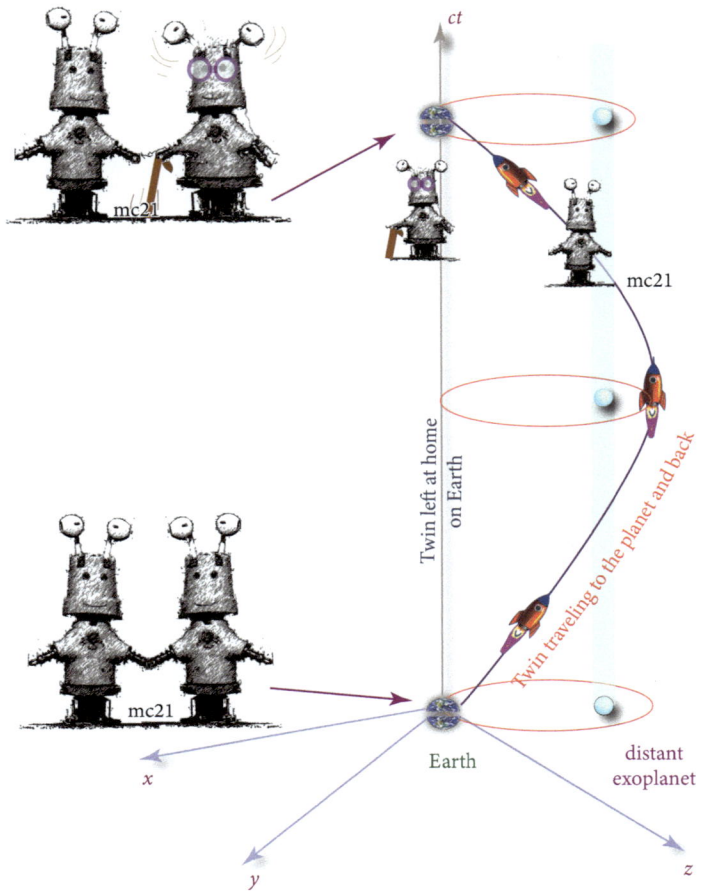

Figure 2.8 The twins paradox. Identical twins involved in exoplanet exploration have different ages when they meet again.

Figure by Mauro Carfora.

This is a foundational re-evaluation of the nature of space and time: the way space-time splits into space and time depends on how an observer moves. But of itself this has no consequences for daily life: we move far too slowly relative to each other for these effects to be discernable. The difference is very small when the speed v of relative motion is small relative to the speed of light $c = 299{,}792{,}458$ metres per second (a figure often approximated as $c = 3 \times 10^8$ m/sec) and so is undetectable in daily life.

The constancy of the speed of light has been verified with extreme accuracy, so that is now used to redefine the centimetre in terms of the second:

$$1 \; centimetre := The \; distance \; travelled \; in \; vacuum \; by \; light \; in \; (1/2.99792458)$$

$$\times \; 10^{-10} \; seconds.$$

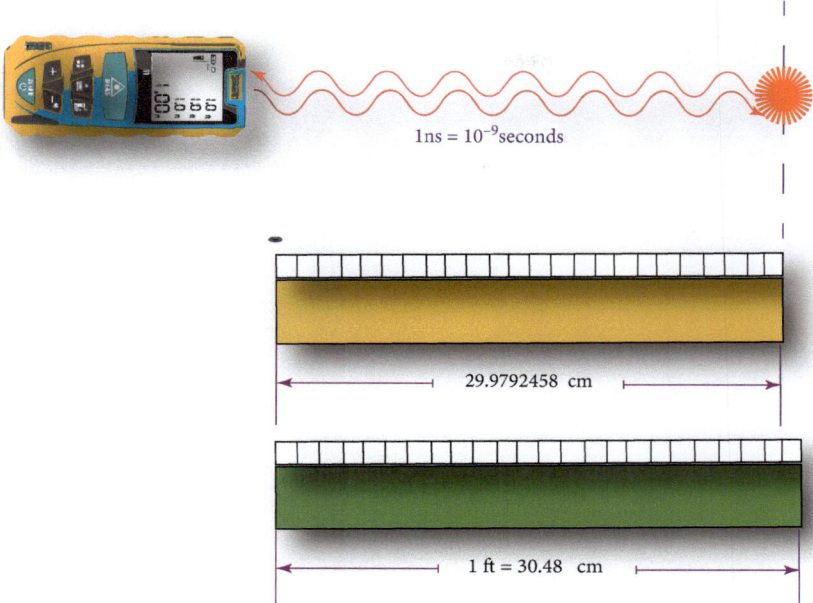

$1\text{ns} = 10^{-9} \text{seconds}$

29.9792458 cm

1 ft = 30.48 cm

Figure 2.9 A laser meter, *used to measure distances at household scales.*

Figure by Mauro Carfora.

Thus, roughly speaking, a light signal travels 1 foot (30.48 cm) in one nanosecond (a billionth of a second). Nowadays we can indirectly experience this fact in daily life by using a laser meter to measure distances (**Figure 2.9**), such as the few feet separating us from a wall to see if a piece of furniture fits into a corner .

A laser meter sends out a finely focused pulse of light to measure small distances. They have electronics based on a high-speed photodiode that measures the time laser pulses take to travel to the target and get reflected back to the photodiode. They can measure time intervals on the order of a nanosecond, so you can experience the speed of light at work in your home. But more than this, all electromagnetic waves travel at the speed of light (see **Figure 2.19**), so this is the way that aircraft use radar at radio or microwave wavelengths[8] to detect other aircraft and hazardous weather.

What matters very much is that Einstein showed that because of special relativity, matter and energy are equivalent. They can be transformed into each other according to the famous equation

$$E = m\ c^2. \tag{2.3}$$

Because of the very large value of the speed of light c, a very small change in mass equals a very large change in energy. This underlies existence of nuclear weapons, nuclear power stations and stellar nucleosynthesis in the Sun. Nuclear reactions such

---

[8] See the Wikipedia article Radar.

Figure 2.10 Gravity and the Solar System. Gravity keeps the Earth and other planets in their orbits around the Sun.

Figure by Mauro Carfora.

as converting hydrogen to helium make the Sun shine, and so provides the energy for life on Earth. We would not be here if that were not the case.

Forces. By Eq. (2.1), the motion of particles is determined by the forces acting on them, or equivalently, by related fields, such as the gravitational field. The four fundamental forces are gravity, electromagnetism, the strong force and the weak force. I discuss them in turn.

Gravity. Isaac Newton's great discovery[9] was that the gravitational force **F** between any two objects is an attractive force of magnitude

$$\boldsymbol{F} = GM_1M_2/R^2, \tag{2.4}$$

where M_1 and M_2 are the masses of the objects, R is the distance between them and G is the gravitational constant. This is an inverse square law because of the $1/R^2$ factor. Combined with Eq. (2.1), this single force is what makes apples fall to Earth, keeps the Moon in its orbit around the Earth (**Figure 2.10**) and keeps the planets in their orbits around the Sun. Understanding that this is the case was the first great scientific unification.

A key point is that the masses M are always positive: gravity in the Solar System, and on Earth, is always attractive. Anti-gravity ideas do not work. If they did, you could use them to generate perpetual motion, which is not physically possible.

As to life on Earth, because of the large mass of the Earth, gravity attracts us toward its centre according to Eqs. (2.1) and (2.4). Thereby it is a dominant force on Earth in architecture and engineering and biology and sport. We experience gravity as pulling us toward the surface of the Earth, underlying standing, sitting and falling, as well as walking, running, climbing and gymnastics. It determines what is locally up and what is down, which is opposite in Australia and England (see **Figure 2.11**).

---

[9] See the Wikipedia article Isaac Newton.

Figure 2.11 Gravity determines what is locally up and down on Earth. That is why up and down can be in opposite directions on opposite sides of the Earth. The force of gravity at any point depends on the mass M of the Earth and the distance R of the place in question from the centre of the Earth. It's a little bit less on the top of a mountain.

Figure by Mauro Carfora.

But more than that, gravity leads to the very existence of the Solar System, firstly holding together each of the Sun, Earth, Moon and other planets (the particles they are made of don't just disperse into space—they are stable objects), and secondly leading to the planets (including Earth) circulating the Sun in stable orbits. It also leads to the Moon circulating the Earth in a stable orbit, causing the tides we experience (**Figure 2.12**). It is a key force shaping the context in which we live.

In 1916, Newton's gravitational theory was replaced by a more accurate theory (**Figure 2.13**): Einstein's theory of general relativity, ascribing gravitational effects to the curvature of spacetime (**Figure 2.14**).[10] It underlies the existence of black holes, with their strange properties, and gravitational radiation, which has recently been detected.

However, these phenomena don't affect our daily lives—Newton's gravitational theory remains an excellent approximation for most engineering and biological contexts. The one significant effect of general relativity in daily life is that it plays a key role in the accurate functioning of GPS navigational systems, because of the effects of spacetime curvature on the passage of time (**Figure 2.15**).[11]

Electromagnetism. Electric and magnetic fields are generated by static and moving electric charges q, which can be positive or negative. Static electric charges attract each other with an inverse square law like Eq. (2.4): the electrostatic force for electric charges q_1, q_2 is

$$F = -k_0\, q_1\, q_2/R^2, \tag{2.5}$$

[10] See Ellis and Williams, *Flat and Curved Spacetimes*.
[11] See the NASA article 'Einstein's Theory of Relativity, Critical for GPS, Seen in Distant Stars' (22 October 2020).

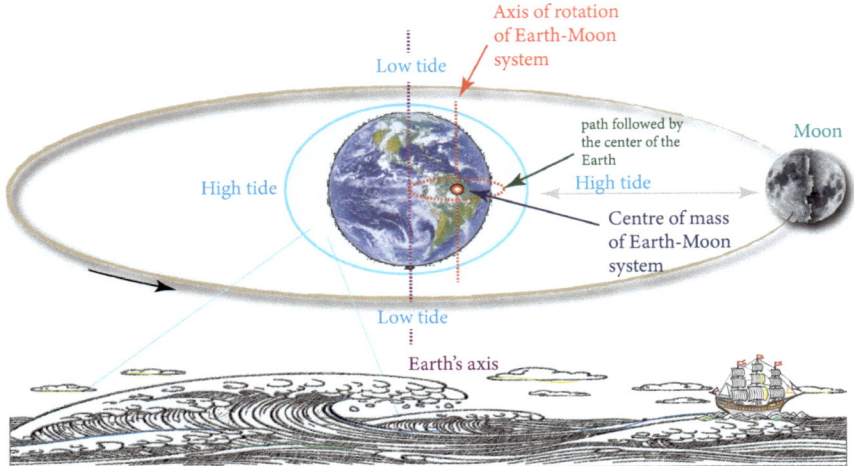

Figure 2.12 Tidal force caused by the Moon. High tides occur on Earth directly under the Moon, but also on the opposite side of the Earth. This is because the Moon attracts not only the water in the oceans, but also the Earth's centre of gravity in accordance with equation (2.4).

Image: Mauro Carfora.

with electrostatic constant k_0. The minus sign is crucial: unlike gravity, like electric charges (with the same sign) repel each other, and unlike charges (with opposite signs) attract each other (**Figure 2.16**).

Electrostatic attraction between oppositely charged particles underlies many engineering and biological effects, for instance underlying the existence of atoms, molecules and chemical bonding, as well as the existence of crystal structures such as metals and rocks. It is the key force underlying chemistry and so the periodic table of elements.

Magnetic fields are generated either by permanent magnets or by solenoids—coils of wire with currents running through them. The magnetic field of the Earth is detected by compasses (magnetized needles), which determine the direction to the magnetic north or south poles. If an electric charge moves relative to a magnetic field, it experiences a Lorentz force at right angles to the direction of motion.

The relativity of motion plays a subtle role in electromagnetism, as illustrated in **Figure 2.17**. On the right, we have a conductor moving in uniform rectilinear motion, with velocity $-u$, relative to a stationary magnet. The resulting Lorentz force causes the free electrons in the conductor to move and thus generate a current. On the left, we have a stationary conductor surrounded by a time-varying magnetic field generated by the magnet moving with velocity u. In this case, the magnetic field generates an electromotive force in the conductor, giving rise to a current. The current is the same in both cases, which is not surprising since what matters is simply the relative motion of the magnet and the conductor.

Yet, the explanations of the origin of the same effect (the current) are quite different. This paradox led Einstein to his revolutionary rethinking of the role of the relativity

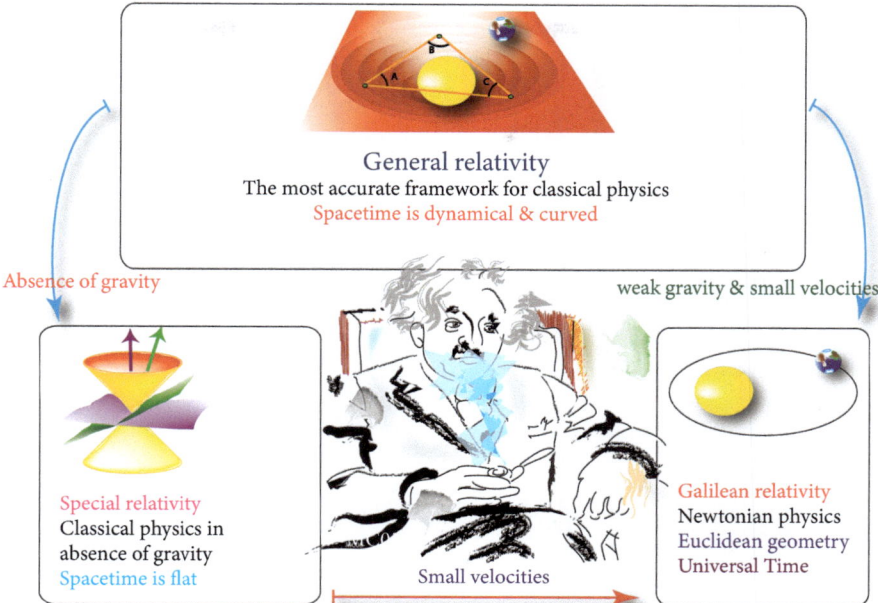

Figure 2.13 The relativity landscape. General relativity, special relativity and Galilean relativity are connected according to the illustrated flow chart. The top box illustrates spacetime curvature as responsible for gravity. The box on the left illustrates special relativity with the lightcone describing the peculiar role played in physics by c, the speed of light in vacuum. The box on the right represents Galilean relativity's role in Newton's gravitation theory.

Figure by Mauro Carfora.

of motion in physics. A sophisticated form of causation is at work here. What are the causes of the above dissymmetry between relative motion and electromagnetic phenomena? The time-honoured Galilean relativity principle, quite effective in Newtonian mechanics, does not hold for Maxwell's electromagnetism since this latter is not Galilean invariant. However, it turns out that the electromagnetic laws are the ones that comply with the correct relativity principle, one that considers all inertial frames (and local inertial frames, in the case of gravity) equivalent for the formulation of the laws of physics (and not just those of Newtonian mechanics).

In nonstatic contexts, electromagnetic fields interact in a complex way: each generates the other when time change takes place, as described by Maxwell's equations.[12] This underlies electric generators and relays and motors, resulting in electric lighting, stoves, refrigerators, washing machines and so on. Nature has discovered this, and used it in our nervous system, where it governs action potential propagation in nerves.

---

[12] See the Wikipedia article Maxwell's equations.

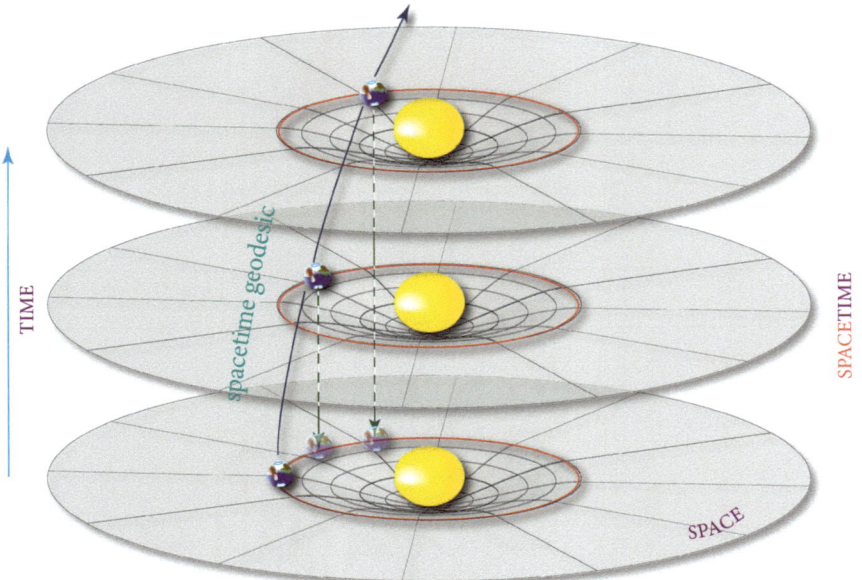

Figure 2.14 The Earth orbiting the Sun according to general relativity. The Sun generates a (tiny) curvature of spacetime. In this curved spacetime, the Earth follows a geodesic worldline slightly bent with respect to a straight worldline. Yet this tiny spacetime curvature gives rise to the elliptical closed orbit that the Earth follows in three-dimensional space.

Figure by Mauro Carfora.

This is what enables us to act via the motions of muscles, and to think via our brain's neural networks (**Figure 2.18**).

Maxwell's momentous discovery was that his equations for electromagnetism predicts that electric and magnetic fields can generate transverse waves moving at the speed of light c (**Figure 2.19**).

Maxwell[13] then realized this meant that light was indeed an electromagnetic wave, with the colour of the light depending on frequency, which is inversely proportional to wavelength. This in turn meant that other wavelengths are possible for electromagnetic radiation: radio waves (long wavelengths), infrared, visible light enabling vision, ultraviolet, X-ray and gamma ray (very short wavelengths), forming the entire electromagnetic spectrum (**Figure 2.20**). These are all forms of electromagnetic radiation. This remarkable realization has transformed human life by enabling radio, television, radar, wifi, iPhones, remotes, CAT scanners and so on. It is a key feature of daily life and technology.

The strong force. The strong force holds atomic nuclei together, despite the repulsive force between the protons in the nuclei. It is an aspect of why matter can be stable, and so is crucial to life.

---

[13] See the Wikipedia article James Clerk Maxwell.

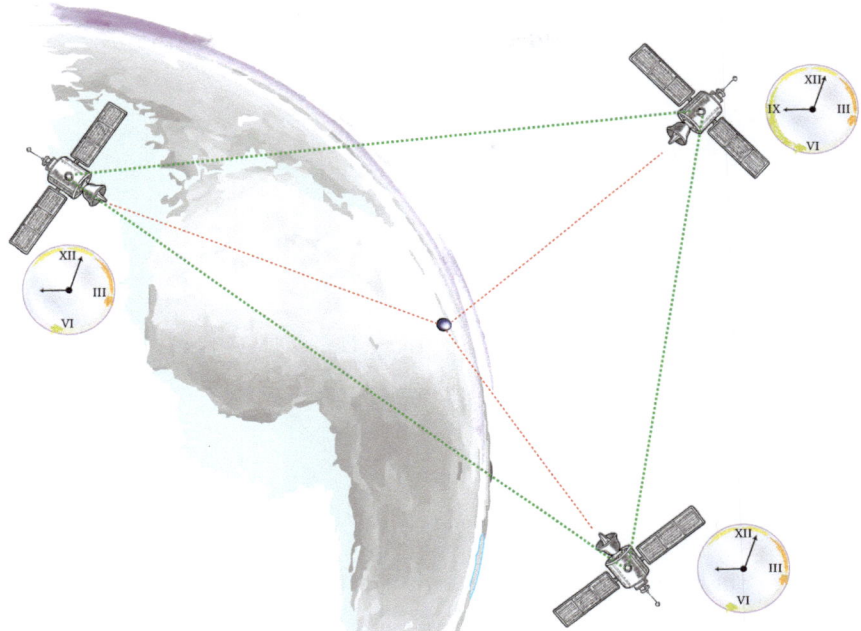

Figure 2.15 The GPS navigational system and relativity theory. The GPS needs accurate clocks to work. Atomic clocks are used. They are so accurate that they are sensible to the special and general relativistic effects, respectively, due to the satellite speed and the extremely tiny spacetime curvature generated by the Earth.
Figure by Mauro Carfora.

The weak force. The weak force governs radioactive decay of unstable elements, and is crucial to internal processes in the Sun. These give off neutrinos as part of nucleosynthesis processes that produce sunlight that is crucial for life. Apart from that, it plays no role in daily life on Earth or biology.

Unification of forces and the standard model. In a tour de force, physicists have shown that the electromagnetic force and the weak force are two aspects of a deeper theory: electroweak theory. This leads on to the well-established Standard Model of particle physics,[14] in which we understand neutrons and protons to be made up of smaller fractionally charged particles called quarks. Even more ambitious attempts to unify the electroweak force with the strong force to obtain **a Grand Unified Theory**, or GUT, have not succeeded so far.

Even more ambitious are plans to unify all these forces with gravity—a so-called **theory of everything** (TOE), of which the most famous is string theory/M-theory. This assumes existence of a symmetry between particles and anti-particles known as

---

[14] R. Oerter, *The Theory of Almost Everything: The Standard Model, the Unsung Triumph of Modern Physics* (London: Penguin, 2006).

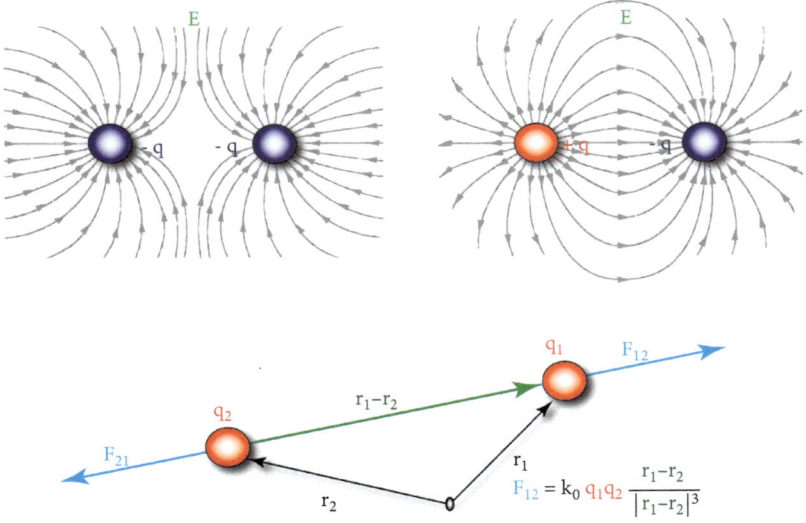

Figure 2.16 **The electric field and the electrostatic force.** The electric field between equal (top left) and opposite (top right) charges. The electrostatic force (in vector form) is illustrated in the lower part of the figure; its intensity is given by the formula (2.5).

Figure by Mauro Carfora.

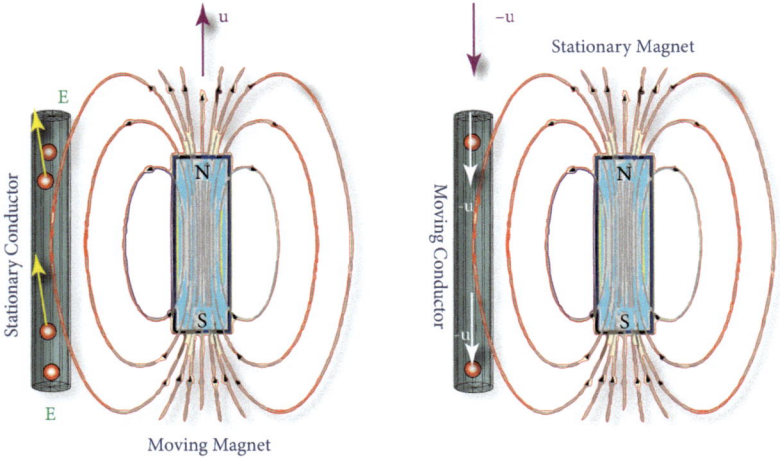

Figure 2.17 **Electromagnetism and the relativity of motion.**

Figure by Mauro Carfora.

supersymmetry, and that spacetime is actually of ten or more dimensions, with all but four dimensions curled back on themselves so tightly so that their existence is indiscernible. No experiment has been able to show that this is actually the case. This

Figure 2.18 The electromagnetic field and nature. Electromagnetic fields are ubiquitous in nature; think of light (the proof that light is an electromagnetic phenomenon is Maxwell's unification of optics and electromagnetism) and of the dramatic atmospheric manifestation in the form of lightning storms. But the most subtle presence of electromagnetic phenomena manifests itself, in an extremely organized aspect, in life processes, here represented by the electric signals propagating in the axion of a brain neuron. [15]

Figure by Mauro Carfora.

theory is not well defined, is not experimentally verified and indeed has not yet been shown to lead to the Standard Model of particle physics, as must be the case if it is a correct theory.

The physics that shapes our lives is not affected by these underlying debates. So if this is all uncertain, how can physics say anything significant about daily life and our existence? The key point is the existence of what Laughlin and Pines[16] label **quantum protectorates:** daily life is independent of the details of quantum physics:

> The crystalline state is the simplest known example of a quantum protectorate, a stable state of matter whose generic low-energy properties are determined by a higher organizing principle and nothing else.

---

[15] The neuron image is the Mauro Carfora's elaboration of an original image from 'Mind the Graph', used under the CC BY-SA license (https://creativecommons.org/licenses/by-sa/4.0/deed.en).

[16] Robert B. Laughlin and David Pines, 'The Theory of Everything', *Proceedings of the National Academy of Science*, 97 (2000): 28–31.

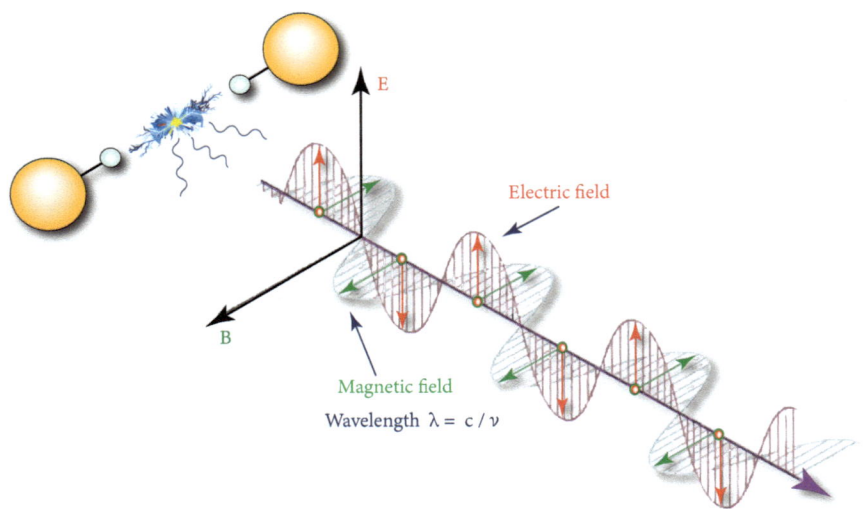

Figure 2.19 Transverse electromagnetic wave. Depicted is a plane electromagnetic wave moving to the right at the speed of light c with wavelength λ and frequency ν (the Greek letter 'nu'). The electric and magnetic components of the wave are at right angles to each other and to the direction of travel, and are in phase with each other in this case: that is, each is maximum at the same time and zero at the same time. A complex wave is made up of combinations of such simple waves.

Figure by Mauro Carfora.

Thus the physics that actually affects us does not demand details of the Standard Model of particle physics, much less details of smaller scales. It just requires existence of positively charged protons, neutral neutrons and much smaller and lighter negatively charged electrons. The interactions between them is characterized by the Schrödinger and Dirac equations of quantum theory (§2.2). What underlies this at a deeper level does not matter, as far as daily life is concerned.

2.2 Quantum Theory and Quantum Field Theory

At a particle and atomic level, quantum physics comes into play that is highly nonintuitive because it is quite unlike classical physics. It was a very surprising discovery, through careful experiments, that physics at the underlying atomic/particle level is quite unlike how it is at the macroscopic everyday level.

Quantum theory. Matter and energy and light are all quantized: they come in discrete units. This is demonstrated in the case of light by the photoelectric effect, showing light behaves as made of particles (photons) when it hits a metal surface and ejects electrons, as in the case of photovoltaic cells, or hits a chlorophyll molecule in a leaf, enabling photosynthesis in plants, or a rhodopsin molecule in an eye, enabling us to see.

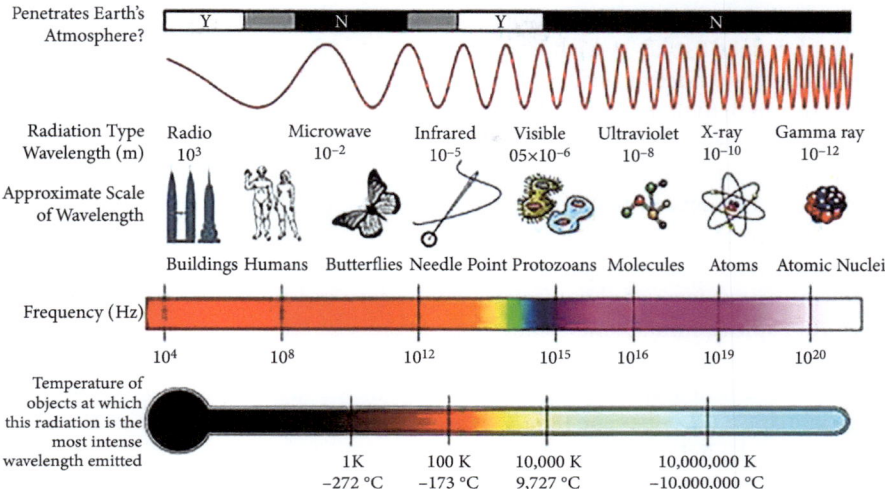

Figure 2.20 The electromagnetic spectrum. Long wavelengths on the left, short on the right; high frequencies (cycles per second) on the right, low frequencies on the left.

Source: Wikipedia, Electromagnetic spectrum.

Quantum states can be described in various ways; a common one is via a wavefunction |Ψ> whose time variation is described by the Schrödinger equation.[17] The uncertainty principle (Heisenberg) states that the position and momentum of a particle cannot be prescribed to arbitrary accuracy. This puts limits on the accuracy of possible predictions.

Superposition of quantum states occurs, where a state can be considered as being partly in each of two or more discrete states. There is no classical correspondence to this situation. This is resolved by **wavefunction collapse**, when a specific classical outcome results—but only the probabilities of such outcomes are determined by quantum theory; the particular outcomes are inherently indeterminate. Thus radioactive decay of excited atoms is unpredictable in principle, as is the position of a particle on a screen after passing through double slits. This uncertainty applies to any quantum process: only probabilities of outcomes can be predicted from the wavefunction, not the result of any particular initial state. In particular this applies to measurements in a laboratory.

How the associated quantum to classical transition takes place is the subject of much contestation. My view is that **contextual wavefunction collapse** occurs.[18] **Entanglement** occurs, where the quantum state of each particle in a group cannot be described independently of the state of the others. But this rapidly decays due to decoherence: it's a very temporary state except under very special conditions.

---

[17] See the Wikipedia article Schrödinger equation.
[18] Barbara Drossel and George Ellis, 'Contextual Wavefunction Collapse: An Integrated Theory of Quantum Measurement', *New Journal of Physics*, 20 (2018), 113025.

Tunnelling occurs: particles with energy E_1 can tunnel through an energy barrier with height $E_2 > E_1$, which is impossible classically. This underlies tunnel diode functioning and nuclear fusion.

Quantum field theory extends quantum theory to a more fundamental theory that is compatible with special relativity. In this case, fields are more fundamental than particles: a particle is a state of a quantum field,[19] and so creation and annihilation of particles can occur when the quantum field changes its state. It famously predicts the existence of **antimatter**: each kind of particle has an antiparticle; for example, the antiparticle of an electron is a positron. They can annihilate each other, giving off energy in the form of photons. Both elementary and composite particles can have **quantum spin**, a fundamental feature of matter.

Integer spin particles (bosons), such as photons, can be in the same state. This leads to **Bose–Einstein statistics**, resulting in **black body radiation**, which to a good approximation describes the light we receive from the early Universe at 2.73 K (see Chapter 1), the light we receive from the Sun at 5500 K and the radiation the Earth re-emits to space at 15°C. It also underlies the functioning of lasers.

Spin-half particles (fermions) such as protons, neutrons and electrons are described by the **Dirac equation**[20] and obey the **Pauli exclusion principle** which states that two particles cannot be in the same state. All known fermions, the particles that constitute ordinary matter, have a spin of 1/2. This leads to **Fermi–Dirac statistics**, which underlies both the stability of matter and the nature of the chemical elements as described in the periodic table of the elements (**Figure 2.22**). Thus it is crucial to our existence. **Quantum electrodynamics** is the form of quantum field theory that extends Maxwell's theory of electromagnetism to the quantum domain.[21]

The domain of quantum theory. When I was a student, it was taken for granted that quantum theory and quantum field theory applied only to very small scales: those of electrons, protons, atoms, and molecules.

A strange thing has happened since then: it has been taken for granted by many practitioners that quantum theory applies also to cats as a whole, brains and even the entire Universe. This leads to the Everett **many-worlds interpretation** of quantum physics with no wavefunction collapse, and all possible outcomes of quantum measurements are physically realized in some universe.[22]

There is even talk of many minds: there are billions of identical copies of you out there, experiencing identical or slightly different events. There is, of course, not the slightest evidence that any of this is true, despite what enthusiasts may say.

Where did things go wrong? The key point is that the Schrödinger and Dirac equations are linear equations in terms of the wavefunction $|\Psi\rangle$—but the real Universe is highly nonlinear. This is demonstrated by the hierarchical structures

---

[19] See David Tong's 2017 Royal Institute lecture, 'Quantum Fields: The Real Building Blocks of the Universe'.
[20] See the Wikipedia article Dirac equation.
[21] See the Wikipedia Quantum Electrodynamics.
[22] See the Wikipedia article Many-Worlds Interpretation.

Figure 2.21 There is no single wavefunction for a cat. Schrödinger himself considered the whole cat matter with irony if not with a sense of the ridiculous. [23]

Figure by Mauro Carfora.

discussed in the Prologue. It is simply not possible for any of these structures to be described in terms of a single linearly evolving wavefunction.

How to fix this? Follow the lead of general relativity theory. Realize that what in fact exists are local wavefunctions $|\Psi>_\alpha$, each restricted to a local domain U_α where the dynamics is linear. The local domains U_α and associated wavefunctions $|\Psi>_\alpha$ cover the entire cat, brain or Universe, as required to make quantum theory apply to them as a whole. But there is no single wavefunction for the whole Universe, a human brain— or even a cat (**Figure 2.21**).[24] This resolves the issue.

This is part of a broader theme: every physical theory has a limited domain of application. When using any physics equations, one must first ask, 'What is its legitimate domain of application?' This domain will always be restricted. Proposed equations

---

[23] Erwin Schrödinger, 'Die gegenwärtige Situation in der Quantenmechanik', *Naturwissenschaften* 23 (1935), 807–12.

[24] George Ellis, 'Quantum Physics, Digital Computers, and Life from a Holistic Perspective', *Foundations of Physics* 54 (2024), 56.

for some claimed physical variable have no physical meaning if that variable simply does not exist. Just assuming that it exists does not make this true.

2.3 Bulk Matter and Thermodynamics

Forces acting on particles at the micro level imply how physical stuff behaves at the macro level, depending on how it is ordered. The four basic kinds of material existence are the familiar solid, liquid and gas, and the less familiar plasma: ionized (electrically charged) matter that occurs under special conditions, such as in space or plasma TV screens.[25] All except the last have been known for a very long time. This relates to how Plato and Aristotle saw the nature of matter: they were not as mistaken has often been suggested.[26]

Kinetic theory of gases. At a macro level, a gas looks like a continuous entity characterized by its density ρ, pressure P, temperature T and entropy S. At a micro level it consists of billions of molecules in rapid motion, banging into each other all the time. The kinetic theory of gases shows how these macro quantities can be expressed in terms of averages of the motions of molecules at the micro level; for example, pressure is related to the averaged square of the speeds of the molecules.

The **first law of thermodynamics** emerges at the macro level from micro level matter and energy conservation: **the total matter and energy are conserved in any isolated system,** although they exist in many forms which can be transformed into each other under suitable conditions through physical and chemical interactions. Energy and matter can be transformed according to Eq. (2.3) at high enough temperatures in nuclear reactors or the interiors of stars, but not under ordinary conditions in daily life. But perpetual motion machines cannot exist, and you cannot power your car with water—as is sometimes claimed.

The **second law of thermodynamics** puts realistic limits on what can be achieved in practice. Entropy—a measure of micro level disorder[27]—grows because of processes such as viscosity and friction which cause loss of useful energy, and result in heat, which cannot generically be accessed for useful purposes. Heat can only be utilized usefully if you have a lower temperature reservoir at hand. You can locally create order by mechanical and chemical and biological processes, but this will inevitably be at the cost of creating some disorder in the environment. It will transform useful resources such as coal and wood and food into unusable stuff such as smoke and ashes and biological waste. Thus a continual process of decay takes place in daily life: heat dissipates, fluids mix, a dropped egg cracks, a dropped glass shatters into pieces. None of these happenings can be undone.

The second law is key in physical and chemical reactions. As famously stated by Arthur Eddington,[28]

---

[25] See the Wikipedia article Plasma (physics).

[26] Carlo Rovelli, 'Aristotle's Physics: A Physicist's Look', *Journal of the American Philosophical Association*, 1 (2015), 23–40; Tim Maudlin, 'Earth, Air, Fire and Water: A Charitable Reading of Ancient Greek Chemistry' (2024) (philsci-archive.pitt.edu).

[27] John Baez, 'What is Entropy?' (2024) https://arxiv.org/abs/2409.09232

[28] Arthur Eddington, *The Nature of the Physical World* (Cambridge, UK: Cambridge University Press, 1928).

The law that entropy always increases holds, I think, the supreme position among the laws of Nature. If someone points out to you that your pet theory of the universe is in disagreement with Maxwell's equations—then so much the worse for Maxwell's equations. If it is found to be contradicted by observation—well, these experimentalists do bungle things sometimes. But if your theory is found to be against the Second Law of Thermodynamics, I can give you no hope; there is nothing for it to collapse in deepest humiliation.

Life has learnt to harness matter and energy in useful ways and to create order despite these limitations via metabolic processes at the micro and macro levels. The costs are that we have to eat and drink on the one hand, and excrete metabolic waste (sweat, urine and faeces) on the other. Analogous processes take place in a society: cities and buildings have similar needs: water and food must be available, waste (solid and liquid) must be removed and dealt with. The arrow of time in the second law (entropy grows in the future direction of time, not the past direction) derives from cosmology (see Chapter 1).

These two laws place fundamental limits on what can occur in biology, technology, society and ecology.

Large scales. Emergent physical laws shape how biology and technology work, and what is possible in both cases. Galilean gravity (emergent because it is due to the mass and radius of the Earth) governs walking, sports, design of buildings and bridges. Bernoulli's theorem for fluid flow governs how birds and aircraft can fly, and so is important in design of aircraft wings. Radiation, diffusion and convection equations govern how fast hot things (coffee, tea, porridge, baths) cool down and how fires keep us warm.

Small scales: Brownian motion and fluctuations. Because a gas is made of billions of particles that are colliding with each other all the time, there is a great deal of randomness in particle motion at the molecular scale, called Brownian motion (the fluctuation in position of small particles suspended in water).[29] The atomic constitution of matter is proved by Brownian motion, and one can determine from it that Avogadro's number—the number of nucleons in one gram of ordinary matter—has the enormous value of 6.0×10^{23}. Emergent matter and life is based in these huge numbers of constituents at lower levels. Molecules in a cell undergo 10^{14} collisions per second, so detailed molecular motions cannot be calculated from initial data. Stochasticity of motion at the molecular level is a fundamental feature exploited by biological mechanisms such as molecular machines[30] to obtain desired emergent outcomes. I return to this theme in later chapters.

Initial data and outcomes. In principle, given suitable initial data for a physical system, the equations of motion and any relevant constraints, the outcomes at later times are uniquely determined by that initial data. However, this is not how it always works out. A key discovery is that in many nonlinear systems, chaotic motion occurs: arbitrarily small changes in initial conditions lead to completely different outcomes after a finite times, so that in practice dynamic outcomes are unpredictable. This was first

---

[29] See the simulation in the Wikipedia article Brownian motion.
[30] See the Wikipedia article on Molecular machines.

discovered by Henri Poincaré for a three-body gravitational system such as the Earth, the Moon and the Sun. Famously, it was shown by Edwin Lorenz that such unpredictability applies to the weather, where 'strange attractors' occur, as described by Tim Palmer and colleagues: the 'butterfly effect' occurs,[31] where in principle a butterfly flapping its wings in Brazil can affect the weather at great distances away.[32] I won't pursue this issue further here—the associated geometry is fascinating—except to say that one should not jump to the conclusion that one can always uniquely predict physical outcomes from the relevant equations of motion.

2.4 Atoms and the Periodic Table of the Elements

Atoms are made of a nucleus consisting of an approximately equal number of positively charged protons and uncharged neutrons of approximately the same mass, surrounded by a cloud of negatively charged, much lighter electrons, orbiting in distinct shells. An element's atomic number N is the number of protons in the nucleus, the same as the number of electrons surrounding the nucleus in a neutral atom. If an atom loses an electron it becomes a positively charged ion; if it gains one, it becomes a negatively charged ion. Ionization can result in lightning and electric sparks as in spark plugs in internal combustion engines.

Atoms have different chemical properties, characterized by their position in the periodic table of the chemical elements[33] (**Figure 2.22**). **This is the key link between physics and chemistry**. All elements in a column have similar physical properties, but size (increasing downward, and decreasing from left to right) and ionization energy (changing in the opposite way) vary both across columns and in columns.

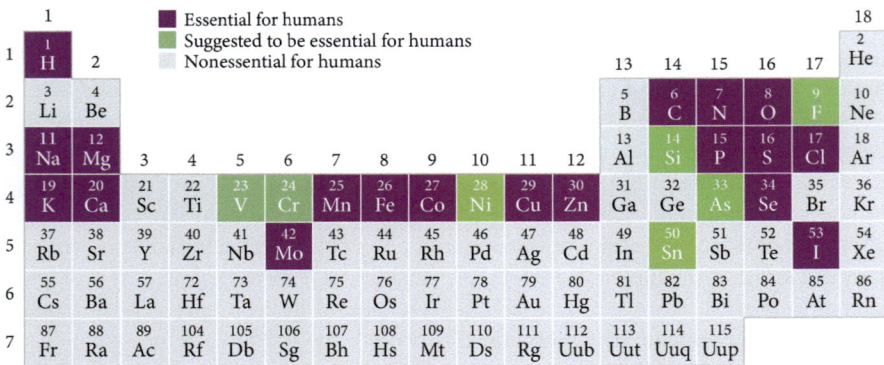

Figure 2.22 The periodic table of the elements, indicating those that are essential for life.

From Libretexts.

[31] See the Wikipedia article Butterfly effect.

[32] Tim N. Palmer *The Primacy of Doubt: From Climate Change to Quantum Physics, How the Science of Uncertainty Can Help Predict and Understand our Chaotic World* (Oxford: Oxford University Press, 2022); Tim N. Palmer, A. Döring and G. Seregin, 'The Real Butterfly Effect', *Nonlinearity*, 27 (2014), R123.

[33] See the Wikipedia page Periodic table of the elements.

The human body. Figure 2.22 indicates the elements essential for life (purple) and maybe so (green). As much as 99% of the mass of the human body is made up of just six elements: oxygen (65%), carbon (18.5%)—the basis of organic chemistry—hydrogen (10%), nitrogen (3.2%), calcium (1.5%) and phosphorus (1%). About 0.85% is composed of sodium, magnesium, potassium, sulphur and chlorine. All these elements are necessary for life, as are the trace elements indicated in Figure 2.22. Physical disorders can arise if we lack them. Water is a key nonorganic molecule enabling cells to function, and so life to exist: an adult body is about 60% water.

How do we acquire these elements? In order of increasing importance, eating, drinking, and breathing. Breathing is crucial: our cells die if we lack oxygen for more than a few minutes, as for example in drowning. Drinking is not required so often, but is vital too; the required water intake is about 3.7 litres per day for an adult male, and 2.7 litres for an adult female. Eating provides the other elements essential to life.

How did these elements come into being? They were generated by stellar nucleosynthesis in the interior of stars of various kinds (massive stars, dying low mass stars, supernova explosions, exploding white dwarfs) and by merging neutron stars, the resulting elements then being spread through space at the end of the star's life (Chapter 1). Hence the phrase, 'We are made of stardust'.

2.5 Molecules, Water, Crystals, Macromolecules

Binding energy and thresholds. A fundamental feature is that when any components—particles, atoms, molecules—are bound together to make larger entities, which may occur spontaneously (self-assembly) or require energy to accomplish (as in metabolism or manufacturing), the resulting emergent entity is held together by binding energy—the energy needed to keep it together. To break it apart requires providing that energy. This applies also to orbital electrons in atoms, resulting in ionization energy required to remove those electrons, and to chemical bonds, holding molecules together. And it applies to atomic nuclei, which weigh just a little bit less than their component parts, by Eq. (2.3).

Bond energy is a measure of the strength of chemical bonds, released in a fire or in catabolic processes in a cell. Bond dissociation energy is a measure of the energy required to break a bond and release the binding energy. It results in threshold energies and temperatures at the macro level required for reactions to take place, such as making tea, boiling an egg or starting a fire. This results in the need for matches or other firelighters so we can generate fire when needed; learning to do so was one of the great cultural transitions in our social evolution. This thermal instability also results in the dangers of forest and bush fires.

Molecules. The basic principle of chemistry is to bind atoms together in a particular order to form molecules[34]—chemically active entities whose three-dimensional shape and associated energies determine what chemical reactions can occur.[35] Binding of atoms to form molecules takes place through chemical bonding enabled by

---

[34] Philip Ball, *Molecules: A Very Short Introduction* (Oxford: Oxford University Press, 2001).
[35] Peter Atkins, *Physical Chemistry: A Very short Introduction* (Oxford: Oxford University Press, 2013).

electrostatic forces, with the nature of allowed bonds being determined by position in columns in the periodic table (which is why elements in the same column have similar chemical properties). Chemical identity is determined by the number of protons in the atomic nucleus, and this is unaffected by chemical bonding which works with the atoms' outermost electrons. It is fair to say that (for example) an oxygen atom in H_2O behaves very differently from a free oxygen atom, but it is still identifiably an oxygen atom. In fact, the atomic composition in organic functional groups such as $-C=O$, $-CO_2H$ and $-NH_2$ determines the distribution of electron charge and hence the specific chemical properties of their host molecules and the chemical reactions these molecules will undergo.[36]

The main types of bonding are the following:

Covalent bonding, when atoms share electron pairs and so are bound together as molecules. True emergence takes place: e.g. as in the case of hydrogen and oxygen bonded to form water; the nature of water is completely different than the nature of either hydrogen or oxygen gases.

Ionic bonding, such as the electrostatic attraction between oppositely charged ions, or between atoms with very different electronegativities. Again true emergence takes place, as in the case of common salt, which has completely different properties than those of its constituents, sodium (a metal of sorts) and chlorine (a poisonous gas).

Much weaker **hydrogen bonding** occurs between H atoms bound (for example) to N atoms ($-NH_2$), which pull the electron distribution away from the H atom, making it slightly electropositive, and O and N atoms, which have 'lone pair' electrons uninvolved in bonding and which are therefore slightly electronegative. Hydrogen bonds between H atoms in $-NH_2$ groups and O atoms in $-C=O$ groups and ring N atoms are responsible for base pairing in RNA and DNA.

Biomolecules[37] bonds form at level **L3** and are key to our physiological functioning: proteins, carbohydrates, lipids and nucleic acids, as well as smaller molecules such as metabolites. Studying how the binding of atoms happens to form these molecules is the topic of quantum chemistry.[38] Chemical reactions take place when two or more molecules interact to produces a new set of molecules made up of the same set of atoms, but arranged in a different way. Matter is conserved in chemical reactions: the number of atoms of each kind on the left- and right-hand sides of the equations describing the interaction must balance, as must energy. Reactions may be exothermic, giving off heat as they take place, or endothermic, requiring input of heat to take place.[39]

I discuss macromolecules in the following. Many smaller kinds of biomolecules are also essential to life:

- 22 amino acids are the small molecules that combine to form proteins.
- Monosaccharides are the basic form of sugars from which carbohydrates are built.

---

[36] A brilliant summary of chemistry is given by Libretexts: General Chemistry, Principles, Patterns, Applications.

[37] See the Wikipedia article Biomolecule.

[38] Martin Karplus, 'Development of Multiscale Models for Complex Chemical Systems: From H+H₂ to Biomolecules (Nobel lecture)', *Angewandte Chemie International Edition*, 38 (2014), 9992–10005.

[39] Baez, 'What is Entropy?'

- 5 Nucleotides (guanine G, adenine A, cytosine C, thymine T and uracil U) are the key coding components of both RNA and DNA .They also play a key role in cell energy processes, particularly via adenine triphosphate (ATP), and in cell signalling processes and enzyme reactions.
- Vitamins are small molecules essential for metabolism.
- Lipids store energy, take part in signalling and self-assemble to form cell membranes. This is essential to cell functioning: they keep active components together so that they can interact with each other.
- Hormones are signalling molecules between organs that regulate digestion, respiration, sleep, metabolism, excretion, growth, movement and so on.
- Neurotransmitters and neuromodulators are key to brain functioning.

Water (H_2O) is crucial to our existence[40] because of its dipole moment (**Figure 2.23**), which makes it an excellent solvent, and because it becomes less dense as it freezes, enabling life to survive underwater in a frozen pond. It plays a key role in metabolic processes in the body through anabolism (helping grow larger molecules) and catabolism (breaking them down).

Crystals are extended ordered arrays of atoms such as snowflakes, diamonds and table salt (**Figure 2.24**).

Crystals fused together form polycrystals, the basis of crystalline solids such as ice, ceramics, rocks and metals, whereby they are key to daily life. Crystals are held together by a range of quantum forces.[41]

Macromolecules. These are long chains of smaller molecules that are the central actors in molecular biology: they are the core of the way physics underlies the existence of life.

The key discoveries made during the twentieth century are those at the molecular level: firstly, DNA is the genetic material storing information coding what proteins can be generated by developmental processes in a cell. Secondly, it is the change of shape of macromolecules (conformational change) that underlies the way biology emerges from physics.[42]

DNA is our genetic inheritance. Its double helix structure (**Figure 2.25**) is determined by evolutionary processes.[43] The two complementary spirals contain the nucleotides thymine (T), adenine (A), cytosine (C) and guanine (G) in a complementary way: T on one side is paired with A on the other side, and C on the one side with G on the other.

---

[40] Lawrence J. Henderson, 'The Fitness of the Environment, an Inquiry into the Biological Significance of the Properties of Matter', *The American Naturalist*, 47 (1913), 105–15; Philip Ball, 'Water is an Active Matrix of Life for Cell and Molecular Biology', *Proceedings of the National Academy of Science*, 114 (2017), 13327–35.

[41] See the article 'The structure of crystals: interatomic forces in crystals', *Crystallography in a nutshell*.

[42] Jean-Marie Lehn, 'Supramolecular Chemistry: From Molecular Information towards Self-organization and Complex Matter', *Reports on Progress in Physics*, 67 (2004), 249. Jean-Marie Lehn, 'From Supramolecular Chemistry towards Constitutional Dynamic Chemistry and Adaptive Chemistry', *Chemical Society Reviews* 36 (2007), 151–60.

[43] Neil A. Campbell and Jane B. Reece, *Biology* (Chennai: Pearson Education India, 2005).

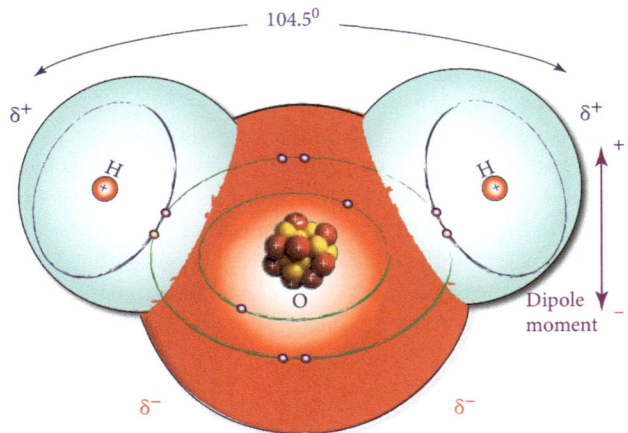

Figure 2.23 The water molecule. Electronic structure, indicating the covalent bonding holding the molecule together, and the dipole moment of water. Each hydrogen atom shares two electrons with oxygen. In this way, both hydrogen and oxygen have full outer shells of electrons, giving rise to a stable configuration.

Figure by Mauro Carfora.

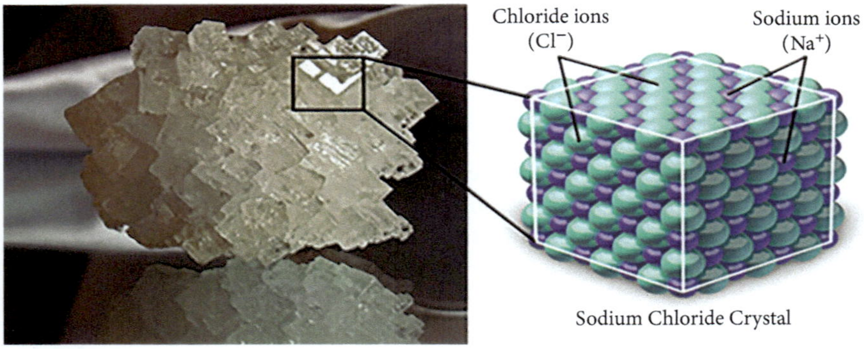

Figure 2.24 A crystal. Sodium chloride (common salt), composed of negatively charge chlorine ions and positively charged sodium ions. Shown are its macroscopic structure (left) and crystalline nature (right).

Source: Labster Theory Pages.

Thus, an arbitrary sequence of nucleotides on the one side, say ACGTCTT, is paired with the complementary sequence TGCAGAA on the other side. The same genetic information is coded in both coils of the DNA. Triplets of these nucleotides

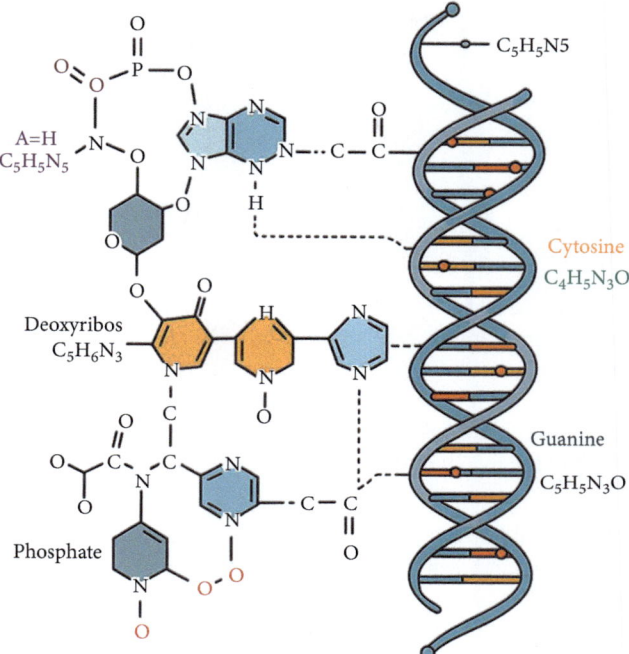

Figure 2.25 The DNA molecule.[44]

Figure by George Ellis.

code for specific amino acids according to the genetic code,[45] and so genes—sequences of triplets, not necessarily consecutively located on the DNA—contain the information to generate needed proteins when decoded by complex cellular machinery.

Proteins. There are a vast variety of proteins, carrying out many different functions.[46]

An important example is voltage-gated ion channels (**Figure 2.26**), underlying the propagation of action potential spike chains in neurons.[47] They open to let the ion through, or not, according to the electrical potential across the cell wall.

This is an example of how logic, in this case

$$\text{IF \{Voltage positive\} THEN \{Closed\} ELSE \{Open\})} \qquad (2.6)$$

emerges from physics, via the alteration of conformation of a protein.

However, proteins have very many other functions: they are the workhorses of biology. A key one is acting as catalysts (enzymes) that play a crucial role in biochemical

---

[44] The link https://www.youtube.com/watch?v=7Hk9jct2ozY provides a very interesting DNA animation.

[45] See the Wikipedia article Genetic code.

[46] Gregory A. Petsko and Dagmar Ringe, *Protein Structure and Function* (London: New Science Press, 2004).

[47] George Ellis and Jonathan Kopel, 'The Dynamical Emergence of Biology from Physics: Branching Causation via Biomolecules', *Frontiers in Physiology*, 9 (2019), 1966.

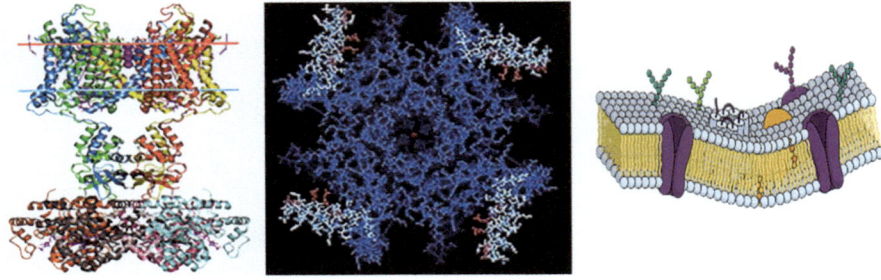

Figure 2.26 A voltage-gated ion channel[48] imbedded in the wall of a neuronal axon. The twisted strands represent the relevant proteins. The ribbons are α-helices. The ion channel has four paddles (at the top) that can open and close. (left) View from the side when they are closed;[49] an ion cannot pass through. (middle) View from the top when they are open. An ion (red dot) is passing through the channel into the neuronal axon.[50] (right) Imbedding of the ion channel in the cell membrane.[51]

functioning by speeding up the rate of metabolic reactions.[52] However, they do much more than this: they carry out structural functions, play a key part in DNA replication, transport molecules in cells, and take part in cell signalling processes. Actin and myosin take part in muscle contraction and cell division, and help maintain cell shape. And the proteins kinesin and dynein—molecular machines—extract energy for useful purposes from the molecular storm.[53]

Overall proteins are the workhorses of biology: DNA is only important because it codes for proteins, albeit in complex ways.[54]

2.6 The Material Basis of Daily Life

The materials that make up everything around us—metals, glass, ceramics, fabrics, paper, plastics, ink, dyes—each have specific physical properties that we need for particular use: being hard, soft, pliable, strong, elastic and so on, based in their underlying specific molecular structure and bonding.

Through a process of adaptive experimentation, we have determined how to make each of these materials that are the background for social life. This is discussed in a marvellous book by Mark Miodownik. He states, after mentioning atoms and quantum mechanics,

---

[48] See the Wikipedia article Voltage-gated ion channels.
[49] *Source*: https://openmembranedatabase.org/) by Andrei Lomeize.
[50] Protein Data Bank:1BL8, https://www.rcsb.org/
[51] The cell membrane image is from Mind the Graph, used under the CC BY-SA license (https://creativecommons.org/licenses/by-sa/4.0/deed.enhttps://creativecommons.org/licenses/by-sa/4.0/deed.en).
[52] See the Wikipedia article Enzymes.
[53] Peter Hoffmann, *Life's Ratchet: How Molecular Machines Extract Order from Chaos* (New York: Basic Books, 2012).
[54] Philip Ball, *How Life Works* (Picador, 2023).

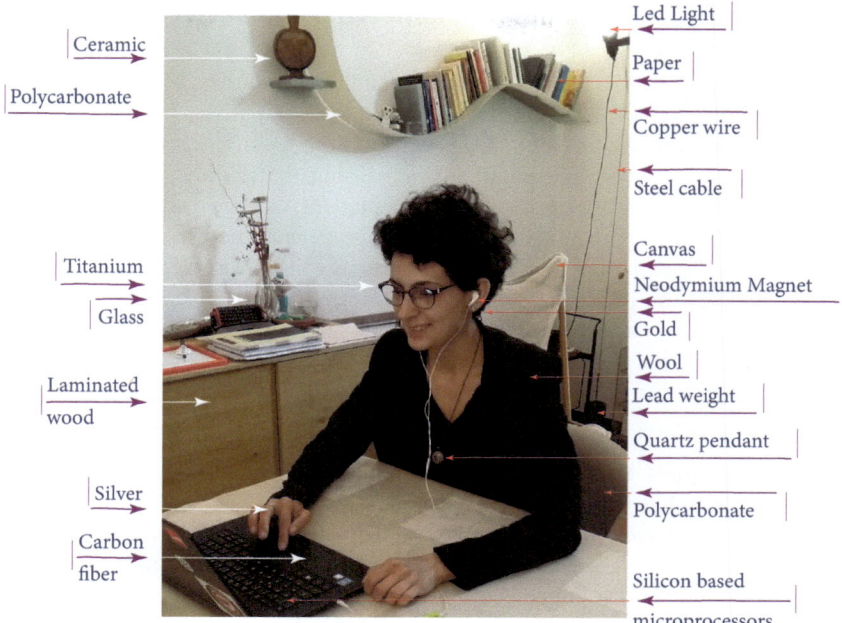

Figure 2.27 Materials in daily life.

Figure by Mauro Carfora, inspired by a figure in Miodownik's book.

The central idea behind materials science is that changes at these invisibly small scales manifest themselves as changes in a material's behaviour at human scales ... It is at these microscopic scales that we discover why some materials smell and others are odourless; why some materials last for a thousand years and others become yellow and crumble in the sun, how it is that some glass can be bullet proof, while a wine glass shatters at the slightest provocation, The journey into this microscopic world reveals the science behind our food, our clothes, our gadgets, our jewellery and of course our bodies.[55]

He then talks about steel and other metals, jelly and aerogels, plastics, glass, graphite and diamonds, plaster casts for broken bones, amalgams for tooth cavities, paper and cardboard, concrete, chocolate and a host of other materials we have created to fulfil our needs (**Figure 2.27**).

At an even higher level, basic design principles, based in what is physically possible, determine what can be achieved in both biology and technology: stresses in trusses, the nature of rigidity, bearings allowing flexible joints, a limited number of ways to achieve propulsion and so on.[56]

[55] Mark Miodownik, *Stuff Matters: The Strange Stories of the Marvellous Materials that Shape Our Manmade World* (London: Houghton Miflin, 2014), 7, 9.

[56] Steven Vogel, *Cat's Paws and Catapults: Mechanical worlds of Nature and People* (London: Penguin, 1998).

2.7 Outcomes in Engineering and Society

Physics and chemistry as discussed in this chapter underlie all engineering outcomes in society, which have transformed our lives. This has, in essence, historically gone through six phases:[57] (a) The simplest technologies; (b) harnessing forces and matter: civil engineering; (c) harnessing motion and heat: mechanical engineering; (d) harnessing electricity and magnetism: electrical and electronic engineering; (e) harnessing chemicals: chemical engineering and pharmaceuticals; and (f) harnessing information: the computational revolution and AI. Underlying all this has been (g) the physical basis for technology: materials science. Later developments have fed back into each of the earlier ones, making them more powerful. The specifics of how they underlie society is discussed informatively by Vaclav Smil.[58]

(a) **The simplest technologies** A chronology of engineering inventions goes something like this:[59] fire probably was tamed about 400,000 years ago; farming (agriculture) introduced 11,000–10,000 BC; the wedge, the key idea of the lever (see **Figure 2.28**) and the harnessing of domestic animals between 10,000 and 8000 BC; the wheel and weaving around 5000 BC; irrigation canals 4000 BC; the pulley 900 BC; water screws 850 BC; winches 500 BC; cranes 450 BC; the catapult 400 BC; gears (wooden) 350 BC, (bronze) 100 BC; paper 150 AD; the loom 300 AD; levers and the three-field system of agriculture 600 AD; the horse stirrup 700 AD; water mills 800 AD; gun powder 900 AD. Nuts and bolts of standardized sizes and ball bearings, both simple but key for engineering came relatively late, about 1794 and 1841, respectively, as they required machine tools to make them. One of the simplest technologies that makes a key difference to the lives of many of us is spectacles, enabling us to see clearly (**Figure 2.29**), probably first introduced in Italy by about 1290.[60]

(b) **Harnessing forces and matter: civil engineering.** A very old human activity is constructing buildings with walls and roofs, doors and windows and internal rooms, providing shelter from the elements. This extended to constructing roads and drains, building bridges and dams and tunnels, and generally shaping the environment to meet our needs. A key development was discovering concrete as a material that would set and retain its shape, and then learning how to reinforce concrete[61] so that it could bear heavy loads and not collapse. Underlying this was a theoretical basis: how to calculate the needed size of foundations, walls and beams so as to carry expected loads safely through understanding stress loads and vibration.

---

[57] David Blockley, *Engineering: A Very Short Introduction* (Oxford: Oxford University Press, 2012).
[58] Vaclav Smil, *How the World Really Works: A Scientists Guide to the Past, Present, and Future* (London: Penguin Random House UK, 2022).
[59] Blockley, *Engineering*.
[60] See the Wikipedia article Glasses.
[61] See the Wikipedia article Reinforced concrete.

Figure 2.28 The lever: the simplest mechanical device. A child playing with his mom on a teeter-totter. Here, the mother takes advantage of the fact that her child is closer to the fulcrum of the teeter-totter than the place where she is applying her pushing effort. This typical situation is a playground example of a first-class lever. She knows that, in this way, less effort is needed to move her child up and down.

Figure by Mauro Cafora.

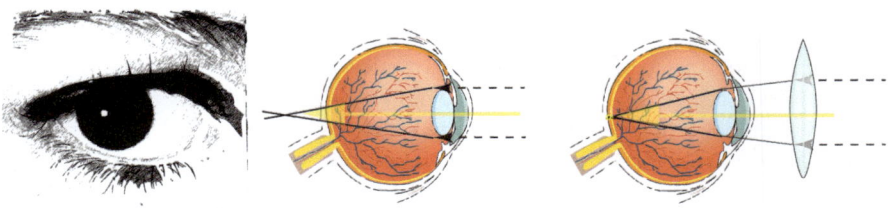

Figure 2.29 Eye glasses Far-sightedness without (left) and with lens correction (right). The simplest technology that makes a major change to human life.

Figure by Mauro Cafora.

Later on mechanical engineering came to the aid of civil engineering through the development of bulldozers, road graders, and other heavy earth-moving machinery, tunnel boring machines and a variety of cranes to assist building. Furthermore surveying instruments developed by electrical engineers ensured accurate placement of all these elements as required: for example, checking levels of irrigation channels, and alignment of tunnels being drilled from both sides.

(c) **Harnessing motion and heat: mechanical engineering.** The earliest way energy was harnessed to do useful work was by wind power, used for travel by sailing ships and for grinding corn or pumping water by windmills. Taming of fire—being able to light a fire when needed for warmth and light and cooking—was a key development in human history. Developing the wheel and axle was another key event enabling transportation of heavy goods. Putting the two together underlay the basic principle of converting heat to do useful work, as allowed by the laws of thermodynamics.

This took place first through the invention of the steam engine powering pumps, trains and ships. Then development of the internal combustion engine (petrol or diesel) transformed life by making rapid long distance transport of people and goods generally available. Next there was the transformative development of aircraft, first powered by piston engines and then utilizing jet engines and pressurized cabins enabling aircraft to travel above the prevailing weather. Finally, there was the development of space travel by rockets, enabling us to travel to the Moon, Mars and beyond.

(d) **Harnessing electricity magnetism: electrical and electronic engineering.** The discovery of the ability to generate and use electrical power to attain desired mechanical purposes via electromagnets, electrical motors and heaters was the next major social revolution, because of how easily controllable electrical power is, and how it can be distributed anywhere from a generating source via wires, or using batteries. This enables electrical heating, lighting, stoves, refrigerators, washing machines, dishwashers, vacuum cleaners and so on to be at our command.

Manufacturing was transformed via arc furnaces, milling machines, drill presses, tools such as electrical drills and screwdrivers and lathes and eventually automated assembly lines. A particular benefit of electricity is electrical street lighting, making it easy to find our way at night and making streets safer. The result is that when viewed from space, cities are marked by their lights (**Figure 2.30**). If there were indeed civilizations on other planets, maybe this would be an easy way to detect that fact.

Electronic engineering allows control of elevators, cranes, traffic lights and railway crossing booms. Radios, phones, TV and then cell phones transformed social life, while radar makes air travel safe.

(e) **Harnessing chemicals: chemical engineering and pharmaceuticals.** The ability to manipulate chemicals to meet human needs was a great transformation in human life. This requires accessing the needed raw materials such as coal and crude oil, refining and separating them to obtain petroleum products such as gasoline (petrol), diesel and jet fuel, which are the

Figure 2.30 Cities in Europe and North Africa seen from space, marked out by their lights.

NASA Earth Observatory image by Robert Simmon, using Suomi NPP VIIRS.

basis of our transport infrastructure and provide asphalt and lubricants that are subsidiary to this, as well as chemical reagents used to make plastics; paint, synthetic rubber and pesticides.

Particularly important are fertilizers, pharmaceuticals, antibiotics and plastics.[62] According to Vaclav Smil, the best return on investment is vaccination,[63] and synthetic ammonia is crucial to feeding the world. Indeed together with cement and steel, plastics and ammonia are the four pillars of modern civilization.[64]

(f) **Harnessing information: the computational revolution and AI.** The invention of the digital computer and its foundation in digital representation of both programs and data enabled the power of abstract algorithms to transform all technology. The key point is that any data—numbers, colour pictures, symbols, music, videos, mathematical equations, quantitative information of any kind, and particularly computer programs—can be represented in digital form: rows of 1's and 0's. These can then be analysed statistically or by multilayer artificial neural networks to make predictions or make decisions such as in automated factories or three-dimensional printing.

The outcome is cell phones, apps able to carry out arbitrary tasks, self-driving cars and more. This transforms social life for better or worse via all the forms of social media enabled by this technology: physically integrated chips containing billions of transistors. Underlying this now is the rise of

---

[62] See the Wikipedia article Plastic.

[63] Vaclav Smil, *Numbers Don't Lie: 71 Things You Need to Know about the World* (London: Random House, 2021).

[64] Smil, *How the World Really Works.*

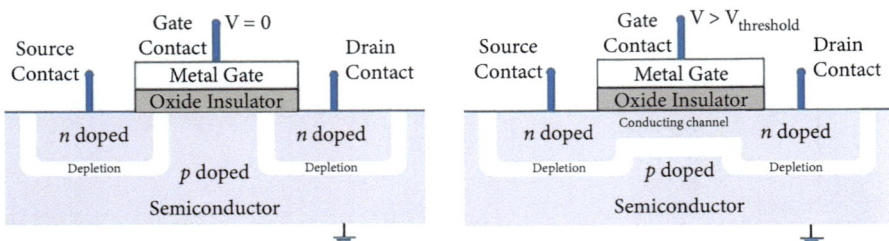

Figure 2.31 A MOSFET transistor. (left) MOSFET with no bias applied to gate. The depletion layer separates the n-doped and p-doped regions, so no current flows. (right) MOSFET with bias applied to gate. The depletion layer shifts, so a continuous channel forms between the source and drain and current flows.

Source: Steven H. Simon, *The Oxford Solid State Basics* (Oxford: Oxford University Press, 2013). (left) p. 204, (right) p. 205. With Permission from OUP and the author.

artificial intelligence and applications such as ChatGPT that potentially has a transformative effect on many aspects of economic and social life.

The crucial element underlying all this is transistors (**Figure 2.31**), which switch on and off according to bias voltage applied to the gate.

(g) **The physical basis for technology: materials science.** Underlying all of this is the way that physics and chemistry underlie the materials that make all technology possible. We have in effect had the following epochs in human life: the Stone Age, the Bronze Age, the Iron Age,[65] the Steel and Concrete Age and the Plastic Age. Developing suitable materials for all our purposes has been a key element in our history.[66]

In the synthesis chapter of his book *Stuff Matters*,[67] Mark Miodownik states,

I want to explore the language of materials science more fully, because it encompasses a unifying concept that encompasses all materials ... This unifying concept is that although all material may look and feel monolithic, although it may appear uniform throughout, this is an illusion: materials are, in fact, composed of many different entities that combine to form the whole, and these entities reveal themselves at different scales. Structurally any material ... is made up of many nested structures, almost all of which are invisible to our eye, each one smaller and fitting exactly into the one before. It is this hierarchical architecture that gives materials their complex identities. ... One of the most fundamental of these material structures is the atom, but it is not the only structure of importance. At the larger scales there are dislocations, crystals, fibres, scaffolds, gels and foams, to name a few ... Taken in isolation, these structures are like characters in a story, each contributing something to its overall shape. Sometimes one character dominates the story, but it is only when they are put back together that they fully explain why materials behave the way they do. As we have seen, the reason why a stainless steel spoon doesn't taste of anything is

---

[65] See the Wikipedia article Stone Age.
[66] Jacob Bronowski, *The Ascent of Man* (BBC, 1973), episodes 3 and 4.
[67] Miodownik, *Stuff Matters*, 237.

Figure 2.32 Civil engineering and technology. The outcome of engineering design and manufacturing processes: cities (left), airports and aircraft (right).

because the chromium atoms within its crystals react with oxygen in the air to form an invisible protective layer of chromium oxide on its surface. . . . A full understanding of why stainless steel behaves the way it does requires you to consider all of the structures of which it is composed.

This is followed by a brilliant description (pp. 238–48) of the hierarchical structure of matter and its emergent nature. Technologies have purposes (**Figure 2.32**). They transform our everyday possibilities and experiences. Every one of these innovations is based in someone's creative thinking.[68] I look at technology from a different perspective in Chapter 6. These are all products of human agency: a main theme of this book. They are definitive proof agency exists.

Further Reading

Physics, chemistry and more

Paul G. Hewitt, *Conceptual Physics* (Boston: Addison Wesley, 1998).
Paul G. Hewitt, Suzanne Lyons, John Suchocki and Jennifer Yeh, *Conceptual Integrated Science* (London: Pearson, 2006).

The periodic table of the elements

Eric Scerri, *The Periodic Table—Its Story and Its Significance* (Oxford: Oxford University Press, 2007).

How physical forces underlie the materials that underlie daily life

Mark Miodownik, *Stuff Matters: The Strange Stories of the Marvellous Materials that Shape Our Man-made World* (London: Houghton Miflin, 2014).

[68] D. H. Cropley, 'Creativity in Engineering', in G. E. Corazza and S. Agnoli, eds, *Multidisciplinary Contributions to the Science of Creative Thinking* (London: Springer, 2015), 155–73. G. E. Dieter and L. C. Schmidt, *Engineering Design* (London: McGraw Hill, 2021).

The nature of technology

W. Brian Arthur, *The Nature of Technology: What it is and How it Evolves* (New York: Simon and Schuster, 2009).

How physics underlies technology

Louis Bloomfield, *How Everything Works* (London: Wiley, 2008).
G. E. Dieter and L. C. Schmidt, *Engineering Design* (London: McGraw Hill, 2021).

Specifically regarding digital computers

David M. Harris and Sarah L. Harris, *Digital Design and Computer Architecture* (Burlington, MA: Morgan Kauffman, 2013).
Subrama Dasgupta, *Computer Science: A Very Short Introduction* (Oxford: Oxford University Press, 2016).

PART II
EMERGENT BIOLOGICAL LEVELS

3

What are the Emergent Biological Levels and What do They do?

Biology is the most extraordinary thing. Life involves function and purpose at the cellular level,[1] and agency at higher levels.[2] Living entities are made of the same chemical elements as non-living things, but the outcome is totally different. The central element underlying life is carbon, which also forms graphite, as in a pencil lead, and diamonds, which can cut glass. Carbon is also the core molecule in flesh and blood. While the hallmark of life is agency, quite unlike anything else, it also includes intelligent beings who can think and plan and understand physics, the Universe, and life, and engage in activities with an ethical dimension.[3] How is this possible? How can life emerge from physics?[4]

The way emergent biology works (see **Figure 3.1**) is briefly summarized in the Prologue. As a reprise, here is a summary from a paper by Patrick McMillen and Michael Levin:[5]

> A defining feature of biology is the use of a multiscale architecture, ranging from molecular networks to cells, tissues, organs, whole bodies, and swarms. Crucially however, biology is not only nested structurally, but also functionally: each level is able to solve problems in distinct problem spaces, such as physiological, morphological and behavioural state space. Percolating adaptive functionality from one level of competent subunits to a higher functional level of organization requires collective dynamics: multiple components must work together to achieve specific outcomes. Here we overview a number of biological examples at different scales which highlight the ability of cellular material to make decisions that implement cooperation toward specific homeodynamic endpoints, and implement collective intelligence by solving problems at the cell, tissue, and whole-organism levels.

Thus we are **adaptive modular hierarchical structures**, for good functional and evolutionary reasons.[6] Each emergent level L in **Figure 3.1** has specific **functions**,

---

[1] Leland Hartwell, John Hopfield, Stanislas Leibler and Andrew Murray, 'From Molecular to Modular Cell Biology', *Nature*, 402(Suppl 6761) (1999), C47–C52.

[2] Philip Ball, *How Life Works* (Basingstoke: Pan Macmillan, 2023).

[3] Philip Ball, *The Book of Minds: Understanding Ourselves and Other Beings* (Chicago: University of Chicago Press, 2022).

[4] George Ellis, *How Can Physics Underlie the Mind: Top-Down Causation in the Human Context* (Berlin: Springer, 2016).

[5] Patrick McMillen and Michael Levin, 'Collective Intelligence: A Unifying Concept for Integrating Biology across Scales and Substrates', *Nature Communications Biology* 7(1) (2024), 378.

[6] Herbert A. Simon, *The Sciences of the Artificial* (Cambridge, MA: MIT Press, 2019); Grady Booch et al., *Object-Oriented Analysis and Design with Applications* (Boston: Addison Wesley, 2007), Chapters 1–3.

How We Come to Be. George F. R. Ellis, Oxford University Press. © George Ellis (2026).
DOI: 10.1093/9780198950189.003.0005

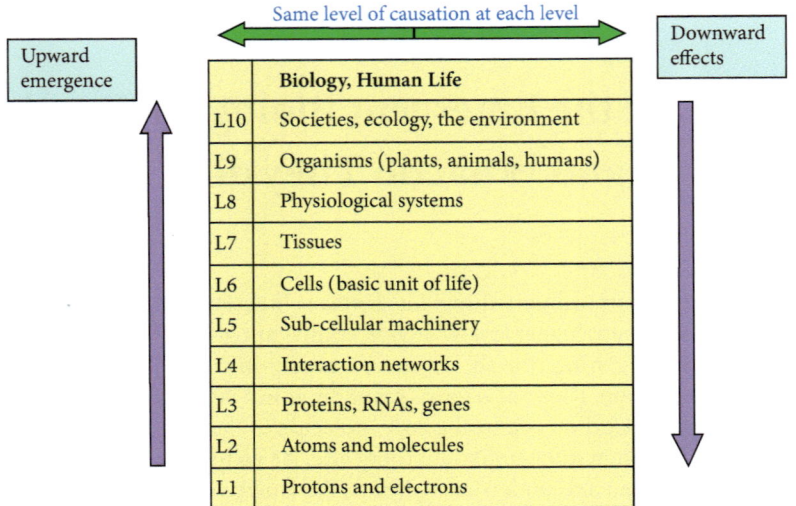

Figure 3.1 Multilevel causation in biology and humans. The hierarchy of emergent levels, with same level functioning occurring at every level. Chaining up of functional processes and down of constraints and adaptive processes creates a hierarchy with multi-level causation underlying emergent complexity. Causal closure only occurs when all the interacting levels are taken into account.

Figure by Mauro Carfora.

expressed as effective laws in terms of appropriate variables and enabled by its structure at that level. This is enabled in turn by the next lower level $L - 1$ (it emerges from that level), and underlies functioning of the next higher level $L + 1$. Each higher level $L + 1$ constrains and controls functioning at the next lower level L in order to fulfil its needs.

The levels in **Figure 3.1** are the biological and social features that make our existence possible. The discovery of the processes of molecular biology at level **L3**, and particularly the nature and function of proteins and the DNA molecule, was a key event in our understanding of ourselves: there was no longer a need for a vital force, something separate from physics, to underlie the processes of life. This could be understood as based purely in biomolecular processes at level **L3**, themselves based in the underlying physics levels **L1** and **L2**, discussed in Chapter 2. But those two levels are independent of biology.

In this chapter, I discuss, first, §3.1 'The Different Biological Levels'. There are four basic principles that apply across all emergent levels in biology and society. They are the following: §3.2 'Open Systems: Metabolism and Control'; §3.3 'Homeostasis/Cybernetics'; §3.4 'Adaptive Selection: Addition, Modification, or Removal of Elements'; and §3.5 'Cycles/oscillations'. These are all emergent interlevel properties that are crucial to emergent adaptive function in biology and society in a specific context. Finally §3.6 considers causal closure in the context of the set of levels overall.

3.1 The Different Biological Levels

Key features of modular hierarchical structures are as follows:[7] a module's internal variables are hidden from outside view; all that is needed is for the module to do what it is required. The system does not mind what internal variables are chosen and what logic is used: the internal variables and even the internal logic of the module can be changed, as long as the externally viewed functionality is maintained. Thus it is multiply realizable: it can be made in many different ways. However, the interface between each module and the rest of the system is crucial: what control variables, parameters and data will be sent to the module from outside, and what will be exported from it to other modules. Taken together, these ensure the module does what the system needs it to do.

I will now explain each level in turn, because they are the heart of our existence as material entities.

3.1.1 Level L3: Macromolecular Chemistry

The foundation of biology at the molecular level (level **L3**) is interaction between macromolecules: large molecules formed of smaller molecular components (see §2.5). The crucial ones are RNA, DNA, and proteins. The discovery of the structure of such molecules transformed biology: the structure of DNA (see Figure 2.25) by Crick and Watson, of haemoglobin by Perutz and of myoglobin by Kendrew (**Figure 3.2**). DNA has segments—genes—that code for RNA that in turn codes for proteins: chains of amino acids folded in complex ways, which are the workhorse of biology.[8] Genes have **promoters**: a sequence of DNA to which proteins bind to initiate transcription of genes that are situated downstream of the promoter. This is how genes are turned on and result in various kinds of RNA generating specific proteins. The reason for the importance of DNA is that it codes for proteins of many kinds.

There are structural proteins such as collagen, keratin and actin; light-sensitive proteins such as rhodopsin in the eye that enables us to see, and chlorophyll in plants that enables them to photosynthesize; ion channels controlling the flow of ions in and out of cells, enabling brain function; quite extraordinary molecular machines such as kinesin and dynein that harvest molecular stochasticity to transport cargoes inside cells;[9] many proteins involved in cell signalling;[10] and crucially, enzymes,[11] which are catalysts that hugely speed up metabolic reactions without themselves being used up. Life would be impossible without them.

---

[7] Booch et al., *Object-Oriented Analysis and Design with Applications.*

[8] Gregory A. Petsko and Dagmar Ringe, *Protein Structure and Function* (London: New Science Press, 2004).

[9] Peter M. Hoffmann, *Life's Ratchet: How Molecular Machines Extract Order from Chaos* (New York: Basic Books, 2012), and see the video here: https://www.youtube.com/watch?v=j11sEzaYCUM/.

[10] Michael J. Berridge, *Cell Signalling Biology* (London: Portland Press, 2014).

[11] See the Wikipedia article Enzyme.

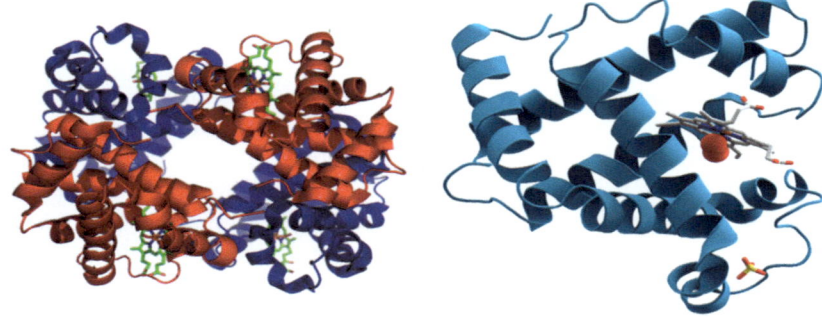

Figure 3.2 The 3D structure of the proteins haemoglobin (left) and myoglobin (right) that transport oxygen in blood cells, hence playing a central role in physiological function. The ribbons are α-helices. Toward the right-centre, a heme group (shown in grey) has a bound oxygen molecule (red).

Acknowledgement: Wikipedia.

The key point is this:

It is the change of shape (conformational change) of macromolecules that enables everything that happens in biology, allowing contextual branching logic to arise from the underlying physics. In this way, macromolecules at level **L3** are the foundation of life, enabling emergent logic to occur.[12]

It is the lowest level where biological activity occurs, underlying all higher level biological activity.

There are two ways this activity happens. One is by the higher-level biological context changing constraints at the macromolecular level, and so changing dynamics at that level. An example is the voltage-gated ion channels shown in Figure 2.26 in axons in the human brain, which change shape to either let ions through the cell membrane or not, according to the voltage across the cell membrane. In this case, the emergent logic is given by Eq. (2.6). The second is by the lock-and-key molecular recognition mechanism, whereby specific macromolecules recognize other specific macromolecules through their shape.[13] This is what underlies cell signalling processes as well as enzyme action (**Figure 3.3**).

Higher-order logic such as AND, OR, NOT can emerge from molecular interactions via allosteric processes discovered by Jacob, Monod and Changeux.[14] They considered a fascinating issue: while bacteria *E. coli* like to eat glucose, they can also digest the sugar lactose if necessary when too little glucose is present, using specific

[12] Jean-Marie M. Lehn, 'Supramolecular Chemistry: From Molecular Information towards Self-organization and Complex Matter', *Reports on Progress in Physics*, 67 (2004), 249. Jean-Marie Lehn, 'From Supramolecular Chemistry towards Constitutional Dynamic Chemistry and Adaptive Chemistry', *Chemical Society Reviews*, 36 (2007), 151–60.

[13] Jean-Paul Behr, ed., *The Lock-and-Key Principle: The State of the Art—100 Years On* (Chichester, UK: Wiley, 2008).

[14] Agnes Ullmann, '*Escherichia coli* Lactose Operon', *eLS* (Chichester, UK: Wiley, 2009).

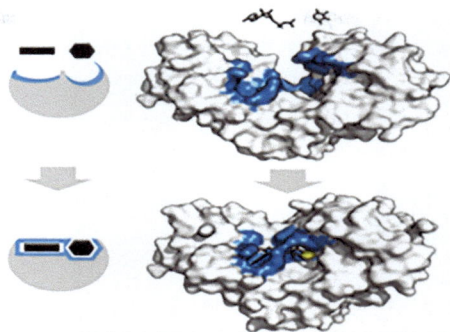

Figure 3.3 Conformational changes of enzymes. Enzyme changes shape by induced fit upon substrate binding to form enzyme–substrate complex. Hexokinase has a large induced-fit change of shape so that it that closes over adenosine triphosphate and xylose. Binding sites in blue, substrates in black and Mg^{2+} cofactor in yellow.

Acknowledgement: from Wikipedia.

enzymes to break down lactose into glucose. They need the gene G_{enz} that produces these enzymes to be turned on when lactose L is present and there is no glucose G, but G_{enz} needs to be off otherwise.

The basic principle to do this is to use a double negative. First, have an entity LR, the lac repressor that prevents G_{enz} from functioning when there is enough glucose G around. Then have a second entity the inducer I that turns off LR when there is too little glucose G. The result is this logic:[15]

$$\text{IF } \{(L) \text{ AND NOT } (I(G))\} \text{ THEN } \{(LR) \text{ ON}\} \text{ ELSE } \{(LR) \text{ OFF}\}. \tag{3.1}$$

The structure of the inducer $I(G)$ and lac repressor LR regulate the level of transcription from lac operon in this way,[16] shown in detail in **Figure 3.4**.

This is a form of downward causation from the local environment at the cellular level **L6** to the macromolecular level **L3**,[17] and then on down to the physics levels **L2** and **L1**, enabled by macromolecular structure. Other examples of logical operations in gene expression are known. For example, double-stranded RNA triggers suppression of gene activity in a homology-dependent manner, a process named RNA interference (RNAi). Discovering this led to the Nobel Prize in Chemistry in 2006.[18]

These are simple examples: more often operations are combinatorial and highly context-dependent,[19] as in the case of micro-RNA function discovered by the 2024 Chemistry Nobel Prize winners.[20] Because these and similar molecular functions

[15] Ullmann, '*Escherichia coli* Lactose Operon'.

[16] The Nobel Prize in Chemistry (2006): Popular Information and Advanced Information.

[17] Ball, *How Life Works*.

[18] Popular Information: The Nobel Prize in Chemistry 2006. Advanced Information: The Nobel Prize in Chemistry 2006.

[19] Ball, *How Life Works*.

[20] Popular Information: The Nobel Prize in Chemistry 2024. Advanced Information: The Nobel Prize in Chemistry 2024.

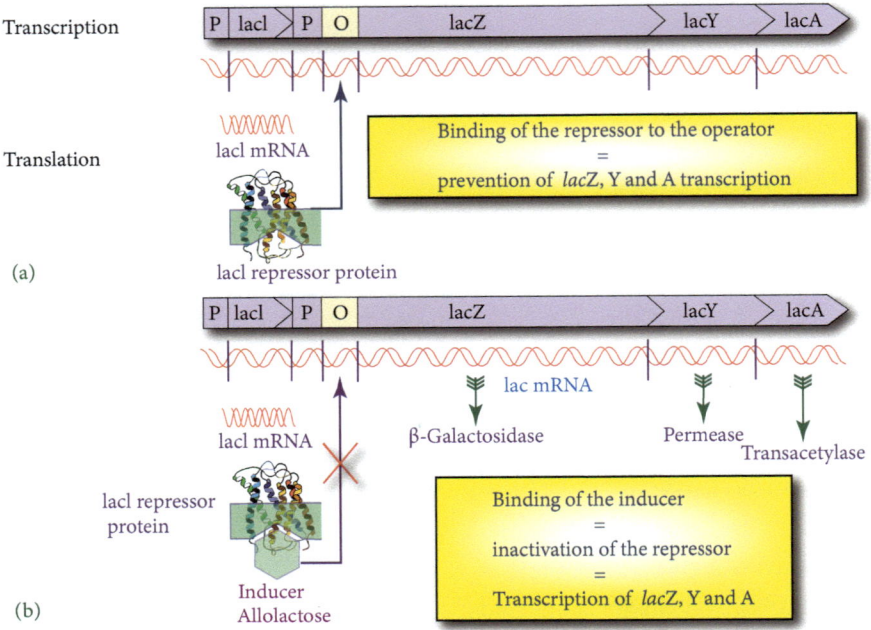

Figure 3.4 Functioning of the lac operon. Diagram of the lactose operon in the repressed (a) and induced (b) states. Synthesis of the lactose operon proteins, genetically determined by the structural genes (lacZ, lacY and lacA), is blocked by the LacI repressor synthesized by the regulator gene, lacI. The operator (O) is the site of specific interaction with the repressor. The repressor can be inactivated by the inducer, thus allowing transcription to take place at the promoter.

Redrawn by Mauro Carfora from Agnes Ullmann, '*Escherichia coli* Lactose Operon', *eLS* (Chichester, UK: Wiley, 2009).

underlie physiological functions, proteins are selected to come into being by the processes of natural selection discussed in Chapter 4; otherwise, these extraordinary structures would not exist.[21]

3.1.2 Level L4: Interaction Networks

Interaction networks are a key feature whereby complexity is built up from simple components. Proteins and RNA and DNA interact via gene regulatory networks, metabolic networks and cell signalling networks, which interact with each other.[22] They are characterized by network motifs: interactions patterns that recur in many

---

[21] Andreas Wagner, *Arrival of the Fittest: Solving Evolution's Greatest Puzzle* (London: Penguin, 2014).

[22] A. Goelzer et al., 'Reconstruction and Analysis of the Genetic and Metabolic Regulatory Networks of the Central Metabolism of *Bacillus subtilis*', *BMC Systems Biology*, 2 (2008), 20.

interaction networks.[23] In particular there are protein interaction networks[24] which inter alia underlie metabolic pathways. There can even be irreducible networks of networks.[25]

3.1.3 Level L5: Subcellular Machinery: Organelles

Cells come in two forms: prokaryotes, such as bacteria, without a nucleus, and eukaryotes with a nucleus, the basis of plants and animals. Eukaryotic cells are modular in nature, being made of many parts. They contain various types of organelles, including mitochondria, the endoplasmic reticulum, the Golgi apparatus and lysosomes (**Figure 3.5**), each carrying out a specific function critical to the cell's function. Lipid bilayer membranes surrounding organelles confine reactions by keeping the inside 'in' and the outside 'out'; this separation permits different kinds of reactions to take place in the different kinds of organelles. Molecular channels made of membrane proteins such as voltage- and ligand-gated ion channels control ingress and egress of molecules and ions through the membranes.

3.1.4 Level L6: Cells (the Basic Unit of Life)

Cells are the lowest level where all the functions of life exist. They are alive when we are alive; indeed they are what keep us alive. We are made up of eukaryotic cells (**Figure 3.5**) with their many organelles, which are complex, including a nucleus which houses our DNA organized as chromosomes.

Cells are replicators—they multiply by a complex process to create two cells where there was one before. This provides the basis of body growth as development takes place, increasing from a single cell to the huge number of 10^{13} cells that make up an adult human being. Note that this replication involves not just duplication of DNA, but also of all the cellular machinery that enables a cell to function. Much of that duplication takes place via metabolic processes that do not involve DNA.

Cell function depends on the continual uptake and conversion of energy and material into useful forms. Metabolism is the complete set of biochemical reactions within a cell that keep it alive,[26] including the key citric acid cycle.[27] We die if our cells do

---

[23] Uri Alon, 'Network Motifs: Theory and Experimental Approaches', *Nature Reviews Genetics*, 8, (2007), 450–61; Uri Alon, *An Introduction to Systems Biology: Design Principles of Biological Circuits*, 2nd edn (London: Chapman and Hall, 2019).

[24] Alexei Vazquez, 'Protein Interaction Networks', in O. Alzate, ed., *Neuroproteomics* (Boca Raton, FL: CRC Press, 2010).

[25] Leonardo R. Gorjão, Arindam Saha, Gerrit Ansmann, Ulrike Feudel and Klaus Lehnertz, 'Complexity and Irreducibility of Dynamics on Networks of Networks', *Chaos: An Interdisciplinary Journal of Nonlinear Science*, 28 (2018):106306.

[26] See the Wikipedia article Metabolism.

[27] See the Wikipedia article Citric acid cycle.

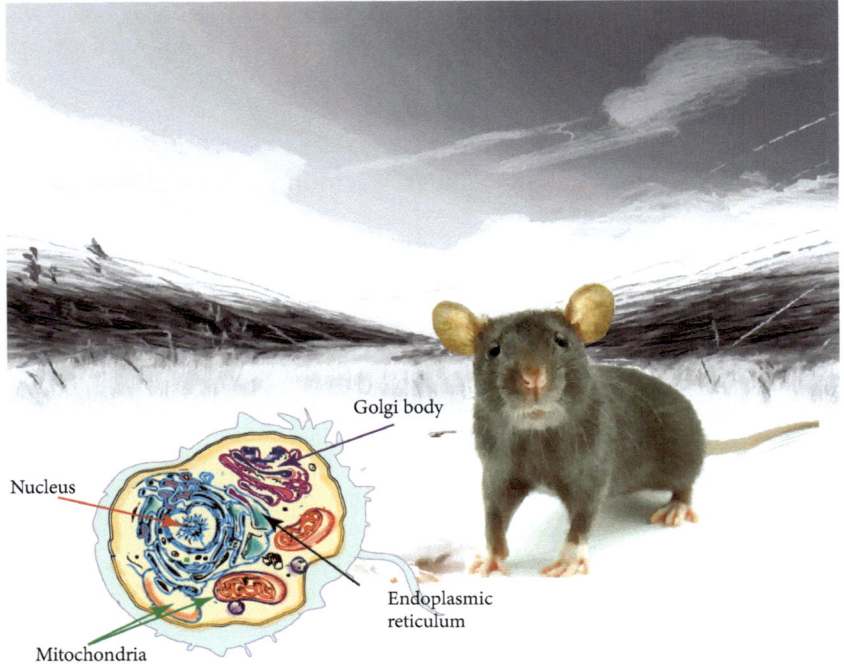

Golgi body

Nucleus

Endoplasmic
reticulum

Mitochondria

Figure 3.5 Eukaryote cell structure. Multicellular animals are made by eukaryotic cells imbedded in an extracellular matrix, and linked via cell adhesion molecules.

Figure by Mauro Carfora.

not get the oxygen, glucose, fats and proteins they need, provided to them by our circulatory system.

3.1.5 Level L7: Tissues

Body tissues are a level of organization between cells and physiological systems. The main kinds include the following:

- **Epithelial tissue**, forming a barrier between the external environment and the organ it covers, providing protection and allowing controlled ingress and egress to the organ. Skin is a key case, covering the body as a whole and so defining our physical boundaries.
- **Connective tissue** is found between other tissues everywhere in the body, including the nervous system. They bind other tissues to each other and form a scaffolding for other cells.
- **Muscular tissue** functions to produce force and cause motion. It is what enables us to speak, see (controlling eye movement), move and act on the world around us.

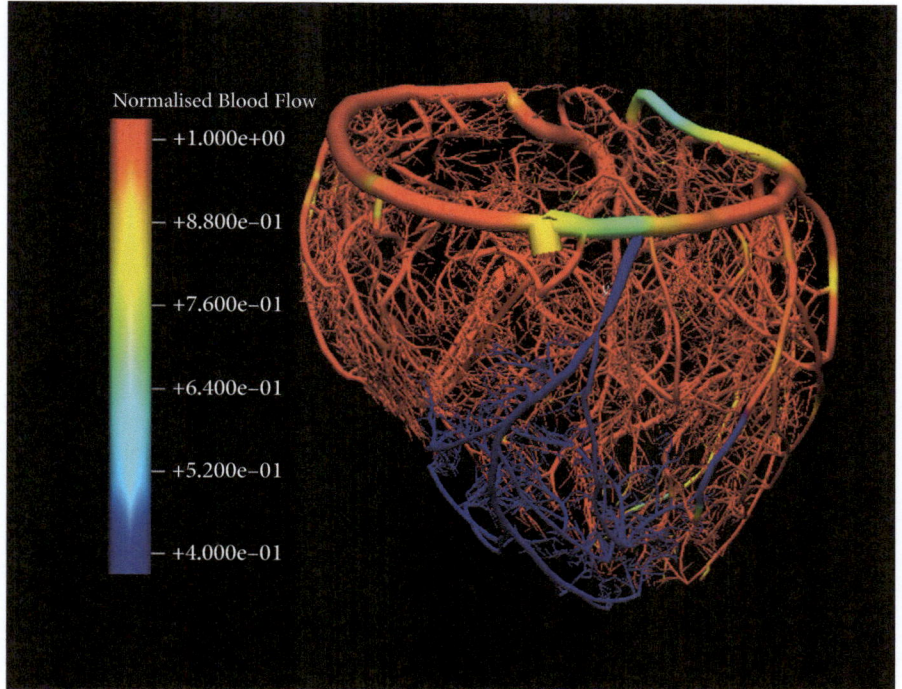

Figure 3.6 The human heart. Hierarchical structure of the heart.

From Denis Noble, 'Modelling the Heart—From Genes to Cells to the Whole Organ', *Science*, 295 (2002), 1678–82, with permission.

- **Nervous tissue** is the main component of the nervous system. It is composed of neurons connected by synapses, which receive and transmit action potential spike chains, and glial cells, which assist propagation of nerve impulses and provide nutrients to the neurons.

Thus they each have different functions in the emergent body, underlying and enabling the emergent physiological level **L8**, which acts down to shape specific outcomes at level **L7**, e.g. specific muscle motions.

3.1.6 Level L8: Physiological Systems

Human physiological systems meet the basic needs of life: each is necessary for our existence.[28] In each case form and function are closely intertwined, as for example in the heart (**Figure 3.6**).[29]

---

[28] See the Wikipedia article https://en.wikipedia.org/wiki/Human_body.

[29] Stephen A. Wainwright, 'Form and Function in Organisms', *American Zoologist* 28 (1988), 671–80.

The physiological systems carry out the following functions:

1. **Establishing our basic structure and identity**: a musculoskeletal system that creates rigidity with flexibility, and the permeable membrane of the skin that establishes the boundary of the body.
2. **Providing the existential needs**: air, water, food, providing energy we need and elimination of waste, met by the digestive system together with the circulatory system powered by the heart.
3. **Giving the ability to act**: arms and hands with fingers, powered by muscles and directed by the brain.
4. **Giving the ability to move**: met by legs and feet with toes, again directed by the brain.
5. **Enabling stability**: homeostatic systems to maintain the internal state within acceptable limits: body temperature, blood pressure, electrolyte levels, calcium levels, blood sugar levels in particular. We are ill if any of them are out of limits. Homeostasis is a deep principle of biological existence (§3.3).
6. **Providing the ability to ward off infectious diseases**: the immune systems (innate and adaptive) exist to fight off both previously encountered and novel infections.
7. **Receiving information**: sensory systems conveying information about the external world and the internal milieu: vision, hearing, taste, smell and somatosensory systems including balance and touch.
8. **Using that information**: brains to analyse that information, predict outcomes and make action choices.
9. **Creating new generations**: reproductive systems that ensure the survival of the species.

Each of these systems is an emergent system with functioning characterized and studied at the physiological level **L8**, enabled by the underlying levels, which this level in turn coordinates to meet these emergent needs, as in the case of the heart, with its hierarchical structure (**Figure 3.6**).

These systems have all evolved to become what they are because they greatly enhance our survival prospects. They work in an integrated way to enable us to function as organisms as a whole, each meeting a separate need. The key point is that, regarding graphs representing physiological processes,[30]

*These diagrams represent the idea of a **system**: a set of components acting in very specific ways to generate properties at the system level that are absent from any one component. The existence of **system-level properties** that are not seen in their components is called **emergence**, and is a very important property of living systems. Indeed life itself can be seen as an emergent property of the physiological mechanisms that underlie it.*

We are in trouble if that emergence fails, as e.g. in the case of heart failure.[31]

---

[30] Jamie A. Davies, *Human Physiology: A Very Short Introduction* (Oxford: Oxford University Press, 2021), 15.

[31] Wouter-Jan Rappel, 'The Physics of Heart Rhythm Disorders', *Physics Reports*, 978 (2022), 1–45.

3.1.7 Level L9: Organisms: Plants, Animals, Humans

This is the integrative level of organisms as a whole, able to react and adapt to the environment.[32] In the case of animals and human beings, this involves agency.[33]

All humans have the same physiology through our common genetic inheritance, but with varied detailed characteristics such as height, weight and colour of hair and eyes and skin. We are very special amongst animals because as well as standing on two feet, we have hands with opposable thumbs enabling fine-scale physical manipulation, together with a larynx, vocal chords, lips and tongue allowing for fine control of speech and song. These physiological adaptations are central to our abilities.

The mind/brain coordinates all this and enables mental causation by humans.[34] Crucially, we have a symbolic brain that uses symbolism to categorize the world and our experiences, underlying our ability to understand and plan and talk.[35] Language enables communicating, planning, logical argumentation, making theories and reasoning via verbalized mental models and metaphors.[36] Written text enables transfer of ideas over time and space. The symbolic structure of mathematics enables science and engineering to emerge, underlying all the varied uses of technology. It is this advanced symbolic ability that above all distinguishes us crucially from all other animals, including the great apes.

3.1.8 Level L10: Societies, Ecology and the Environment

Taken together, we form populations made of social groups of various sizes and degrees of coherence. We are social animals relying on our communities for life opportunities, leading to the existence of structured societies with varied cultures and a variety of organizations that fulfil specific functions. Societies interact with the people in them in upward and downward ways.[37] They emerge from the people making them up, but they do so through organizations that are more than the sum of the parts:[38] as organizations, they have agency. We could not exist without these social structures: they support us, provide us with social, emotional and material needs, and shape the way we think and act.

Societies in turn are imbedded in hierarchically structured ecologies and natural environments that provide their larger context and underlie their existence by

---

[32] See the Wikipedia article https://en.wikipedia.org/wiki/Organism

[33] Ball, *How Life Works*; Kevin Mitchell, *Free Agents: How Evolution Gave Us Free Will* (Princeton: Princeton University Press, 2022).

[34] David Robb, John Heil and Sophie Gibb, 'Mental Causation', in Edward N. Zalta, ed., *The Stanford Encyclopedia of Philosophy* (2021). https://plato.stanford.edu/archives/spr2023/entries/mental-causation/.

[35] Terence Deacon, *The Symbolic Species: The Co-evolution of Language and the Brain* (New York: Norton, 1998).

[36] George Lakoff and Mark Johnson, *Metaphors We Live By* (Chicago: University of Chicago Press, 2008).

[37] Peter L. Berger, *Invitation to Sociology: A Humanistic Perspective* (New York: Anchor Books, 1963).

[38] David Elder-Vass, *The Causal Power of Social Structures: Emergence, Structure and Agency* (Cambridge, UK: Cambridge University Press, 2010).

providing water and many other natural resources, some renewable and some not. This is discussed further in Chapter 6.

3.2 Open Systems: Metabolism and Control

All living beings are open systems.[39] They and the environment reciprocally exert forces on each other, and they exchange matter, liquids, energy, heat and information with the environment. The fact that they are open systems follows because they are finite: they have a boundary separating them from the rest of the Universe. These interactions inevitably take place because that boundary exists.

Because of the laws of thermodynamics (§2.3), we need energy to function which we derive from food, and we need to get rid of waste material and heat. Hierarchically structured metabolic systems handle these issues.[40] A key feature is how enzymes speed up these reactions at the molecular level, enabling the digestive system to function at the physiological level.

Additionally, being open systems means that living beings must be able to react appropriately to influences from the environment which can only partially be predicted from data currently available. Full knowledge of the visible current state of external affairs cannot reliably predict everything that will happen in the near future. Rain or snow may start to fall, a threatening figure may emerge from behind a wall or a motor car may crash in front of you. Two key features have been developed to handle these issues. The first is homeostatic systems that can routinely respond to classes of events likely to recur often (§3.3). The second is agency and an analytic brain that enables us to respond appropriately to situations that have never been encountered before. This is discussed in Chapter 5.

The same metabolic issues arise for buildings, factories, villages, towns and cities. They must have access to needed resources such as water and food, and must be able to get rid of the waste products after use. This is why they all need lavatories (toilets), waste water systems and rubbish storage and disposal facilities. They also need access to petrol (gasoline) and electricity for energy, and the Internet for information.

3.3 Homeostasis/Cybernetics

Homeostasis is the basic principle whereby stability of biological systems is achieved in the face of internal and external perturbations that threaten its stability.[41]

---

[39] George Ellis, 'Biological Emergence: a Key Exemplar of the Open Systems View', in Michael E. Cuffaro and Stephan Hartmann, eds, *Open Systems: Physics, Metaphysics, and Methodology* (Oxford: Oxford University Press, 2026). https://philpapers.org/archive/ELLBEA-4.pdf.

[40] Wainwright, 'Form and Function in Organisms'. David Fell, *Understanding the Control of Metabolism* (London: Portland Press, 2004).

[41] Walter B. Cannon, 'Organization for Physiological Homeostasis', *Physiological Reviews*, 9 (1929), 399–431; Harold Modell, William Cliff, Joel Michael, Jenny McFarland, Mary P. Wenderoth and Ann Wright, 'A Physiologist's View of Homeostasis', *Advances in Physiology Education*, 39 (2015), 259–66.

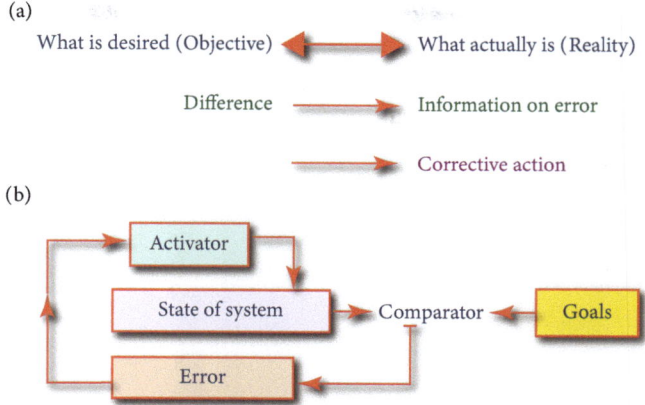

Figure 3.7 Homeostasis/cybernetics. (a) The basic principle. (b) The basic mechanism.
Figure by Mauro Carfora.

In engineering, the same principle is embodied in cybernetic systems.[42] The basic idea is shown in **Figure 3.7(a)**. The basic underlying mechanism is shown in **Figure 3.7(b)**. A goal G or setpoint is chosen.

A sensor determines the state of the system S, and a comparator compares it with the set goal G and so determines the error $E = S - G$. The activator changes the state of the system if $|E|$ is bigger than the threshold Δ to reduce this discrepancy between the actual and desired state. The point is that the specific disruptive effects, whether internal (heart rate too high, temperature too high) or external (too hot or cold, a wind threatening to blow you off the cliff), cannot be predicted, but the classes of such effects can, and can be protected against by homeostatic mechanisms (physical, biological, engineering, social, mental) to counteract them reliably.

An example is a thermostat that changes room temperature by controlling a heater (an emergent system) so as to attain the desired temperature T_0 at a macro level L. If the temperature is too low, the system turns the heater on, and hence changes the mean speed of motion $<v>^2$ of molecules (a level $L - 1$ variable). This is an example of emergence: the corrective action is due to the closure of the feedback control loop indicated in **Figure 3.8**: the thermostat that measures the state of the system is connected via the comparator to the activator which heats the air. If the loop is broken by disconnecting the heater, all the constituent elements (and the atoms out of which they are made) will still be there but the topology is changed (there is no longer a closed loop). It is also an example of downward causation: the setting on the thermostat is changed by a dial or buttons (macro-level objects) and results in the average temperature T of the room changing, which results in the mean square velocity $<v>^2$ of atomic molecules in the room increasing.

---

[42] Norbert Wiener, *Cybernetics: Or Control and Communication in the Animal and the Machine*, 2nd rev edn (Cambridge, MA: MIT Press, 1961).

Figure 3.8 A thermostat and heating system. The control sets the desired temperature T_0. A sensor measures the actual temperature T. A circuit at the macro level turns a heater on if $E = T - T_0$ exceeds a threshold Δ. In this way the average mean square velocity $\langle v \rangle^2$ of molecules at the micro level gets controlled in a downward way by the heater. Thus it demonstrates the downward causation involved.

Figure by Mauro Carfora.

Branching logic. The key point then is that the operation of any feedback control system enables the emergence of a form of contextual branching logic. In the case of the thermostat, the logic is

$$\text{IF } \{E > \Delta\} \text{ THEN } \{\text{turn on heater}\} \text{ ELSE } \{\text{not}\}. \tag{3.2}$$

This emergence of an **(IF . . .THEN . . . ELSE)** logic is in sharp contrast to the underlying laws of physics, which by themselves entail no such branching logic.

This principle underlies the way that engineering control systems work and living organisms function. Indeed, all purposive activity that reliably attains goals works this way. Immensely complex feedback control systems occur in plant and animal physiology, resulting in goal seeking and apparently purposeful activity at all levels of the biological hierarchy. The processes in each living cell are controlled by numerous feedback control loops associated with metabolism and the activities of enzymes at that level; the temperature of the human body is maintained accurately through physiological control loops at the macroscopic scale; heartbeats are automatically regulated at the physiological level; metabolite levels are controlled within precise limits;

you are ill if any of these systems is not functioning properly. In all these cases, the mechanism and setpoints have been determined by Darwinian evolution.

In engineering cybernetic systems, the key difference is that the setpoint is under our control, as in the case of a thermostat. They are central in many engineering contexts, such as a chemical plant, manufacturing, and aircraft autopilots. In advanced cases it can include predictive and adaptive elements.

3.4 Adaptative Selection: Addition, Modification, Removal

Addition, modification and removal of elements takes place at all levels. In particular, lower-level elements are shaped by higher levels so as to fulfil higher-level needs by these processes. This is crucially different to cases where the existence and nature of lower-level elements is independent of context, as in the kinetic theory of gases. In Aristotelian terms, this is material causation.[43]

Addition of lower-level elements can take place either by some kind of creation process or by allowing entry into the system of selected kinds of elements.

Creation happens at the molecular level via metabolic networks that create needed molecules such as ATP, and via gene regulatory networks that determine what proteins will be created in the body by reading specific genes in a particular context. At the cellular level, it happens by cell division, which is the basis of human growth, and includes the processes whereby DNA is duplicated. At the physiological level, it happens via developmental systems that create hearts, eyes, lungs, bones and all the rest in a developing embryo. At the organism and population level, this creates new kinds of animals that were not there before.

Entry into the system is controlled at the molecular level by ion channels such as voltage- and ligand-gated ion channels, determining what ions and molecules are allowed to enter a cell. At the physiological level it is controlled by the digestive system and respiratory system, both under mental control to some extent (we can usually choose what we eat and drink, avoiding substances we don't like). At the ecosystem level, it takes place by migration.

Modification of lower-level elements at the molecular level includes the amazing proofreading systems whereby errors in DNA replication are detected and corrected. At the cellular level this is via the developmental systems that determine what type of cell will be created where in the human body, out of pluripotent cells (Chapter 4). At the physiological level it is via the digestive system that digests incoming food into usable form as nutrients, and discards the waste. At the ecosystem level it takes place by the processes of natural selection.

Removal can take place by deleting or exporting unwanted elements. At the molecular level, ion channels selectively remove ions from cell interiors in a controlled way, such as voltage-gated proton channels. At the cellular level this takes place by

[43] George Ellis, 'Efficient, Formal, Material, and Final Causes in Biology and Technology', *Entropy*, 25 (2023), 1301.

apoptosis—programmed cell death. At the physiological levels this is via the excretory systems.

At the ecosystem level this is both the central process that underlies Darwinian natural selection, with less fit species being driven to extinction, and a central process in ecosystem dynamics.

There are analogues to these three processes in all social organizations; indeed they are central to their functioning. For example, a firm can recruit people, thus letting them into the firm; train them, thus altering them so as to fit a specific job; and fire them if they fail at what they are supposed to do.

They are also key to chemical engineering and various manufacturing processes. Indeed purification processes are crucial to all manufacturing, and particularly the pharmaceutical industry and chemical engineering, for example fractional distillation of petroleum products[44] and water treatment plants.[45]

3.5 Cycles/Oscillations

These are pervasive in biological, engineering and social systems. They are based in one of two major principles: deriving from some orbital motion or rotation, or caused by filling up and then emptying a reservoir of some kind. They are also of two basic types: externally imposed, or internally generated.

Externally imposed. There are two key cycles shaping our lives, related to each other, based on the fact that we live in the Solar System with the axis of rotation of the Earth tilted relative to our orbit round the Sun (see §1.7). This results in yearly cycles—the annual progression of summer, autumn, winter and spring—superimposed on the daily cycles caused by the rotation of the Earth—the daily progression of day followed by evening, night and dawn. The former modulates the latter in a latitude-dependent way: the day is longer in summer and shorter in winter, even failing to occur at all in extreme latitudes in the heart of winter.

These cycles provide the basic temporal framework for life on Earth, in particular through the influence of the seasons on weather, thereby dominating daily life, social life, many forms of work, farming, gardening, sports, leisure and travel. These weather effects reach down to shape biological development in vegetation, adapted by natural selection to have annual growth patterns according to the seasons, and to affect patterns of behaviour of animals who hibernate and birds who carry out extraordinary large distance annual migrations to get to better weather.

In this way the astronomical nature of the Solar System reaches down to affect biomolecules and electrons in living plants and animals and birds, thereby effecting these seasonal patterns of behaviour in the biological world.

Additionally, monthly cycles generated by the motion of the Moon around the Earth have an important effect through the way they modulate ocean tides, which are important in coastal contexts, affecting both marine life on the one hand and human

---

[44] See Wikipedia article Fractional distillation.
[45] See Wikipedia article Water purification.

activities—swimming, sailing, canoeing and so on, and maritime transport—in a predictable way.

Internally generated. There are many internally generated cycles at all scales in biological, technological and social systems. Two key physiological cycles are those of breathing, generated by the lungs, and the pumping of the heart, circulating blood; these are what keep us alive. The oscillations of the heart are generated by protein interactions within the context of subcellular, cellular, tissue, organ and system structures, including pacemaker cells. The lungs work via cycles of inhalation and exhalation controlled by the lower brain in combination with a variety of other receptor and reflex centres. These cycles are both central to our survival.

The brain has many basic rhythms: they occur in delta (0–4 Hz), theta (4–8 Hz), alpha (8–14 Hz), beta (14–30 Hz) and gamma (30–80 Hz) bands, the lower frequency ones being related to sensory processing. The brain also forms temporary adaptive resonant circuits that can be claimed to underlie autonomous adaptive intelligence.[46]

The daily cycles of life—waking and sleeping, daily meals, work and play and rest— are due to a combination of internal clocks and the imposed astronomically based rhythm of day and night (remember that our bodies have adapted to this daily rhythm over millennia of adaptive selection). In industrial societies, timekeeping via watches and clocks plays a key role in regulating activities in a socially agreed way: opening and closing of shops and offices, working hours and school hours, times of opening of entertainment venues and restaurants and so on. The daily cycle of time as monitored by clocks is a central feature of daily life.

Four major cyclical processes that result from internal physiological needs are crucial to our continued well-being, although the times involved can be very variable. They are, breathing in and breathing out; ingesting liquids and disposing of waste; eating food and disposing of waste; being awake and sleeping.

An intriguing issue is where does the universally adapted rhythm of a seven-day week, with a one- or two-day weekend, come from? It is not imposed astronomically, as the daily, monthly and yearly rhythms are. I hypothesize that it is a natural internal timescale in the human brain—we need rest after five days of continual activity (work, school, whatever) in order to recover and recharge for the next cycle of regular activity. It seems basic to human life across the world.

3.6 The Set of Levels Overall, Causal Closure, and the Varieties of Existence

Each of the emergent levels **L3** to **L10** has been brought into being by evolutionary processes (Chapter 4) because each of them fulfils a vital function enabling life to exist. Each of these levels has its own characteristic behaviours different from the

---

[46] Stephen Grossberg, 'Adaptive Resonance Theory: How a Brain Learns to Consciously Attend, Learn, and Recognize a Changing World', *Neural Networks*, 37 (2013), 1–47. Stephen Grossberg, *Conscious Mind, Resonant Brain: How Each Brain Makes a Mind* (Oxford: Oxford University Press, 2021).

L10	Sociology/economics	Smoking advertising, social pressure
L9	Individual psychology	"I enjoy smoking": rationality, emotion
L8	Human physiology	Smoking causes cancer: causal relation (Pearl)
L7	Tissues	Inter alia affects lymph nodes
L6	Cells biology	Uncontrolled cell duplication
L5	Sub-cellular machinery	Enables cell duplication
L4	Interaction networks	Protein production based on DNA code changes
L3	Proteins, RNAs, genes	DNA altered, macromolecule interactions change
L2	Atomic physics	Atoms bind to give molecules
L1	Particles	Electron/proton interactions

Figure 3.9 Causal closure: smoking causes lung cancer.

others, characterized by effective laws and variables at that level. They are irreducible to any lower level, although they are able to exist and function only because of those lower levels.

Causal closure only occurs via the upward and downward interactions between all the levels **L1** to **L10**: specific real-world outcomes are not determined if we omit any of these levels.[47] An example is a well-established causal relation: at the physiological level, **smoking causes lung cancer (Figure 3.9)**. This has been analysed by Judea Pearl,[48] carefully distinguishing causation and correlation. It is driven in a downward way, but enabled in an upward way.

Efficient causation takes place at each level **L1** to **L10**, characterized by effectives laws at that level. This starts with pressures at the societal level **L10** where major advertising campaigns, carefully researched, encourage smoking. It may also be encouraged at social events. At the psychological level **L9** a choice has to be made as to whether to smoke: this is where emotion and rationality battle it out. If smoking is chosen, smoke particles will enter the lungs and greatly enhance the probability of catching cancer at level **L8**,[49] as demonstrated by Judea Pearl. It will be caused by changes to gene regulatory networks at Levels **L5** and **L4** due to the smoke particles altering DNA at level **L3**,[50] changing cell division processes and

[47] George Ellis, 'The Causal Closure of Physics in Real World Contexts', *Foundations of Physics*, 50 (2020), 1057–97.

[48] Judea Pearl and Dana Mackenzie, *The Book Of Why: The New Science of Cause and Effect* (London: Penguin Books, 2018).

[49] Stephen Hecht, 'Lung Carcinogenesis by Tobacco Smoke', *International Journal of Cancer*, 131 (2012), 2724–32; Wendy Cooper, David C. L. Lam, Sandra A. O'Toole and John D. Minna, 'Molecular Biology of Lung Cancer', *Journal of Thoracic Disease*, 5(Suppl 5) (2013), S479.

[50] Hecht, 'Lung Carcinogenesis by Tobacco S'; Cooper et al., '"Molecular Biology of Lung Cancer'.

Figure 3.10 **A variety of landscapes, partly natural and partly man made.**

hence causing uncontrolled cell proliferation at level **L6** that affect the lymph nodes at level **L7**. This is all enabled by molecular processes based in physical chemistry and physics at levels **L2** and **L1**. Specific outcomes that occur are determined by this entire set of interactions as a whole. This process chains down from level **L10** to level **L1**, which enables the whole to take place. The physics level **L1** by itself is not causally closed, contra claims often made. It is the set **L1** to **L10** as a whole that is closed.[51]

The varieties of existence. I have summarized a great deal of biological understanding in this chapter. What is the point of doing so? It is to fill out the claims in the Prologue in a solid way. Needs chain up through the levels discussed here: cells need nutrition and oxygen (**L6**), which is why related physiological systems exist (**L8**), and why we need to eat and drink (**L9**), and therefore why agriculture is needed in society (**L10**). Outcomes chain down: agriculture (**L10**) provides food which is used by physiological systems (**L8**) to ensure cells (**L6**) have the energy they need to continue functioning and stay alive via levels **L4** and **L3**, which shape outcomes at **L1** and **L2** that would otherwise be undetermined. This all taken together leads to the varieties of existence indicated in **Figure 3.10**. The three essential levels underlying functioning of life are **L4** (macromolecular chemistry), **L6** (cells) and **L9** (the organism as a whole).

---

[51] Matteo Mossio, 'Causal Closure', in Dubitzky et al., eds, *Encyclopedia of Systems Biology* (Berlin: Springer, 2013), 415–18. Matteo Mossio, Leonardo Bich and Alvaro Moreno, 'Emergence, Closure and Inter-level Causation in Biological Systems', *Erkenntnis*, 78 (2013), 153–78. Maël Montévil and Matteo Mossio, 'Biological Organisation as Closure of Constraints', *Journal of Theoretical Biology*, 372 (2015), 179–91.

Further Reading

For biology in general: two classic books

N. A. Campbell and J. B. Reece, *Biology* (Chennai: Pearson Education India, 2005).

P. Ball, *How Life Works: A User's Guide to the New Biology* (Chicago: University of Chicago Press, 2023).

For the macro-molecular level

L. H. Hartwell, J. J. Hopfield, S. Leibler and A. W. Murray, 'From Molecular to Modular Cell Biology', *Nature*, 402(Suppl.) (1999), C47–C52.

G. A. Petsko and D. Ringe, *Protein Structure and Function* (London: New Science Press, 2004).

M. J. Berridge, *Cell Signalling Biology* (London: Portland Press, 2014).

P. M. Hoffmann, *Life's Ratchet: How Molecular Machines Extract Order from Chaos* (New York: Basic Books, 2012).

For the Cellular Level

S. Mukherjee, *The Song of the Cell* (London: Vintage, 2023).

For the interactions between Levels

D. Noble, 'Modelling the Heart—From Genes to Cells to the Whole Organ', Science, 295 (2002), 1678–82.

R. Noble and D. Noble, *Understanding Living Systems* (Cambridge, UK: Cambridge University Press, 2023).

For the social level: another classic book

P. L. Berger, *Invitation to Sociology: A Humanistic Perspective* (New York: Anchor Books, 1963).

4

How Did We Get Here?

Natural Selection and Developmental Processes

The properties of life, and the huge range of its manifestations, are truly remarkable (**Figure 4.1**). How did the various forms of life come into existence? How does each individual living entity grow to its mature form? Both processes are fundamental to our existence: we would not be here without both of them—and the way they interact with each other. These are the topics of this chapter. The context of the discussion is the hierarchy of existence shown in **Figure 4.2**.[1]

Two different timescales are involved in our coming into being: evolutionary timescales (millions of years to decades in some cases) and developmental timescales (decades to hours), depending on the organism. Quite different processes are involved in these two cases. In the illuminating book *Endless Forms Most Beautiful* by Sean B Carroll,[2] hereafter referred to as EFMB, Carroll expresses it this way (EFMB:4):

> Every animal form is the product of two processes: development from an egg and evolution from its ancestors. . . . Development is the process that transforms an egg into a growing embryo and eventually an adult form. The evolution of form occurs through changes in development.

But they depend on a further process and timescale: the functioning of our minds and bodies, and the timescale on which they react. This after all determines whether we survive some critical situation.

The attempts to understand the developmental processes by which each of us come into existence stretch back at least to Aristotle. They have been revolutionized by the understanding of the processes of molecular biology, as presented in Chapter 3. The discovery of natural selection by Wallace and Darwin[3] in 1870s was an absolutely central event in our understanding of our existence. It removed the need for a divine creator to bring into being birds, giraffes, zebras and ourselves.[4] This could, from then on, be regarded as the result of a natural process, taking place over very long times—much longer than any individual life. This again was transformed by the molecular biology revolution, because evolution could then be understood as a

---

[1] The same as **Figure 3.1**, which applies to physiological functioning, but here applies to developmental and evolutionary processes. Levels **L3** to **L6** come into being by abiogenesis: the creation of life (§4.7).

[2] Sean B. Carroll, *Endless Forms Most Beautiful: The New Science of Evo Devo and the Making of the Animal Kingdom* (London: Wiedenfeld, and Nicholson, 2005).

[3] Charles Darwin and Alfred R Wallace, *Evolution by Natural Selection (Memorial Volume)* (Cambridge, UK: Cambridge University Press, 1958).

[4] William Paley, *Natural Theology or Evidences of the Existence and Attributes of the Deity* (London: J. Faulder, 1802).

How We Come to Be. George F. R. Ellis, Oxford University Press. © George Ellis (2026).
DOI: 10.1093/9780198950189.003.0006

Figure 4.1 The variety of life.

Figure by Mauro Carfora.

process of selection of genes. However, natural selection does not only operate at the genetic level: it acts at all emergent levels, in an interactive way. This is a key feature I discuss here.

The sections of this chapter are, §4.1, 'Evolution by Natural Selection'; §4.2, 'Debates Regarding Evolution'; §4.3, 'Developmental Processes and the Life Cycle'; §4.4, 'EVO–DEVO Interactions'; §4.5, 'Including Physiological Processes: EVO–DEVO–Physio'; §4.6, 'The Whole Shebang: The Multilevel Nature of Evolution'; and §4.7, 'Abiogenesis: How Did Life Start?'

4.1 Evolution by Natural Selection

Evolutionary processes of natural selection are the origin of the DNA that makes up the genome. This is a key foundation of our existence,[5] as DNA shapes what kind of animal each species becomes. Human DNA is different than that of plants, mice and great apes, even though many genes are shared with them. Each individual has slightly different genes, playing a key role in shaping who we are as persons.

The evolutionary process is this:[6] in a given population (sharing the same DNA pool), when reproduction takes place, as a result of mutation, genetic recombination

---

[5] Theodore Dobzhansky, 'Nothing in Biology Makes Sense Except in the Light of Evolution', *The American Biology Teacher*, 75 (2013), 87–91.

[6] Ernst Mayr, *What Evolution is* (London: Phoenix, 2002); Andy Gardner, 'Adaptation as Organism Design', *Biology Letters*, 5 (2009), 861–4.

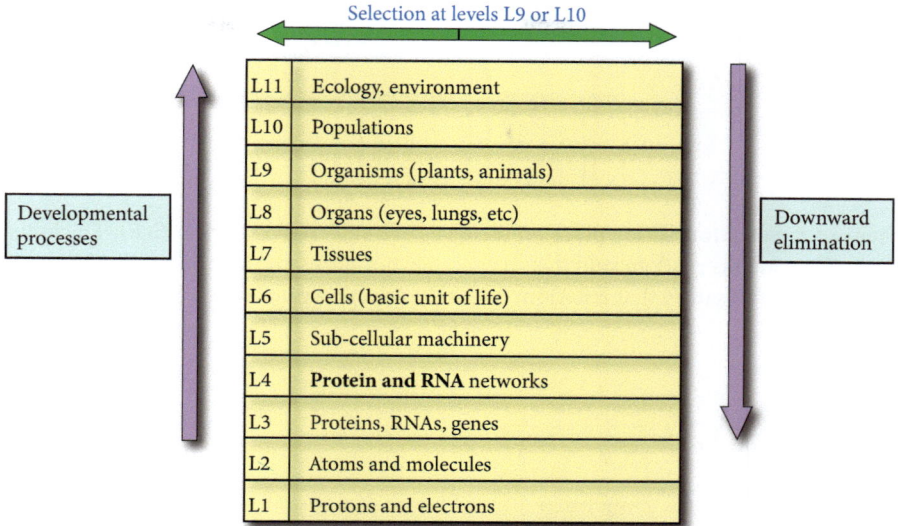

Figure 4.2 **The chaining up of developmental processes and down of selective processes** in the hierarchy of emergent levels. As this takes place, every emergent level must adapt in a consonant way, else the whole will not work. Selection works right down to the macromolecular level **L3**, altering the nature of components and the effective equations at this level. It thereby shapes what occurs at levels **L2** and **L1**, but without altering the components present or effective equations at those levels. A crucial feature is the creation of higher emergent levels (**L7** to **L11**) through both developmental and evolutionary processes.

and other sources of change, variations in DNA in the offspring will take place. Developmental processes will result in a varied population of adults based on these varied genes. They will compete with each other for resources and survival until they themselves produce progeny. Those adults that have a relatively better reproduction rate because of their physiological and ecological interactions will become relatively dominant in the shaping the next generation gene pool, because the genes of other less successful adults will be subdominant, or eliminated. Selection chains down from either the organism or the group level, depending on which is more dominant in determining survival and reproduction, to the gene level. Major transitions in evolution occurred as new levels of structure emerged through this process,[7] involving new levels of information processing.[8] After life started—cells existed at level **L6** based in existence of levels **L3** to **L5**—the higher levels **L7** to **L11** did not yet exist. They had to be added on by one by one over evolutionary timescales, with two key transitions: the coming into being of eukaryotic cells from prokaryotic cells, and the coming into being of the first multicellular organisms, both at level **L6**.

[7] Eörs Szathmáry and John Maynard Smith, 'The Major Evolutionary Transitions', *Nature*, 374 (1995), 227–32.

[8] Eva Jablonka and Marion J. Lamb, 'The Evolution of Information in the Major Transitions', *Journal of Theoretical Biology*, 239 (2006), 236–46.

Figure 4.3 Selection taking place on the basis of a selection criterion *C(E)*. The initial ensemble contains random stuff that may or may not satisfy the criterion *C(E)*. The chosen group all satisfy that criterion; hence the selection gate creates order from disorder.

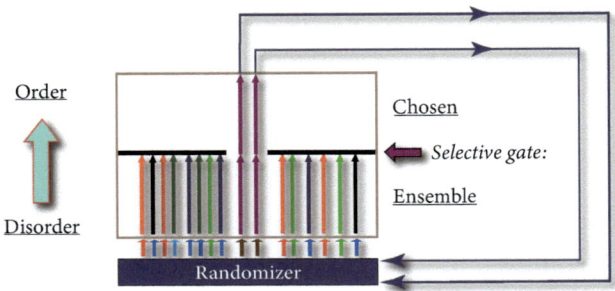

Figure 4.4 Repetition of randomization and selection. The true power of a selective process results from imbedding it in a larger system that repeatedly involves randomization before the next round of selection. This is how Darwinian natural selection works.

In more detail, this process can be regarded as having two stages. The first is selection of members of an ensemble on the basis of a selection criterion *C(E)* which depends on the environment *E*. Meaningful information is gained by discarding all the information/stuff received that is not meaningful in the sense that it does not obey the selection criterion *C(E)* (see **Figure 4.3**).

In this case, the logic of the process is

$$\text{IF } (X \text{ in } C(E)) \textbf{ THEN } (\text{keep } X) \textbf{ ELSE } (\text{delete } X). \tag{4.1}$$

There can be multiple selection stages of this kind, some of which may have multiple selection criteria.

A simple case is sorting out incoming emails, selecting those that are meaningful and deleting the rest. One can then apply a multiple-criterion selection process to those that are kept, sorting them into categories such as personal, work, financial, travel, and so on by using multiple criteria $C_N(E)$ to sort the emails into N distinct categories. This is a very general process, for example underlying the selective nature of vision via predictive processing, and how learning selects specific neural connections (as discussed in Chapter 5).

As shown in **Figure 4.4**, the key point about natural selection in biology is the way that selection is repeated over and over, with randomization processes creating new varied ensembles that can be selected from each time. It is the relentless repetition of the selection process in this way that gives it its awesome power, evidenced by the great variety of species it has produced (**Figure 4.1**). Evolutionary selection extended to the evolution of consciousness, because this enabled causal thinking and predictive powers that enabled better adaption to the ecological environment, and then development of technology that enabled humans to dominate the world. Cecelia Heyes states,[9]

Humans are animals that specialize in thinking and knowing, and our extraordinary cognitive abilities have transformed every aspect of our lives. In contrast to our chimpanzee cousins and Stone Age ancestors, we are complex political, economic, scientific and artistic creatures, living in a vast range of habitats, many of which are our own creation. Research on the evolution of human cognition asks what types of thinking make us such peculiar animals, and how they have been generated by evolutionary processes. New research in this field looks deeper into the evolutionary history of human cognition, and adopts a more multi-disciplinary approach than earlier 'Evolutionary Psychology'. It is informed by comparisons between humans and a range of primate and non-primate species, and integrates findings from anthropology, archaeology, economics, evolutionary biology, neuroscience, philosophy and psychology. Using these methods, recent research reveals profound commonalities, as well striking differences, between human and non-human minds, and suggests that the evolution of human cognition has been much more gradual and incremental than previously assumed. It accords crucial roles to cultural evolution, techno-social co-evolution and gene–culture co-evolution. These have produced domain-general developmental processes with extraordinary power—power that makes human cognition, and human lives, unique.

Downward causation. Because evolution depends on the ecological context **L11**, it is a form of downward causation from level **L11** to animal structure and function[10] (see Figure P.3 in the Prologue and **Figure 4.2**).

A key issue is whether this takes place due to organism changes at Level **L9**, leading to more useful individual traits (stronger limbs, better eyes, etc), in turn leading to a more successful group; or due to group level changes at Level **L10**, due to the existence of cooperative traits that enhance group survival, therefore leading to organism survival (Level **L9**). Some claim that group level selection does not take place, and others that it does,[11] with selection of groups that are genetically related leading to altruistic behaviour ('kin selection').[12]

I will not enter this technical argument here, but just state that from the viewpoint of this book, it is quite clear that either process is possible, depending on context.

[9] Cecelia Heyes, 'New Thinking: The Evolution of Human Cognition', *Philosophical Transactions of the Royal Society*, B367 (2012), 2091–6.
[10] Donald Campbell, 'Downward Causation in Hierarchically Organised Biological Systems', in F. J. Ayala and T. Dobhzansky, eds, *Studies in the Philosophy of Biology: Reduction and Related Problems* (London: Palgrave, 1974), 179–86.
[11] Jonathan Birch, 'The Inclusive Fitness Controversy: Finding a Way Forward', *Royal Society Open Science*, 4 (2017), 170335. Jonathan Birch and Samual Okasha, 'Kin Selection and Its Critics', *BioScience*, 65 (2015), 22–32.
[12] See the Wikipedia articles Kin selection and Inclusive fitness.

Indeed both have happened in the past. In some cases the group survives better because the individuals making up the group are better equipped to survive. They are stronger, can run faster, see better and so on, so a group (at level **L10**) made up of such individuals (at level **L9**) is better equipped to survive. That is the driver of group survival.

In other cases it is the fact that living together in groups at level **L10** enables survival more likely, as the group can hunt together in a coordinated way, make weapons through manufacturing processes only made possible by group cooperation and so on. Group survival at level **L10** chains down to ensure survival of cooperative individuals in the group at level **L9**. A central example of this process is the development of a symbolic brain[13] allowing for the evolution of language[14] and hence development of cooperation and culture. Language would not exist if group-level selection did not take place.

There are three fundamental principles regarding how evolution works.

The first is that selection does not just involve the group level **L10** and organism **L9**: it chains down to all the other lower levels **L8** to **L3**, selecting entities at each of these levels that will enable desirable changes at level **L9** to occur, including selecting DNA, RNA and proteins at level **L3**. Each level is adapted to provide the services needed by the higher levels. During this process, multiple realizability occurs: there are ever more alternative ways of realizing each lower level while delivering the outcomes desired by higher levels in order that the higher levels can function. By the time one reaches the macromolecular level **L3** there are literally billions of alternative ways the desired outcome can be realized. That is why higher-level purposes—functions that are desired, such as creating an eye that can see—are effectively invisible at the molecular level, where it is only in very few cases that a single gene or small number of genes determine higher-level outcomes. This multiple realizability underlies some of the differences between the way population geneticists and others such as ecologists and physiologists see evolution (see §4.2).

Second, the principles of assembly theory[15] hold:

You can only get to something new on the basis of what already exists.

Things can't just arise from nowhere: you have to modify what is available. This ties evolutionary processes into historical contingency and the concept of **the adjacent possible**,[16] central to the evolutionary processes discussed in Wagner's important book *Arrival of the Fittest*.[17]

Third, the higher levels shown in **Figure 4.2** did not exist at early stages of the evolutionary process, when only single-cell organisms existed (at Level **L6**). Evolution led to the higher levels coming into existence. Multicellar organisms had to arise from

---

[13] Terence Deacon, *The Symbolic Species: The Co-evolution of Language and the Brain* (New York: WW Norton, 1998).

[14] W. Tecumseh Fitch, 'The Evolution of Language: A Comparative Review', *Biology and Philosophy*, 20 (2005), 193–203.

[15] Abishek Sharma et al., 'Assembly Theory Explains and Quantifies Selection and Evolution', *Nature*, 622 (2023), 321–8.

[16] Lennart Björneborn, 'Adjacent Possible', in V. P. Glăveanu, ed, *The Palgrave Encyclopedia of the Possible* (Cham: Palgrave Macmillan, (2022), 16–28.

[17] Andreas Wagner, *Arrival of the Fittest: Solving Evolution's Greatest Puzzle* (London: Penguin, 2014).

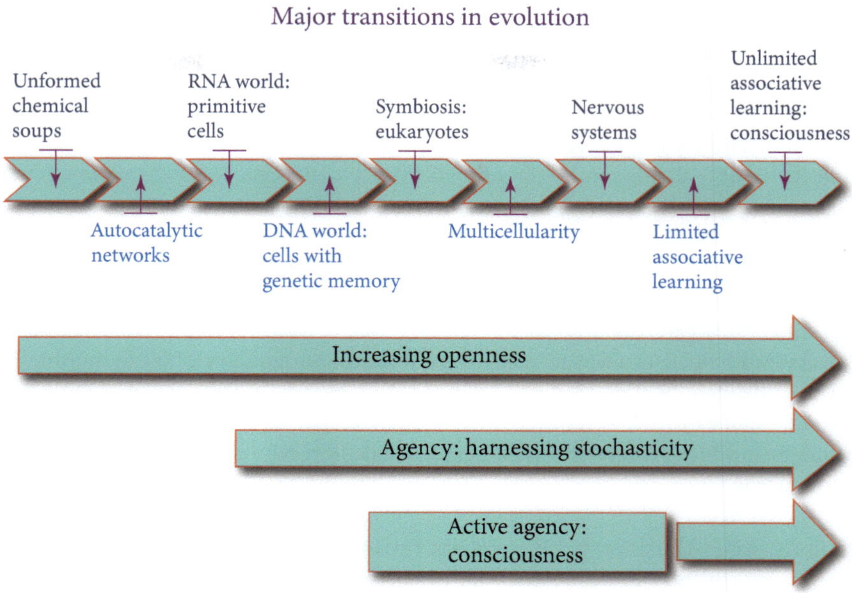

Major transitions in evolution

Figure 4.5 **The major transitions in evolution.**

Adapted by Mauro Carfora from Raymond Noble and Deni Noble, *Understanding Living Systems* (Cambridge, UK: Cambridge University Press, 2023).

single-cell organisms, and then tissues enabled physiological systems to arise. These changes are some of the major transitions in evolution (**Figure 4.5**).[18] Assembly theory applies in these cases: the higher levels **L7** to **L11** could only come into existence when the next lower level already existed, although they could modify that lower level as needed, once they existed.

4.2 Debates Regarding Evolution

While there is agreement on the overall process, there are many vigorous debates about specifics. Darwin laid down the general basis as discussed earlier—a competition to survive, the relatively fittest win, resulting in apparent design. But there are major disputes in evolutionary theory between those with a strongly reductionist viewpoint and those with a more holistic viewpoint.

This book presents an integrative view of evolution that I believe is correct. However, the reader should be aware that much of it is denied by the reductionist approach of the modern synthesis. Andreas Wagner puts it like this:[19]

[18] Szathmáry and Maynard Smith, 'The Major Evolutionary Transitions'.
[19] Wagner, *Arrival of the Fittest*, 19–21. See also Scott F. Gilbert, John M. Opitz and Rudolf A. Raff, 'Resynthesizing Evolutionary and Developmental Biology', *Developmental Biology*, 173 (1996), 357–72 for a similar analysis.

The modern synthesis . . . is a grand achievement of the human mind. There is, however, a dirty secret behind its success. The architects of the modern synthesis focused on the genotype at the expense of the organism and its phenotype. They neglected the marvellous complexity of organisms with their trillions of cells, each inhabited by billions of molecules whose functions are themselves incredibly complex. And they neglected how all this complexity unfolds from a single fertilised cell, and how genes contribute to this unfolding. By neglecting this complexity, the architects of the modern synthesis effectively ignored its product: the organism itself. . . .

To say that all evolutionists had thrown the organism under the bus, however, would be unfair to a minority of them, those who compared how the complexity of different organisms unfolds in embryos. But these embryologists, whose forebears had helped Darwin to recognize the common ancestry of all living things, were sidelined by the modern synthesis and its advocates, who had no need for the embryo. . . . Evo-devo, however, has taught us an important lesson. To understand innovability, we cannot ignore the complexity of phenotypes. We must embrace it.

Similarly, Andy Gardner in his excellent review of evolutionary theory[20] contrasts in an insightful way the difference between the view of evolution held by population geneticists and the view held by cell biologists, physiologists and ecologists. He characterizes Darwin's three major contributions to biology as follows.

First, by amassing diverse evidence from a range of different disciplines, Darwin firmly established the fact of biological evolution.

Second, Darwin proposed natural selection as a mechanism to explain evolutionary change: 'We now understand that other factors also contribute to the evolutionary process and that natural selection may not even be the main driver of evolutionary change, particularly at the level of DNA sequence evolution. Nevertheless, natural selection is undoubtedly a major driver of phenotypic evolution.'

Third, Darwin explained how natural selection could give rise to the apparent design of organisms. 'This may be the greatest of his three contributions; certainly it is the most far reaching. The theory of Darwinian adaptation knocks the foundation out of the Teleological Argument for the existence of God, or the "argument from design".'

Then Gardner comments,

Despite the importance of this third contribution, and despite the fact that it underlies a great deal of biological research—from functional anatomy to behavioural ecology—the link between natural selection and organism design has been almost completely neglected by population geneticists, who are charged with the task of formalizing all aspects of evolutionary theory. Indeed, many researchers in this discipline regard the notion that natural selection leads to optimization of the phenotype as naive. I suggest that this paradoxical situation—whereby the core of evolutionary theorists reject the core result of evolutionary theory—is owing to a

---

[20] Gardner, 'Adaptation as Organism Design'.

misunderstanding of what biological design (adaptation) really entails, and I suggest that getting the definition right resolves controversies such as the long-running 'group-selection' debate.

I agree with this, and adopt his proposal of **weak adaptationism**:

Organisms appear designed to maximize their inclusive fitness, without implying that they are optimally designed in this respect. This property of organisms is attributed to the action of natural selection, which need not be the sole (or even the main) driver of evolutionary change. The usefulness of weak adaptationism is that it permits biologists to describe organisms and their traits in purposeful, functional, intentional terms.

The selfish gene. What is simply wrong is Richard Dawkins' claim that genes are replicators, shaped by natural selection, that control human beings.[21] Firstly, DNA by itself cannot replicate; this is a basic biological error. The lowest-level replicators are cells. When dividing, they duplicate all the structures making up the cell, including the cell nucleus and the DNA it contains. Secondly, selection cannot act directly on genes. In adult organisms it acts at either level **L9** or **L10** as discussed earlier, then chaining down to level **L3** to select suitable genes. As stated by Stephen J Gould,

No matter how much power Dawkins wishes to assign to genes, there is one thing that he cannot give them—direct visibility to natural selection.[22]

Thirdly, genes by themselves do not control unique biological outcomes. Developmental processes do (§4.3).

Genetic drift. A specific debate is about to what degree selection shapes outcomes, or whether it is random genetic drift that does so: many features are not, in fact, the result of adaptive processes. The analogy used to make the point is the existence of the spandrels of San Marco cathedral—decorative features with no structural significance, so their existence is not explained by selection of function.[23] The theory of neutral evolution asserts that most evolutionary outcomes at the molecular level are due to genetic drift, not adaptive selection.[24] It has even been stated that this is the case at macro levels. It is indeed true that drift plays a key role in molecular evolution at level **L3**, because of the multiple realizability of higher needs. However, arguably it plays a minor role at the physiological level **L8**. For example, it underlies details of

[21] Richard Dawkins, *The Selfish Gene* (Oxford: Oxford University Press, 1976).
[22] Quoted in the Wikipedia article Gene-centered view of evolution.
[23] Stephen J. Gould and Richard C. Lewontin, 'The Spandrels of San Marco and the Panglossian Paradigm: A Critique of the Adaptationist Programme', *Proceedings of the. Royal Society of London B*, 205 (1979), 581–98.
[24] Motoo Kimura, *The Neutral Theory of Molecular Evolution* (Cambridge, UK: Cambridge University Press, 1985).

skull shape, but not the overall features of the skull which enable eyes, ears, and the nose to function.[25]

We are indeed a product of our evolutionary past, and some aspects of that past may remain with us as useless appendages, such as male nipples and the human tailbone (the coccyx). This has also in the past been claimed to be true for the appendix and tonsils; however, these are now believed to play important roles in maintaining gut flora and in the immune system. There are some evolutionary outcomes that can certainly be regarded as negative, such as the pain and problems experienced by women while giving birth.

Genes and culture. Once consciousness evolved, evolution extended to gene-culture co-evolution with genes and culture both evolving in a symbiotic way,[26] and cultural evolution itself evolving.[27] This leads to a key feature: according to Hinton and Nowlan, learning can guide evolution, because learning alters the shape of the search space in which evolution operates.[28] They state,

> This effect allows learning organisms to evolve much faster than their non-learning equivalents, even though the characteristics acquired by the phenotype are not communicated to the genotype.

4.3 Developmental Processes and the Life Cycle

Development of a human being from a fertilized cell to an adult human being is an extraordinary process,[29] enabled by developmental systems[30] that have arisen through evolution. The problem is that every cell contains identical genetic information, but different cell types must be generated in different positions in the body so as to create bone, skin, legs, eyes, neurons and so on. Different proteins are needed to create the required different cell types, so different genes must be read in different cells. How does that occur?[31]

I look in turn at the macro-level processes occurring, the underlying micro-level processes, and ECO–DEVO: the effect of the ecological context.

---

[25] Rebecca R. Ackermann and James M. Cheverud, 'Detecting Genetic Drift versus Selection in Human Evolution', *Proceedings of the National Academy of Science*, 101 (2004), 17946–51.

[26] Heyes, 'New Thinking'. Kevin Laland, John Odling-Smee, and Sean Myles, 'How Culture Shaped the Human Genome: Bringing Genetics and the Human Sciences Together', *Nature Reviews Genetics*, 11 (2010), 137–48.

[27] Jonathan Birch and Cecilia Heyes, 'The Cultural evolution of Cultural Evolution', *Philosophical Transactions of the Royal Society B*, **376** (2021), 20200051

[28] Geoffrey Hinton and Stephen Nowlan, 'How Learning Can Guide Evolution', *Complex Systems*, 1 (1987), 495–502.

[29] Lewis Wolpert, Cheryll Tickle and Alfonso Arias, *Principles of Development* (Oxford: Oxford University Press, 2002).

[30] Russell Gray, Paul E Griffiths and Susan Oyama, eds, *Cycles of Contingency: Developmental Systems and Evolution* (Cambridge, MA: MIT Press, 2001).

[31] See the Wikipedia article Development of the human body.

4.3.1 The Macro-Level Processes

The start of an organism is a fertilized cell—a zygote—at the cellular level **L6**, which contains all the sub-cellular machinery at levels **L5** to **L3** needed for the zygote to replicate, composed of the physical stuff at levels **L2** and **L1**. At this point of embryonic development, none of the levels **L6**–**L9** exist, although the developing organism is imbedded in the already existing levels **L10** and **L11**.

A complex series of events occurs (**Figure 4.6**). The zygote divides multiple times, forming a blastocyte that in the case of human beings is embedded in the uterus. As further development of the embryo takes place, each of the levels **L7** to **L9** is built up sequentially. Crucially, as each new level **L** is generated and modified (being constructed in an upward way), they each act downwardly on the levels below them. They alter levels **L6** to **L3** to shape cells and proteins as needed so that the higher levels can be constructed. The particles at Levels **L2** and **L1** do not change their nature as this happens—physics is what it is—but rather are rearranged so as to constitute the needed entities at levels **L3** and up. This is the process of emergence.

When the entire organism (level **L9**) is ready, eggs hatch or birth takes place. Further developmental process take place as a baby grows into a juvenile and then transitions to an adult state. This may involve major transformations of form (a tadpole to a frog, a caterpillar to a butterfly and so on), or lesser but still crucial ones as during the adolescence of human beings.[32] There are critical stages where the environment must provide necessary stimuli in order that development proceed normally; e.g. vision will not develop if an animal is kept in the dark or otherwise receives no visual stimuli while the visual cortex is developing.[33]

Developmental programmes shape our lives as we progress through these various life stages: zygote to embryo to birth to infant to youth to puberty, to adult with responsibilities, to aging, to death. We share all these stages with all mammals. The emergence of the brain and consciousness is part of the process, with the preschool years being developmentally crucial.[34] Both informal and formal learning processes are key.

The final stage is the inevitability of death: each life comes to an end.[35] Despite desperate attempts by some billionaires to avoid it, it will forever be a profound unavoidable aspect of the human condition. Prime causes of death are different in the developed world and in the developing world. In the latter case, malnutrition and TB remain major causes of death; in the former, heart disease and cancer are dominant.

---

[32] See the Wikipedia article Development of the human body.
[33] Bryan Hooks and Chinfei Chen, 'Critical Periods in the Visual System: Changing Views for a Model of Experience-Dependent Plasticity', Neuron, 56(2) (2007), 312–26.
[34] Hugo Lagercrantz and Jean-Pierre P. Changeux, 'The Emergence of Human Consciousness: From Foetal to Neonatal Life', Pediatric Research, 65(3) (2009), 255–60.
[35] William Breitbart, 'On the Inevitability of Death', Palliative & Supportive Care, 15 (2017), 276–8.

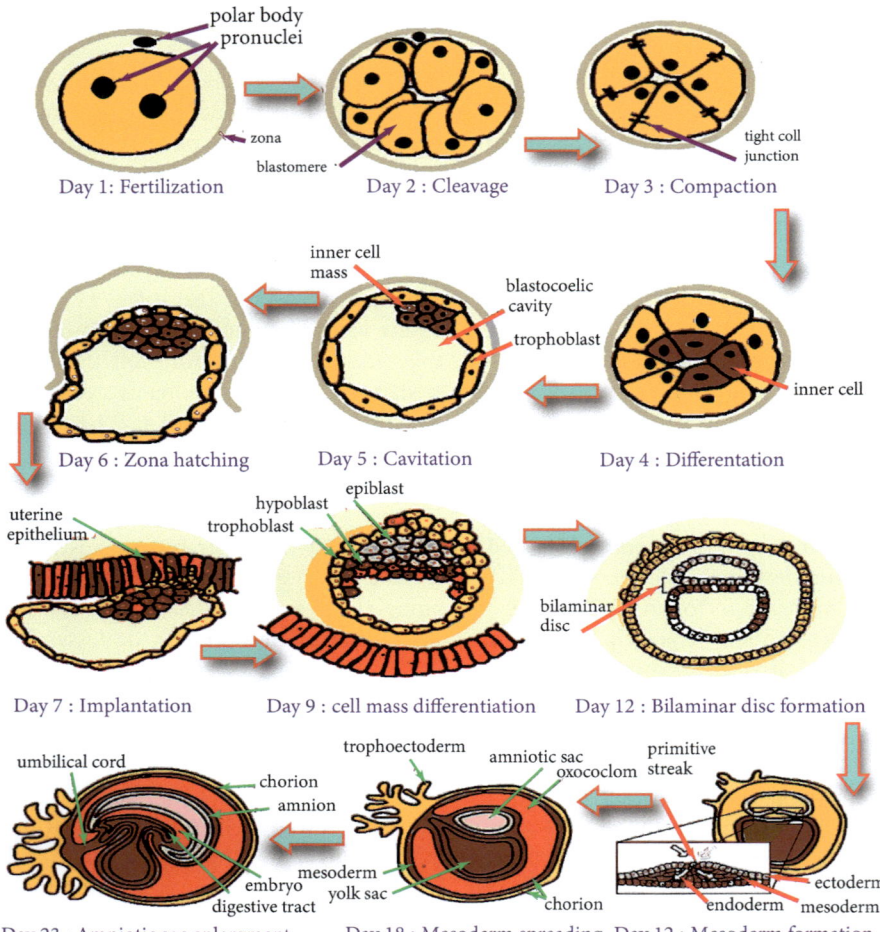

Figure 4.6 The initial stages of human embryonic development. [36] Gastrulation [37] is a key event occurring around day 17, involving symmetry breaking and setting up the basic body axes and fundamental structure. [38]

[36] From the Wikipedia article Human embryonic development.

[37] See the Wikipedia article Gastrulation.

[38] https://commons.wikimedia.org/wiki/File:HumanEmbryogenesis.svg

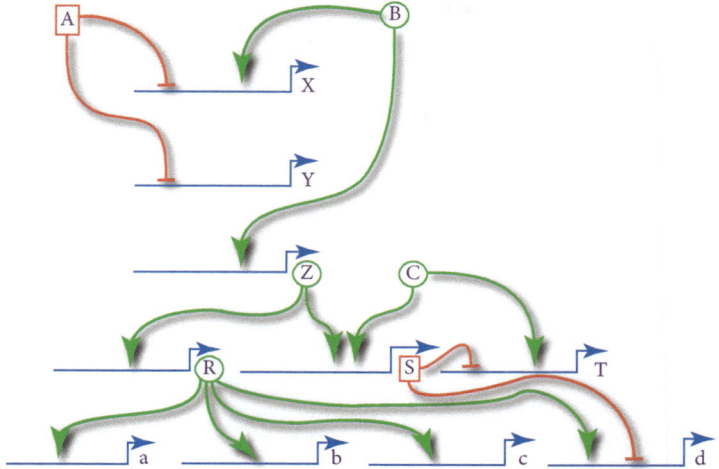

Figure 4.7 Higher-level logic involved in reading genes. Activators (circled letters) and repressors (square letters) act on switches (solid lines). Arrows indicate activation, horizontal endings repression. The outcome is a complex logic determining what genes are read.

Adapted by Mauro Carfora from Figure 5.7 of EFMB.

4.3.2 The Underlying Micro-Level Processes

Developmental processes are based on positional information derived from morphogens[39]—signalling factors diffusing from organizing centres (Spemnan–Mangold organizers).[40] The morphogens control which genes are turned on and off in which places and at which times in a developing organism, and so shape embryonic development (Figure P.7 in the Prologue). This process relies on the interaction of gene regulatory networks (GRNs) and metabolic networks, where the GRNs involve allosteric processes as described in Chapter 3. The kind of logic that emerges was explained there (see Figure 3.4) and is illustrated in **Figure 4.7**, showing activation and repression of genes.

The key feature is this: our bodies are hierarchically organized, so higher levels of logic shape what happens as developmental processes take place. A hierarchical structure of gene expression underlies these processes, involving downward causation due to spatial location.

EFMB describes in depth how this works. Unsurprisingly, the genes that lead to the hierarchical structures of our bodies are themselves controlled in a hierarchical way. What is surprising is that broadly the same sets of genes are shared by so many animals, including ourselves. The same segmentation processes involving similar sets

---

[39] Wolpert, et al, *Principles of Development*. See also EFMB.
[40] See the Wikipedia article Speman–Mangold organizer.

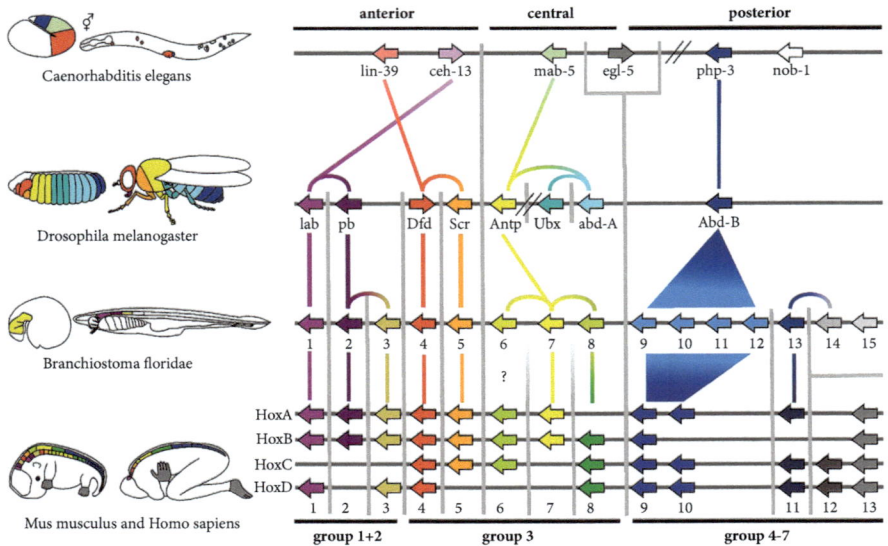

Figure 4.8 Gene homology. Clusters of Hox genes pattern different regions in different kinds of embryos: C. elegans, Drosophila, a fish, a mouse and humans. Broadly the same pattern of genes is involved in all cases.

Source: Stefanie D. Hueber, Georg F. Weiller, Michael A. Djordjevic and Tancred Frickey; From the Wikipedia article Evolutionary developmental biology.

of genes result in the *Caenorhabditis elegans* body, fruit fly segments and vertebrate backbones (**Figure 4.8**), including backbones of fish and mice and humans.

The key actors in the developmental process are Homeobox genes (Hox genes being the most common subset),[41] involving a regulatory sequence of base pairs TATA (called a TATA box[42]) that hierarchically control gene expression as shown in **Figure 4.8**. Crucially, as in the case of evolution, developmental processes generate the higher levels that were not there initially, starting through bifurcation at the cellular level **L6** but using information stored at the macromolecular level **L3**. These processes are cases of downward causation,[43] because body position (a macro variable) determines the reading of genes at the molecular level, and thereby underlies cellular differentiation. Cell processes correct errors that occur in DNA duplication;[44] otherwise, cells would cease to function in short order.

The GRNs and the entire process itself have evolved over time through evolutionary processes,[45] and shaped the evolution of animal body plans.[46] Aging results from cellular and genetic changes, such as genome instability

---

[41] See the Wikipedia articles Homeobox.

[42] See the Wikipedia articles TATA box.

[43] Giovanni Pezzulo and Michael Levin, 'Top-down Models in Biology: Explanation and Control of Complex Living Systems above the Molecular Level', *Journal of the Royal Society Interface*, 13 (2016), 20160555.

[44] Leslie A. Pray, 'DNA Replication and Causes of Mutation', *Nature Education*, 1 (2008), 214.

[45] Gray, et al, *Cycles of Contingency: Developmental Systems and Evolution*.

[46] Eric H. Davidson and Douglas H. Erwin, 'Gene Regulatory Networks and the Evolution of Animal Body Plans', *Science*, 311 (2006), 796–800.

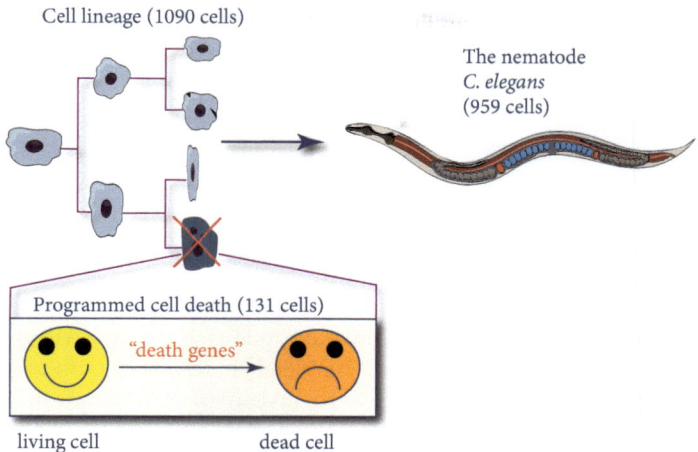

Figure 4.9 Death genes and the nematode.

Figure by Mauro Carfora, adapted from the Nobel Prize press release: Carlos López-Otín et al., 'The Hallmarks of Aging', *Cell*, 153 (2013), 1194–217.

and telomere shortening.[47] On smaller scales, developmental processes involve programmed cell death. Proving this is true in the case of the nematode earned the Nobel Prize in Medicine in 2002, as explained in the press release for the prize[48] and the related diagram given in that press release (**Figure 4.9**). Underlying this, one should remember that in the end it is the cell at level **L6** that is the active agent driving all these dynamics.[49]

Overall, these complex developmental processes are a classic example of emergence. As stated by Sean B. Caroll,[50]

> The term 'reductionism' refers to the biologist's quest to understand life's processes at the molecular level, often by breaking down—'reducing'—processes and structures into their molecular components. The process has been enormously successful over the past half century . . . The frequent objection to reductionist thinking is that many important biological entities—cells, individuals, populations, ecological communities—are organised at levels above that of the molecules, such that knowledge of molecules alone does not explain the properties of the levels above. . . . having an inventory of toolkit genes still leaves us well short of an understanding of how an animal is put together during development.

The way this emergence occurs is a central theme of this book.

[47] Carlos López-Otín et al., 'The Hallmarks of Aging', *Cell*, 153 (2013), 1194–217.

[48] https://www.nobelprize.org/prizes/medicine/2002/press-release/.

[49] Brian Ford, 'The Cell as Secret Agent—Autonomy and Intelligence of the Living Cell: Driving Force of Development', *Academia Biology*, 1(3) (2023).

[50] Carroll, *Endless Forms Most Beautiful*, 82–3.

4.3.3 ECO–DEVO: The Effect of the Ecological Context on Developmental Processes

The ecological level **L11** acts down to the lower biological levels so as to crucially affect developmental outcomes. This is discussed in depth by Scott Gilbert and David Epel in their classic book *Ecological Developmental Biology*.[51] The book gives numerous examples of such effects at the macro level (see the book's Table 1.1), discussing the epigenetic mechanisms that underlie them at the micro level. I will mention two of the examples they give, of quite different natures.

The first is the blue-headed wrasse (a fish).[52] A single blue-headed male lives with some less colourful females. Should the male die, one of the females will grow testes to become a male. The phenotype changes completely in response to the social context. If a juvenile approaches a reef where a single male lives and defends his territory, it will become a female; otherwise it will become a male.

The second example is how baby rats being licked by their mothers changes the number of glucocorticoid receptors (GRs) in the brain's hippocampus when it is an adult. The more GRs it has then, the better it is able to deal with stress. And what determines that number? It depends on the quality of grooming and licking the rat pup receives during its first week after birth.[53] The mechanism underlying this is histone methylation, a key epigenetic mechanism (**Figure 4.10**).

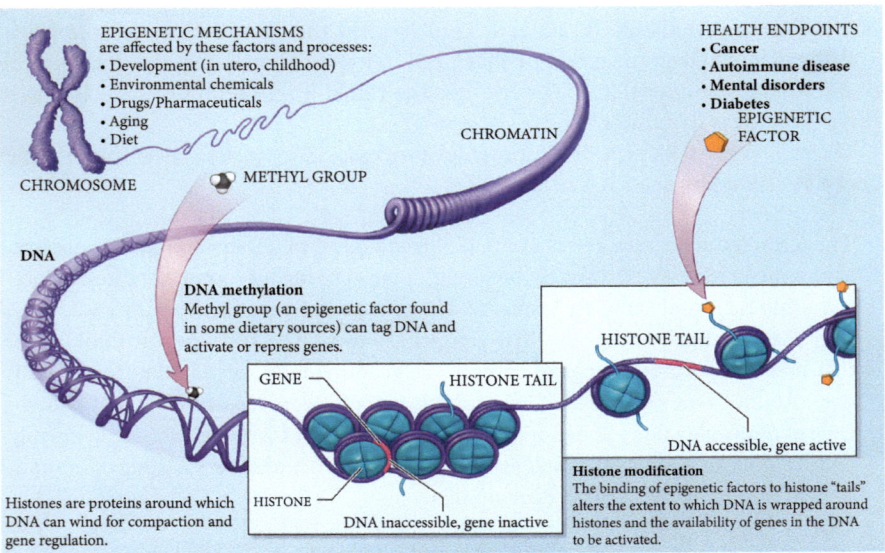

Figure 4.10 Histone methylation.[54] The mechanism and its endpoints. DNA is wound round histones. Methyl groups activate or repress genes by binding to histone tails, altering the accessibility of genes.

Source: https://upload.wikimedia.org/wikipedia/commons/f/fc/Epigenetic_mechanisms.png

---

[51] Scott F. Gilbert and David Epel, *Ecological Developmental Biology: Integrating Epigenetics, Medicine, and Evolution* (Sunderland, MA: Sinauer, 1949).
[52] See the Wikipedia article *Thalassoma bifasciatum*.
[53] Gilbert and Epel, *Ecological Developmental Biology*, 44–5.
[54] See the Wikipedia article Histone methylation.

All development is shaped by the interaction of our internal states with social and cultural influences. A negative example is foetal alcohol syndrome,[55] which has social roots. If the mother drinks a lot while pregnant, her foetus gets severely damaged. It is prevalent amongst farm workers in my home country of South Africa, because it is customary to partly pay farm workers in terms of alcohol.

4.4 EVO–DEVO Interactions

Developmental processes and evolutionary processes affect each other reciprocally, because organisms that survive are the outcome of developmental processes, so adaptive selection selects developmental systems that lead to preferred organismal outcomes.

The EVO–DEVO integrated view[56] arises by combining the effects of these two processes that occur on different timescales.[57] This takes for granted the existence of physiological processes that enable the organism to function (§4.5), and interact crucially with both evolutionary and developmental processes.[58] The approach has five key elements that I will briefly mention.

4.4.1 Emphasizing the Existence of Developmental Systems

- They are ignored in the modern synthesis. The EVO–DEVO view emphasizes that they exist, and claims they have a central role in evolutionary processes.[59]

4.4.2 Emphasizing the Key Role of Developmental Processes in Evolution

Developmental system outcomes are the physiological systems and organisms at levels **L8** and **L9** that are selected for by evolutionary processes. Without developmental processes, they would not exist. The modern synthesis misses this, because it claims that selection acts directly down to the levels of genes—which cannot happen, as emphasized earlier. Evolutionary processes as such cannot 'see' genes. They can see and select outcomes at all levels from the level of cells **L6** and up, and these outcomes then 'reach down', by rejection or inclusion, to select outcomes at the lower levels **L5** to **L3**—the level of genes and proteins.

---

[55] See the Wikipedia article Fetal alcohol spectrum disorder.
[56] See the Wikipedia articles Evolutionary developmental biology and Evo-devo gene toolkit.
[57] Carroll, *Endless Forms Most Beautiful*. Gilbert et al., 'Resynthesizing Evolutionary and Developmental Biology'.
[58] Lewis Wolpert, Cheryll Tickle, and Alfonso Arias (2002) *Principles of development* (Oxford: Oxford University Press;
[59] Carroll, *Endless Forms Most Beautiful*.

4.4.3 Emphasizing the Evolutionary Selection of Developmental Systems

Because of those processes, developmental systems themselves have evolved over time as evolutionary processes took place.[60] Consequently, the evolutionary process itself has evolved over time, and continues to evolve.[61]

4.4.4 Emphasizing the Consequent Homologous Gene Systems between Species

The existence of homologous families of genes between species (**Figure 4.8**) is a consequence of EVO–DEVO interactions. They are simply unaccounted for if you do not take these interactions into account.

4.4.5 Emphasizing Niche Construction

Evolution also has an upward component from level **L10** to **L11**. Due to their agency, living beings can engage in ecological niche construction,[62] altering the environment.[63] This can be represented by Richard Dawkin's concept of the extended phenotype.[64] Examples are ants creating anthills, beavers creating dams (**Figure 4.11**) and humans engineering the environment to suit their needs (**Figure 3.10**: bottom left).

Figure 4.11 Niche construction. Anthills (left), beavers constructing dams (right).

Photo by David Parkyn, from Beaver Trust.

---

[60] Russell Gray et al., *Cycles of Contingency*.
[61] Kevin N. Lala, Tobias Uller, Nathalie Feiner, Marcus Feldman and Scott F. Gilbert, *Evolution Evolving: The Developmental Origins of Adaptation and Biodiversity* (Princeton: Princeton University Press, 2024).
[62] See the Wikipedia article Niche construction.
[63] Richard Lewontin, 'Gene, Organism and Environment', in D. S. Bendall, ed., *Evolution from Molecules to Men* (Cambridge, UK: Cambridge University Press, 1983). F. J. Odling-Smee, 'Niche Constructing Phenotypes', in H. C. Plotkin, ed., *The Role of Behavior in Evolution* (Cambridge, MA: MIT Press, 1988), 73–132.
[64] Richard Dawkins, *The Extended Phenotype* (Oxford: Oxford University Press, 1982).

4.5 Including Physiological Processes: EVO–DEVO–Physio

Charles Darwin provided convincing evidence of adaptation to the local ecological environment through his study of the variety of Galápagos finches on various Galápagos islands.[65] Because the islands are somewhat isolated from each other, fruits vary on the different islands, so slightly different beaks (**Figure 4.12(a)**) are better adapted to the ecological contexts on the different islands. This is clear evidence that physiological outcomes (level **L8**) matter in evolution. Indeed they drive it, because physiology is key to survival.

Consider, for example, the zebras shown in **Figure 4.12(b)**. They have eyes to see, ears to hear, a mouth to eat, indeed all the physiological systems discussed in Chapter 3 (§3.1). These are all there because they are key to enabling the organism as a whole to function and survive. They are physiologically necessary. Therefore it is crucial that selection in the case of mature organisms includes selection at the physiological levels **L7** and **L8,** as this underlies selection at the organism-level **L9** and group-level **L10**.

This has rightly been emphasized by the distinguished physiologist Denis Noble and his brother Raymond Noble,[66] resulting in ridicule from some reductionist evolutionary theorists whose focus is on outcomes at the molecular level. But, I submit, they are mistaken.

(a)   (b)

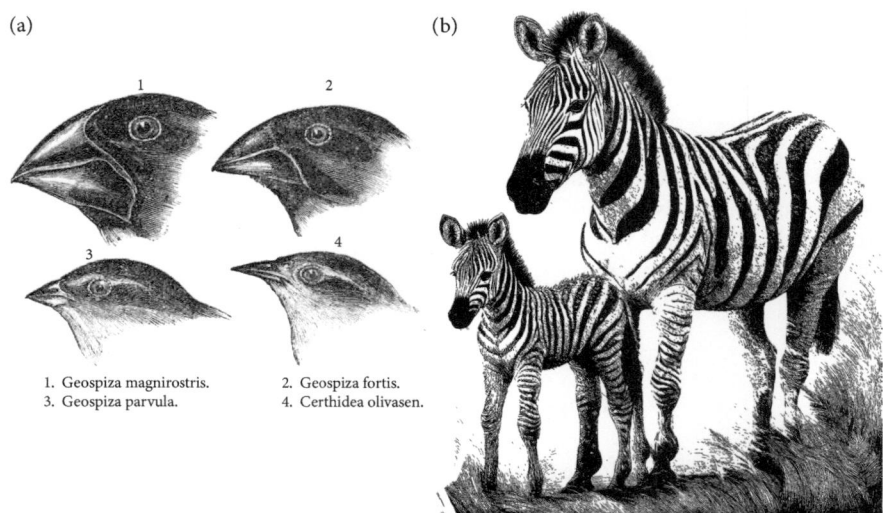

1. Geospiza magnirostris.
3. Geospiza parvula.
2. Geospiza fortis.
4. Certhidea olivasen.

Figure 4.12 Physiological adaptation to the environment. (a) Darwin's finches; (b) zebra mother and foal.

Figure by Mauro Carfora.

[65] Jefrey Podos and Stephen Nowiki, 'Beaks, Adaptation, and Vocal Evolution in Darwin's Finches', *BioScience*, 54(6) (2004), 501–10. Also see the Wikipedia article Darwin's finches.

[66] Denis Noble, 'Physiology Is Rocking the Foundations of Evolutionary Biology', *Experimental Physiology*, 98 (2013), 1235–43. Denis Noble, 'Evolution Viewed from Physics, Physiology and Medicine', *Interface focus*, 7 (2017), 20160159. Raymond Noble and Denis Noble, 'Physiology Restores Purpose to Evolutionary Biology', *Biological Journal of the Linnean Society*, 139 (2023), 357–69.

As argued earlier, the causal relationships emphasized by evolutionary physiologists and ecologists—the latter including Charles Darwin—are hidden at the
macromolecular level because of the huge multiple realizability linking these different levels. This does not mean those relations do not exist. A special issue of the
Journal of Physiology looks at this issue in depth.[67]

4.6 The Whole Shebang: The Multilevel Nature of Evolution

When you consider all the above taken together, it amounts to the statement that
evolution, development and physiology are an integrated package where all the levels
L3 to **L11** are subject to selection, with a strong linkage between these processes at
levels **L3** to **L9**. Evolutionary processes as a whole depend on an interaction of development and function (physiology, ecology) with evolutionary genetics. All levels are
involved.[68] Consequently,

Noble's principle of biological relativity[69] for biological functioning applies to devel-
opmental/evolutionary processes as well. Every emergent level **L3** to **L9** takes
part both in development as a basis for emergence of the next highest level and
in the selective evolutionary processes as a constraint on the next lower level
(**Figure 4.2**). Evolution thereby tunes every emergent level so they all function
together consistently.

Evolution is a multilevel affair. Not only can selection can take place at either the
group or organism level, it takes place at every emergent level from **L3** up in a consistent and coordinated way. Claiming it only happens at the genetic level **L3** misses
out most of what is happening.

4.7 Abiogenesis: How Did Life Start?

What was the origin of life? We do not know how abiogenesis took place: that is, how
life came into being from non-living matter.[70] This was an extraordinary step in the
evolution of our planet: a quite new form of matter, with quite different properties
than those of inanimate matter, came into being (**Figure 4.13**).

---

[67] *Journal of Physiology*, Vol. 592, no. 11.
[68] Richard Lewontin, 'The Units of Selection', *Annual Review of Ecology and Systematics*, 1 (1970), 1–18.
[69] Denis Noble, 'A Theory of Biological Relativity: No Privileged Level of Causation', *Interface Focus*,
2(1) (2012), 55–64.
[70] See the Wikipedia article Abiogenesis.

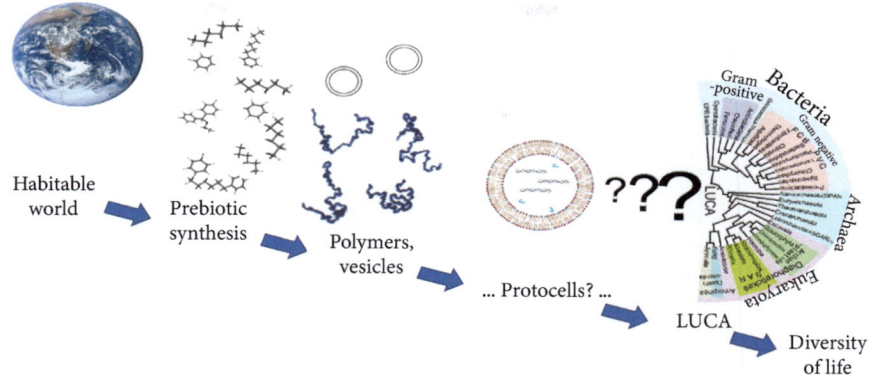

Figure 4.13 Abiogenesis. A conceptual scheme.[71]

Source: Chiswick Chap.[72]

There are competing theories,[73] particularly as to whether metabolism or genetic information came first, or both evolved together simultaneously, maybe via an RNA world,[74] and in what environment this took place. That does not mean that it cannot be explained scientifically: just that we are not yet in a position to decide unambiguously.

To create life, you need a cell that is able to duplicate itself. In order for this to happen, you have five key things to contend with. The cell must be able to find the energy to do so (it must get metabolism going). It needs a store of genetic information that can be replicated (it must have the needed DNA and various RNAs available). It must have the internal mechanisms needed to enable duplication of the cell itself—all its organelles—and duplication of the DNA. It must have a membrane separating the interior of the cell from the exterior, so that all its internal processes take place in a confined and protected space with ingress and egress controlled by ion channels.[75] These membranes are made of lipids with hydrophilic and hydrophobic ends that spontaneously form vesicles (**Figure 4.14**) in an aqueous environment.[76] As for metabolism, it seems quite likely that initially the energy needed for metabolism to take place was provided by deep-sea hydrothermal vents (**Figure 4.15**).

All of this has to be put together in such a way that replication and selection can start to happen. Assembly theory is a way of characterizing when this key change

---

[71] Denis Noble, 'Physiology Is Rocking the Foundations of Evolutionary Biology', *Experimental Physiology*, 98 (2013), 1235–43. Denis Noble, 'Evolution Viewed from Physics, Physiology and Medicine', *Interface focus*, 7 (2017), 20160159. Raymond Noble and Denis Noble, 'Physiology Restores Purpose to Evolutionary Biology', *Biological Journal of the Linnean Society*, 139 (2023), 357–69.

[72] https://commons.wikimedia.org/wiki/File:Origin_of_life_stages.svg

[73] Sara Walker and Paul C. W. Davies, 'The Algorithmic Origins of Life', *Journal of the Royal Society Interface*, 10 (2013), 20120869. Sara Walker, N. Packard and G. D. Cody, 'Re-conceptualizing the Origins of Life', *Philosophical Transactions of the Royal Society A*, 375(2109) (2017), 20160337.

[74] See the Wikipedia article RNA world.

[75] See the Wikipedia article Biological membranes.

[76] See the Wikipedia article Lipid bilayer.

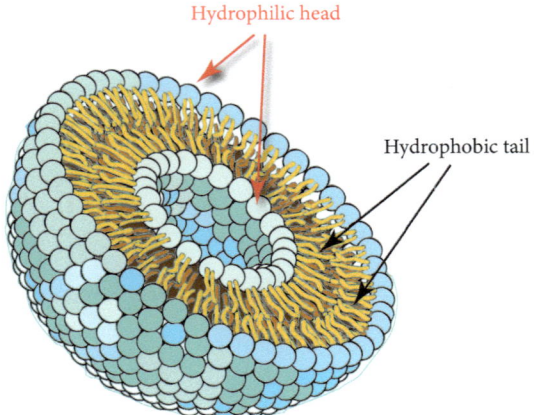

Figure 4.14 A vesicle made of lipid bilayers. The hydrophilic head is on the outside and the hydrophobic tail on the inside.

An elaboration by Mauro Carfora of a Wikipedia image (*Source*: Mariana Ruiz Villarreal).

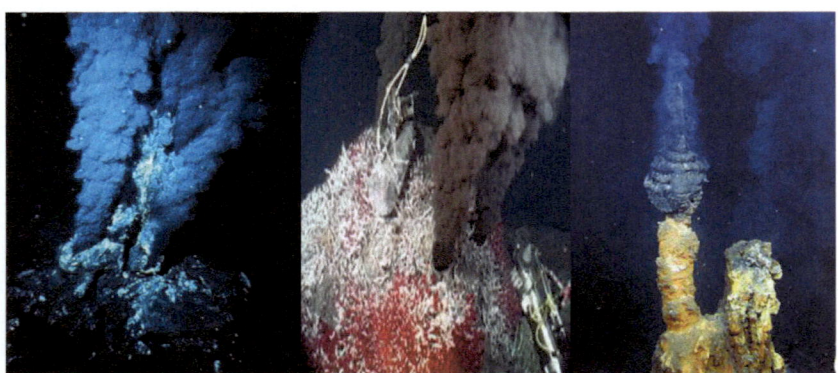

Figure 4.15 Hydrothermal vents.[77]

Figure from Wikipedia.[78]

takes place.[79] There are two proposals at present that may be on the right track as to how this took place. Vesicles spontaneously assemble themselves once the lipids are available. Note that as they are not proteins, lipids are not coded for by DNA. They are rather produced by metabolic processes in a cell[80]—once the cell exists. As to the issue of how information storage and reproduction initially took place, there are plausible proposals for an RNA world initially doing this,[81] with DNA being a later entrant on the scene.

---

[77] See the Wikipedia article Hydrothermal vent.
[78] *Source*: W.R. Normark, Dudley Foster, Public domain, via Wikimedia Commons.
[79] Sharma et al., 'Assembly Theory Explains and Quantifies Selection and Evolution'.
[80] See the Wikipedia article Lipid metabolism.
[81] See the Wikipedia article RNA world.

Consideration of all the elements that have to be in place for self-replicating life in the form of prokaryotic cells to first exist shows just how difficult it must have been to get this all in place—and therefore why it took so long after the formation of the Earth. I am certainly not an expert on this topic, which is not yet clarified. But what's been discussed in this chapter should give some idea of the magnitude of complexity involved in making it happen.

The coming into being of the first unicellular living organisms was an astonishing event. It already embodied agency at that level. It was the foundation of the later coming into existence of higher-level agency, then consciousness, and the ability to form theories of the coming into existence of life.

Further Reading

On evolution

Ernst Mayr, *What Evolution is* (London: Phoenix, 2002).

John Maynard Smith and Eors Szathmary, *The Major Transitions in Evolution* (Oxford: Oxford University Press, 1997).

Massimo Pigliucci and Gerd B. Müller, *Evolution: The Extended Synthesis* (Cambridge, MA: MIT Press, 2010).

On developmental biology

Scott F. Gilbert, ed., *Developmental Biology*, 8th edn (Sunderland, MA: Sinauer Associates, 2006).

Lewis Wolpert, Cheryll Tickle and Alfonso Martinez Arias, *Principles of Development* (New York: Oxford University Press, 2015).

On ECO–DEVO

Scott F. Gilbert and David Epel, *Ecological Developmental Biology: Integrating Epigenetics, Medicine, and Evolution* (Sunderland, MA: Sinauer Associates, 1949).

Sean B. Carroll, *Endless Forms Most Beautiful: The New Science of evo devo and the Making of the Aanimal Kingdom* (New York: Norton, 2005).

Russell D. Gray, Paul E. Griffiths and Susan Oyama, eds, *Cycles of Contingency: Developmental Systems and Evolution* (Cambridge, MA: MIT Press, 2003).

On the influence of physiology on evolution

Denis Noble, 'Physiology Is Rocking the Foundations of Evolutionary Biology', *Experimental Physiology*, 98 (2013), 1235–43.

Raymond Noble and Denis Noble, 'Physiology Restores Purpose to Evolutionary Biology', *Biological Journal of the Linnean Society*, 139(4) (2023), 357–69.

Overall integrative views

The *Third Way of Evolution* website gives many references to relevant books and authors giving integrative views of evolutionary processes: https://www.thethirdwayofevolution.com/

5

Perception, Rationality, Emotions, Free Will

The mind/brain integrates all the systems of an organism through the physical central nervous system and the phenomenal experience of consciousness: what it is like to be a person, feeling alive. Decisions made change outcomes in the physical world around us. Brains are networks made of particular kinds of cells (neurons) connected together via synapses, and supported by other cells called astrocytes. We collect information, store it in symbolic form, predict outcomes[1] and make action choices that change the world.[2]

Excellent big picture views of the mind/brain are given by Eric Kandel[3] and Chris Frith.[4] Humans can be regarded as having a hierarchy of needs which extends from basic physical needs to higher levels of needs, including self-actualization and transcendence.[5] The mind/brain underlies why we have those higher needs, and how we try to handle all of these needs. In short, we have agency.[6]

It is an extraordinary thing that this all arises out of the underlying physics—interactions of electrons and protons—leading to higher-level thoughts and contemplations: how to understand the world and the Universe, economics and society; how to handle love and joy, grief and loss, and yet keep hope; and above all, to have worthy values that lead to a life well lived. An example is this:

> As President Trump sat in the first pew of the National Cathedral on Tuesday 21 January 2025 during a traditional prayer service, Bishop Budde asked Trump 'to have mercy upon the people in our country who are scared now.' Trump glared and shifted uncomfortably as she spoke. 'There are gay, lesbian and transgender children in Democratic, Republican, and Independent families, some who fear for their lives.' In his first hours as President, Trump ordered immigration officers to ramp up deportations of people in the country without authorization. Budde said in her sermon that those being targeted for deportation 'may not be citizens or have the proper documentation, but the vast majority of immigrants are not criminals. They pay taxes

---

[1] Jeff Hawkins and Sandra Blakeslee, *On Intelligence* (London: Macmillan, 2004).

[2] Rainer Feistel, 'Self-Organisation of Prediction Models', *Entropy*, 25 (2023), 1596.

[3] Eric R. Kandel, *The Age of Insight: The Quest to Understand the Unconscious in Art, Mind, and Brain, from Vienna 1900 to the Present* (London: Random House, 2012).

[4] Chris Frith, *Making up the Mind: How the Brain Creates Our Mental World* (Malden, MA: Blackwell Publishing, 2013).

[5] See the Wikipedia article Maslow's hierarchy of needs, and Francis Heylighen, 'A Cognitive-Systemic Reconstruction of Maslow's Theory of Self-actualization', *Behavioral Science*, 37 (1992), 39–58.

[6] Michael Gazzaniga, *Who's in Charge?: Free Will and the Science of the Brain* (London: Constable and Robinson, 2012). Kevin Mitchell, *Free Agents: How Evolution Gave Us Free Will* (Princeton, NJ: Princeton University Press, 2023).

How We Come to Be. George F. R. Ellis, Oxford University Press. © George Ellis (2026).
DOI: 10.1093/9780198950189.003.0007

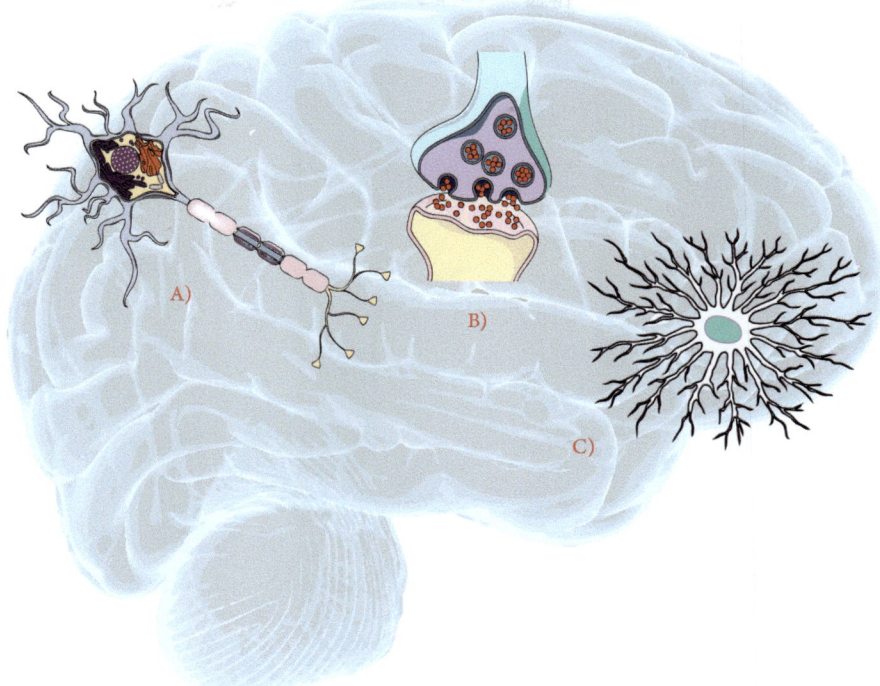

Figure 5.1 The microstructure of the brain: neurons (A), synapses (B), astrocytes (C).

Figure by Mauro Carfora.

and are good neighbours. They are faithful members of our churches and mosques, synagogues, gurudwaras and temples.' She added, 'I ask you to have mercy, Mr. President, on those in our communities whose children fear that their parents will be taken away.'[7]

One thing is now certain: our extraordinary gift of consciousness, the central feature of our existence, is enabled by the properties and activities of the physical brain, which is an immensely complex neural network. This simply was not understood in the past. There are three aspects that deserve comment.

Firstly, we learnt how the brain is made up of a specific kind of cell—neurons—connected to other neurons by synapses, forming a network, and supported by other cells called astrocytes[8] (**Figure 5.1**). Electrical signals (action potential spike chains) from neurons lead to outgoing signals in the next neuron if the total incoming signal exceeds a threshold. This is the micro-level logic underlying brain function: it is what enables us to think.

---

[7] See the Wikipedia article Mariann Budde. As a result of saying this, she has been subject to death threats.

[8] See the Wikipedia articles Neuron, Synapse, Astrocyte and Action potential.

Figure 5.2 The physical brain and qualia. We do not know how the extraordinary complex neural network structure of the brain enables qualia to come into existence.
Figure by Mauro Carfora.

But despite all this progress, we do not know how this extraordinary complex neural network structure enables qualia—our felt experiences such as seeing red and feeling pain—to come into existence (**Figure 5.2**). We understand a huge amount about the neural correlates of consciousness, but not how this leads to consciousness itself. The brain imaging revolution, however, is able to localize specific conscious effects to particular brain regions, and we can even use brain implants to enable direct manipulation of mechanical devices by thought.

Secondly, our present understanding of the molecular basis of brain function is in essence an extension of the molecular biology revolution (Chapter 2) to the brain, for it reveals how change of shape in a contextual way of macromolecules—specifically DNA, RNA and proteins—enables brain functioning.

Thirdly, we now understand the embodied nature of brain function, as expressed by Roberto Maffei:[9]

At the core of it, cognitivism conceives the brain as a data processing machine, the Central Nervous System (CNS) as the absolute command and control centre of the whole organism, and the body as a mere instrument of the CNS, like a mechanistic effector only. Oppositely, the many different versions of embodiment share the idea that the body, and its interactions with the CNS and with the environment, have a

[9] Roberto Maffei, 'Between Instincts and Reason: Understanding a Critical Relationship', *Academia Letters*, 3459 (September 2021).

crucial role in human behaviour, knowledge and cognitive processes. . . . For cogni-tivists, brain is a real command and control centre which works through algorithms and information processing (perception as input, motor answers as output); the brain/body relationship is one-way from perceptions to brain processing to motor answers. Embodiment upholders have a different concept: brain processes are fed through the 'mediation' of the body and the relationship between the two is cir-cular, with continuous reciprocal influence. In other words: cognitivism proposes an expressly divided vision, with the brain (and/or mind) rigidly separated from the body and in dominant position. For embodiment theories, the organism is an 'integrated system'.[10]

This chapter discusses, §5.1, 'The Central Nervous System and the Brain'; §5.2, 'Consciousness is Real'; §5.3, 'A Complex Set of Interactions Take Place in the Mind/Brain'; §5.4, 'The Nature and Function of Emotions'; §5.5, 'What Is Innate in Our Brains and What Is Not'; §5.6, 'The Existence of Agency and Free Will'; and 5.7, 'Values, Purpose and Meaning Shape Outcomes'.

5.1 The Central Nervous System and the Brain

The human central nervous system (CNS) is a network of 10^{11} neurons connected by synapses to form an immensely complex neural network.[11] For all the reasons discussed in this book, the CNS has a modular hierarchical structure, discussed by Churchland and Sejnowski[12] and shown in **Figure 5.3**. It connects the brain—the cerebral cortex—to all the muscles and sensors in the body in the following way. The somatic nervous system sends signals received from the external environment by sen-sory neurons to the CNS, and acts on the external environment via signals sent from the CNS to motor neurons (**Figure 5.4**). The autonomic nervous system does the same as regards the internal environment. These together form the peripheral nervous system (PNS).

I now discuss in turn (§5.1.1) 'Neurons and Synapses'; (§5.1.2) 'The Cerebral Cor-tex'; (§5.1.3) 'Subcortical Structures and the Ascending Systems'; and (§5.1.4) 'The Interaction between the Nervous System and the Immune System'.

5.1.1 Neurons and Synapses

At the cellular level, the CNS is made of neurons (**Figure 5.5**), consisting of many branching dendrites connected to a nucleus and then on to many branching axons.[13]

---

[10] He continues 'as much as to say it is composed by different parts which are just assembled to work together, not composing a true unique reality'. Here I part company with him: I see them as integrated.
[11] Eric Kandel et al., eds, *Principles of Neural Science* (New York: McGraw-Hill, 2021).
[12] Patricia S. Churchland and Terence J. Sejnowski, *The Computational Brain* (Cambridge, MA: MIT Press, 2016).
[13] See the Wikipedia articles Neuron, Synapse, Astrocyte and Action potential.

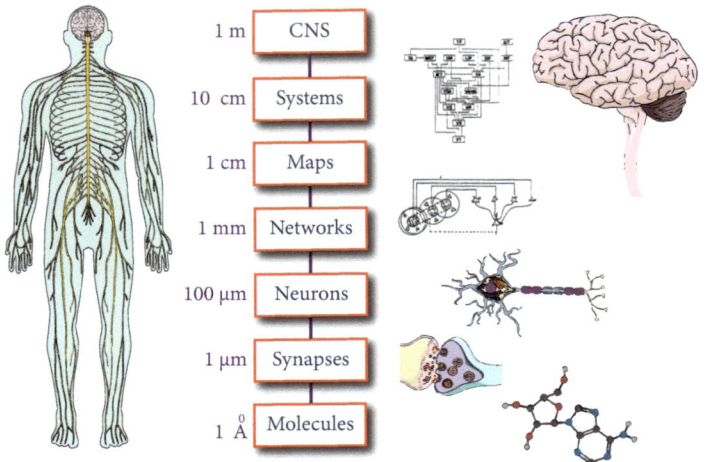

Figure 5.3 The hierarchical structure of the central nervous system (CNS).

Figure adapted by Mauro Carfora from Patricia S. Churchland and Terence J. Sejnowski, *The Computational Brain* (Cambridge, MA: MIT Press, 2016).

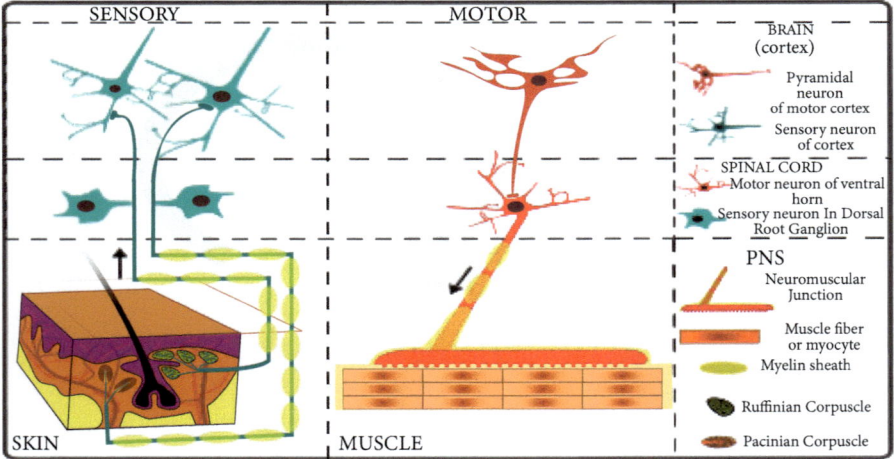

Figure 5.4 Basic organization of the somatic nervous system. Sensory neurons send signals to the brain via the spinal cord. The brain sends signals back to muscles.

Wikipedia image by Looie496; from Wikimedia Commons (https://commons.wikimedia.org/wiki/File:Nervous_system_organization_en.svg)

The axons usually connect to many other neurons via neurotransmitters crossing synapses linking one neuron to another. A neurotransmitter can either help (excite) or hinder (inhibit) the next neuron from firing its own action potential.[14]

---

[14] See the Wikipedia articles Neuron, Synapse, Astrocyte and Action potential.

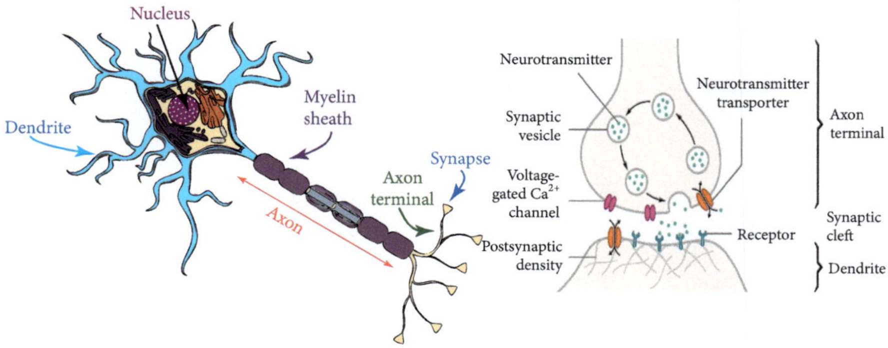

STRUCTURE OF NEURONS

Figure 5.5 Neurons connected together via synapses. A neuron (left) made of dendrites, a nucleus and axons, joined to other neurons via synapses; (right) the joining of an axon to a dendrite. The dendrites convey incoming signals and axons convey outgoing signals in the form of action potential spike chains. At the synapses, information is conveyed across the synaptic cleft from one neuron to another by neurotransmitters.

Figure adapted by Mauro Carfora from University of Queensland Brain Institute: Action Potentials and Synapses.

There are also electrical synapses involving direct electrical connection between neurons.[15]

Information is conveyed down dendrites and axons via action potential spike chains,[16] enabled similarly to how voltage-gated ion channels control passage of ions through a cell wall.[17] A neuron spikes when a combination of all the excitation and inhibition it receives causes it to reach a threshold. This is how logic arises from the underlying molecular biology in this case. Neurons connected together in this way form a network.

5.1.2 The Cerebral Cortex

The cortex has two cerebral hemispheres separated by a fissure, joined by the corpus callosum. It has a folded structure, crucial for the efficiency of brain circuitry and for its functional organization. It consists mainly of six layers, organized into columns (**Figure 5.6**).

Various regions in the cerebral cortex are dedicated to specific functions, for example motor and sensory regions such as the visual cortex and auditory cortex (**Figure 5.7**). However, the brain is plastic, and if some exceptional circumstance occurs, this can be changed. For example, if a person loses their sight through damage

[15] See the Wikipedia article Electrical synapse.

[16] See the Wikipedia articles Neuron, Synapse, Astrocyte and Action potential.

[17] George Ellis and Jonathan Kopel, 'The Dynamical Emergence of Biology from Physics: Branching Causation via Biomolecules', *Frontiers in Physiology*, 9 (2019), 1966.

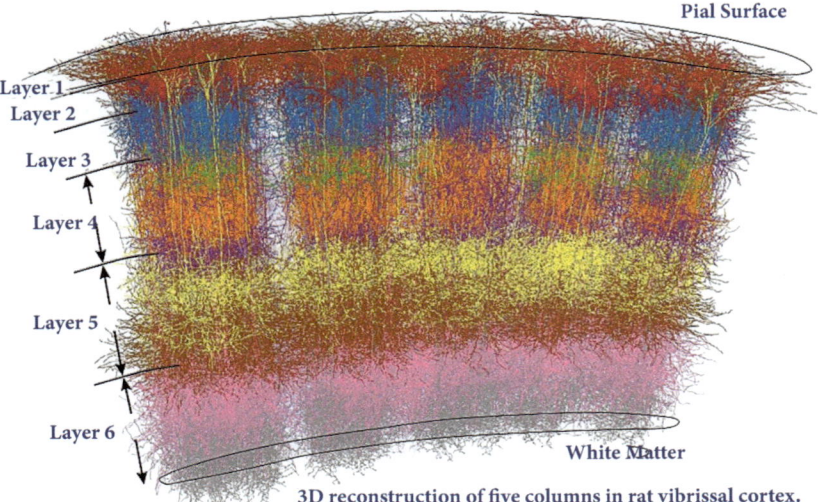

3D reconstruction of five columns in rat vibrissal cortex.

Figure 5.6 Neural networks in the cortex are structured into columns and six layers. Higher-level processing takes place in levels 1 to 3. Figure 5.13 shows the relation to incoming signals in the case of the visual and auditory cortices.

Wikipedia image by Marcel Oberlaender (see the Wikipedia article Cortical column).

Motor and Sensory Regions of the Cerebral Cortex

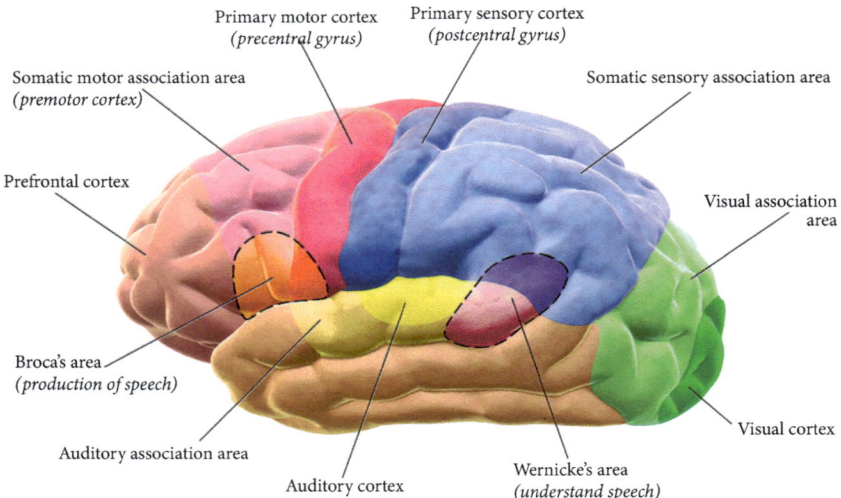

Figure 5.7 Motor and sensory regions of the cerebral cortex.

Source: Blausen.com staff, 'Medical Gallery of Blausen Medical 2014', *WikiJournal of Medicine*, 1(2) (2014).

to eyes or to the optic nerve, then the visual area can be rededicated to auditory processing so as to make hearing much more sensitive.

5.1.3 Subcortical Structures: The Ascending Systems and Neuromodulators

A key feature of the brain is the role of subcortical structures and their interaction with the cortex. A very simplified model of this is Paul McLean's model of the tri- une brain[18] (**Figure 5.8**). In this view, the brain is divided into three main areas: the neocortex (thinking brain), limbic brain (the mammalian brain that is the source of emotions) and the reptilian brain (the source of instincts). The neocortex enables thinking rationally, verbal expression and memories of facts and events. The mam- malian brain enables experiences of feelings and emotions. The reptilian brain is the oldest part of the brain and enables survival by controlling heart rate and breathing, and underlies instinctual (unthought) reactions.

A key feature is that emotions are not an unnecessary diversion from the logical thinking necessary for our survival: they are crucial guides to that thinking. This was forcefully pointed out by Antonio Damasio,[19] who used the case of Phineas Gage[20] to provide solid evidence that this is the case. The physiological basis of this effect is that the limbic brain is the source of neuromodulators such as dopamine and norepinephrine that shape neural connections in the brain through a process that

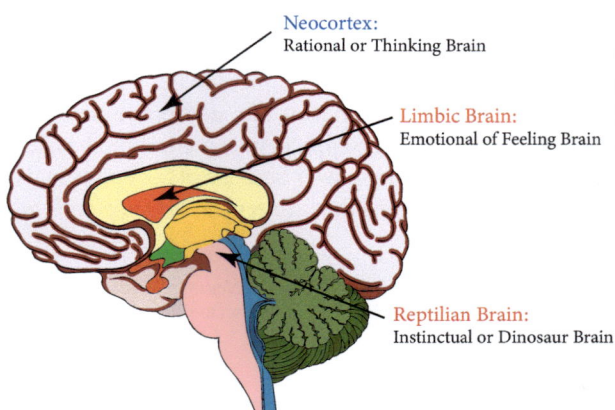

Figure 5.8 The triune brain. This model broadly splits the brain into the instinctual reptilian brain, the mammalian or limbic brain and the thinking neocortex. It is a useful approximation to how the brain functions.

Figure adapted by Mauro Carfora from Simple Overview of the Brain. Creator: @InvaderStich.

---

[18] See the Wikipedia article Triune brain.
[19] Antonio R. Damasio, *Descartes' Error: Emotion, Reason, and the Human Brain* (New York: Putnam, 1994).
[20] See the Wikipedia article Phineas Gage.

Gerald Edelman called **neural Darwinism** (later, **neuronal group selection**).[21] This is a sophisticated theory[22] based on selectionist principles and developing from Edelman's work on the adaptive immune system,[23] for which he won the Nobel Prize. It has been criticized from various directions, particularly as to whether it is really Darwinian (just a terminological issue) and for being different from the dominant computational paradigm for brain function: adjustment of neural network weights through learning processes.[24] However, it does not contradict the latter: it adds to it a second extraordinarily effective mechanism for shaping neural connections in the neocortex. This has also, in effect, been proposed by others[25] and has solid support.[26]

The point is this. Unlike ordinary neural network connections, which connect specific neurons to specific other neurons via synapses, the ascending systems spread neuromodulators such as dopamine diffusely from nuclei in the limbic brain to whole areas in the neocortex (**Figure 5.9**, left). This enables the dopamine to strengthen or weaken entire patterns of connections in the cortex at once, depending on whether specific neurons are firing when the dopamine arrives. The release of these neuromodulators from subcortical nuclei is controlled by what Edelman calls a **value system**, which has been shaped to be what it is because this mechanism has enhanced survival rates over evolutionary timescales.

However, there is a key feature that Edelman did not comment on. Neuroscientist Jaak Panksepp spent many decades studying innate emotional systems in the brains of mammals and humans. They are our evolutionary heritage: they have been selected for by the evolutionary processes discussed in Chapter 4, because they enhance survival rates by guiding our cognitive processes appropriately. This has to be the case; otherwise, they would not exist. Panksepp identified seven such primary emotional systems[27] and investigated the neuromodulators associated with them and the reasons they promoted survival.[28]

What neither Edelman nor Panksepp commented on is that they were both referring to the same set of neurological systems and neuromodulators. Edelman's value system is, in fact, the same as Panksepp's primary emotional systems.[29] Edelman's neural Darwinism attains its driving power because of the emotional force created by neurotransmitters such as dopamine that are central to Panksepp's work.

---

[21] Gerald Edelman, *Neural Darwinism: The Theory of Neuronal Group Selection* (New York: Basic Books, 1987). Gerald Edelman, 'Neural Darwinism: Selection and Re-entrant Signalling in Higher Brain Function', *Neuron*, 10 (1993), 115–25

[22] See the Wikipedia article Neural Darwinism.

[23] See the Wikipedia article Adaptive immune system.

[24] Jean-Pierre Changeux, Philippe Courrège and Antoine Danchin, 'A Theory of the Epigenesis of Neural Networks by Selective Stabilization of Synapses', *Proceedings National Academy of Science USA*, 70(10) (1973), 2974–8.

[25] Edelman collaborated with Mountcastle and Tononi.

[26] Anil Seth and Bernard Baars, 'Neural Darwinism and Consciousness', *Consciousness and Cognition*, 14 (2005), 140–68.

[27] Jaak Panksepp, *Affective Neuroscience: The Foundations of Human and Animal Emotions* (Oxford: Oxford University Press, 2004).

[28] Jaak Panksepp and Lucy Biven, *The Archaeology of Mind: Neuroevolutionary Origins of Human Emotions* (London: W. W. Norton, 2012).

[29] George Ellis and Judith Toronchuk, 'Neural Development: Affective and Immune System Influences', in N. Newton and R. Ellis, eds, *Consciousness and Emotion: Agency, Conscious Choice and Selective Perception* (Philadelphia, PA: John Benjamins, 2005), 81–119.

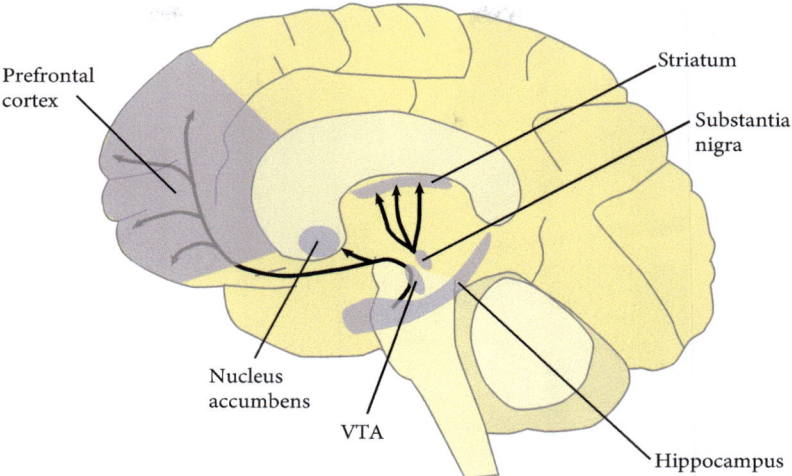

Figure 5.9 Dopamine pathways. As part of the reward pathway (paths indicated on the left), dopamine is created in nerve cell bodies within the ventral tegmental area (VTA) and released in the nucleus accumbens and prefrontal cortex.

From the Wikipedia article Dopamine and National Institutes of Health.

A key issue then is what triggers these emotional systems. I return to their nature in §5.4 below.

5.1.4 The Interaction between the Nervous System and the Immune System

I mentioned in §5.1.3 that Edelman developed his ideas of neural Darwinism from his understanding of the adaptive immune system. So one can ask the question: is it just a coincidence that similar mechanisms operate in these two cases? Or is there a causal relationship?

Indeed there is. Esther Sternberg, in her book *The Balance Within: The Science Connecting Health and Emotions*,[30] makes the case that there is a real causal relationship between the nervous system and the adaptive immune system,[31] namely that some immune system antigens are also neuromodulators. This gives a direct link between the brain and the adaptive immune system: each can influence the other.[32]

---

[30] Esther Sternberg, *The Balance Within: The Science Connecting Health and Emotions* (New York: W. H. Freeman, 2001).
[31] See the Wikipedia articles Adaptive immune system and Antigen.
[32] Diana Kwon, 'Your Brain Could Be Controlling How Sick You Get—and How You Recover', *Nature News Feature*, 22 February 2023. Hao Jin et al., 'A Body–Brain Circuit That Regulates Body Inflammatory Responses', *Nature*, 630 (2024), 695–703. Giorgia Guglielmi, 'Found: The Dial in the Brain That Controls the Immune System', *Nature News*, article 01, May 2024.

PLACEBO DIET SUPPLEMENTS

Figure 5.10 The causal power of placebos without deception.

Figure by Mauro Carfora.

As stated by Sternberg, 'The immune system talks to the brain and the brain talks back'. This leads to two key questions she considers: can stress make you sick? And can believing make you well?

The answer in both cases appears to be 'Yes'. Indeed evidence that the latter occurs is that this plausibly underlies the demonstrable power of placebos to improve well-being.[33] Placebos are pills or injections with no active pharmacological component: there is no physiological basis for them affecting bodily health. Just the fact that a pill has been taken can improve health, even if the pill has no active ingredient. This has the consequence that clinical trials of any new drug must be compared with the effect of taking placebos, to show that it is not simply the taking of the drug that has improved health.[34] The effect is so strong that it can occur even if the person affected knows that they pills they are taking are placebos (**Figure 5.10**).[35]

5.2 Consciousness is Real

We experience qualia and our own consciousness, being aware of the passing of time and our own existence in an external world.[36] Indeed denying this is so, as some do

---

[33] See the Wikipedia article Placebo
[34] See the Wikipedia articles Clinical trial and Placebo-controlled study.
[35] Ted Kaptchuk et al., 'Placebos without Deception: A Randomized Controlled Trial in Irritable Bowel Syndrome', *PloS One* 5(12) (2010), e15591.
[36] Massimo Pigliucci, 'Consciousness is Real', *Aeon*, 16 December 2019.

('Consciousness is an illusion'), doesn't even make sense: you can't have any illusion whatever if you are not conscious.[37]

This is one of the most extraordinary features of our existence, characterizing the nature of our being. It is also the foundation of our ability to make the choices that shape our lives and aspects of the world around us. Being awake and sleeping are different states of brain activity. Consciousness largely recedes during sleep but dreams take place and can be remembered to some extent when we wake.

Our perceptions are shaped by our senses (sight, hearing, touch, taste and smell, and balance/proprioception), which means we experience the world differently than animals with other senses (bats must perceive the world differently[38]) or extended sensitivity such as dogs' sense of smell and bird's ultraviolet vision.[39] Our sensory perceptions are not direct reports of what is out there in the external world: they are what our brain interprets to be out there, and may sometimes result in sensory illusions. Nevertheless they provide by and large a good representation of the external world: that must be the case, or we would not survive. These representations are strongly influenced by culture: some cultures can make fine distinctions others do not. An example is the many Eskimo words for different kinds of snow, revealing how culture and language are shaped by physical conditions and experience.

We experience both rational thought and emotions, which interact with each other to shape our thought processes and action choices. We make a difference to the world by our actions, at smaller (personal) and larger (social, environmental) scales; we make choices and act. Some can be undone, but some cannot. Time passes, and some actions make irreparable changes to our physical or social situation.

> The Moving Finger writes;
> And having writ,
> Moves on: nor all thy Piety nor Wit
> Shall lure it back to cancel half a Line,
> Nor all thy Tears wash out a Word of it.
>
> —Omar Khayyám

This is a key feature of our individual and communal lives: a fundamental aspect of the human condition, with the arrow of time ultimately originating in cosmology (§1.1).

We have no idea how consciousness is possible. A great many proposals have been made;[40] however, four main theories have emerged as prime contenders. **Integrated information theory** is a serious attempt to characterize modular hierarchical information processing in the brain. However, firstly, you can't actually do the needed calculations for truly complex systems; a combinatorial explosion prevents this. It's certainly not possible to carry them out for the human brain, although approximate calculations are possible. Secondly, because of this, you can't use it to determine

---

[37] Galen Strawson, 'The Consciousness Deniers', *The New York Review*, 13 March 2018.
[38] Thomas Nagel, 'What Is It Like to Be a Bat?', *The Philosophical Review* 83 (1974), 435–50.
[39] https://en.wikipedia.org/wiki/Dog_sense_of_smell
[40] Robert Lawrence Kuhn, 'A Landscape of Consciousness: Toward a Taxonomy of Explanations and Implications', *Progress in Biophysics and Molecular Biology*, 190 (2024): 28–169.

whether an arbitrary hierarchical system is conscious. However, it has been claimed that this is possible, and even that it supports panpsychist theories. This has led to a considerable backlash against the theory, claiming it is unscientific.[41]

Nevertheless it's an interesting approach that is informative in many ways. **Re-entry and predictive coding** is undoubtedly a central aspect of how perception works (see §5.3.3). **Global workspace theory** is a plausible proposal related to how downward causation takes place in the brain.[42] **Higher-order theories** relate to the way that understanding meaning is a central feature brain function (see §5.3.6 and §5.7).[43]

A problem with all of those proposals is that they are only cortical, whereas subcortical structures play a key role in brain function, as discussed earlier. In particular, Mark Solms' book *The Hidden Spring* claims that the essence of consciousness lies in the brain stem.[44] He gives solid evidence that this is indeed the case.

5.3 A Complex Set of Interactions Take Place in the Mind/Brain

The most fundamental features arising out of the physical structure of the brain are, firstly, the existence of qualia and consciousness, and, secondly, the brain's abilities to learn, to predict and to make action choices. These are guided by our emotions, on the one hand, and our understanding of meaning on the other.

In this section I look at (§5.3.1) 'The Big Picture'; (§5.3.2) 'Automation and Intuition'; (§5.3.3) 'Perceptions and Expectations: Predictive Processing'; (§5.3.4) 'Taste and Smell'; (§5.3.5) 'Rationality, Action Choices and Abstract Causation'; (§5.3.6) 'Possibilities and Imagination'; (§5.3.7) 'Stories and Narratives'; (§5.3.8) 'Shortcuts: Emotions, Heuristics, Satisficing and Lock-in'; (§5.3.9) 'The Social Brain and Mind Reading'; and (§5.3.10) 'Emergence and Agency'.

5.3.1 The Big Picture

This complex interaction of rationality, perception, emotion and values guiding thought and action is summarized in **Figure 5.11**. The brain makes action choices on the basis of rational predictions of future situations informed by past experience,[45] and on the basis of narratives: stories that shape how we think. These are all remembered through the key attribute of brain plasticity: we learn by altering

---

[41] See the Wikipedia article Integrated information theory.

[42] George Ellis, 'Top-down Effects in the Brain', *Physics of Life Reviews*, 31 (2019), 11–27.

[43] Anil K. Seth and Tim Bayne, 'Theories of Consciousness', *Nature Reviews Neuroscience*, 23 (2022), 439–52; Mariana Lenharo, 'The Consciousness Wars: Can Scientists Ever Agree on How the Mind Works?' *Nature*, 17 January 2024.

[44] Mark Solms, *The Hidden Spring: A Journey to the Source of Consciousness* (London: Profile Books, 2021). For a useful review see Leon Hoffman and Stevan Obradovic, 'The Hidden Spring: A Journey to the Source of Consciousness', *jaPa Book Reviews* (2023), 151–160.

[45] Hawkins and Blakeslee, *On Intelligence*. Donald Davidson, 'Actions, Reasons, and Causes', in *Essays on Actions and Events* (Oxford: Clarendon Press, 2001), 3–19.

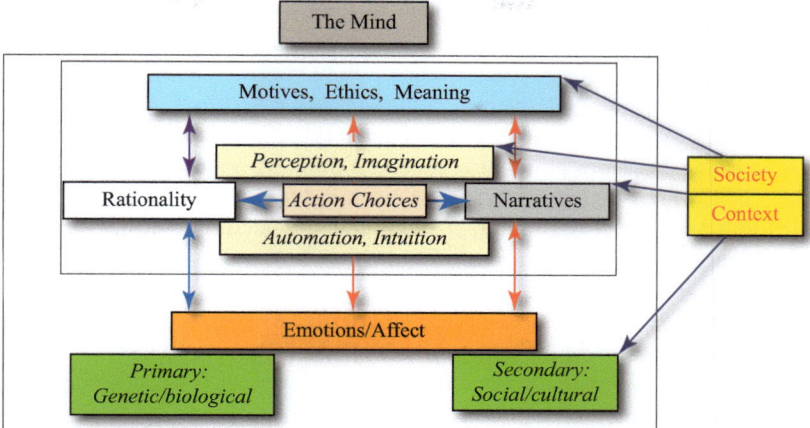

Figure 5.11 The complexity of interacting factors shaping the choices we make.

Figure by Mauro Carfora.

neural connection weights in response to experiences.[46] Learning mathematics or how to play the piano, for example, alters your brain: that is how memories are formed.[47] Automation of learnt thoughts and actions occurs, with intuition being a fast form of subconscious thinking based on past experience. Our choices are influenced by our perception of the current situation and resulting expectations of what will happen in the future. Imagination is the framework of our understanding of possibilities when we make rational choice. All of this is influenced by the society we live in.

Primary (genetically based) emotional systems, built in by evolution because they enhance survival probabilities, guide rationality as discussed in §5.1.3. The nature and function of these systems is discussed in §5.5. Secondary (culturally based) emotions are also key factors in mental life, piggybacking off the functioning of the primary ones. How emotion and cognition work together is discussed by Pessoa.[48] The way that motives, ethics and meaning shape outcomes is discussed in §5.7. Global workspace theory is a model that tries to put this all together, but it does not deal with the role of emotions.[49] A more comprehensive view is the LIDA cognitive architecture, representing many of the interactions in **Figure 5.11**.[50]

---

[46] Feistel, 'Self-Organisation of Prediction Models'.

[47] Eric R. Kandel, 'The Molecular Biology of Memory Storage: A Dialogue between Genes and Synapses', *Science*, 294 (2001), 1030–8.

[48] Luiz Pessoa, 'Précis on the Cognitive-Emotional Brain', *Behavioral and Brain Sciences*, 38 (2015), e71.

[49] Bernard Baars, 'Global Workspace Theory of Consciousness: Toward a Cognitive Neuroscience of Human Experience', *Progress in Brain Research*, 150 (2005), 45–53.

[50] Stan Franklin et al., 'LIDA: A Systems-Level Architecture for Cognition, Emotion, and Learning', *IEEE Transactions on Autonomous Mental Development*, 6 (2013), 19–41. Stan Franklin et al., 'A LIDA Cognitive Model Tutorial', *Biologically Inspired Cognitive Architectures*, 16 (2016), 105–30.

5.3.2 Automation and Intuition

It has been known for a long time that subconscious processes play a key role in our actions and thoughts.

Automation occurs when one has carried out a specific task, such as driving a car, for so long that it becomes automatic: it can be done without any thought.[51] It has become a habit which enables us to free up our mind for other tasks.[52]

Intuition is a way of fast thinking, enabling the understanding of situations without a slow process of rational analysis.[53] Intuitive understandings are not the same as instinct: they have been learned, arising on the basis of past experiences. They have become internalized, but can be rationalized later if need be.[54] Examples are why a doctor concludes you might have pneumonia after a cursory inspection ('You need an X-ray!'), or how a person understands what their partner is about to do without specific communication taking place between them.

5.3.3 Perceptions and Expectations: Predictive Processing

The mind has evolved so as to predict the future on the basis of incoming data taken together with expectations shaped by past experience.[55] This is a key understanding regarding brain function in general,[56] which applies in particular to perception. What we see is what the brain expects us to see, modulated by incoming data that corrects expectation in the light of incoming information. This is illustrated by the examples in **Figure 5.12.**

What I if told you
You the read first line wrong?
Same the with second line
And also the third ...

Figure 5.12 Three visual illusions. Left: the triangle illusion. There are no triangles in this picture. Middle: the duck/rabbit duality. You can interpret the picture as either one or the other, but not both at the same time. Right: Reading what you believe is there (because it makes sense), not what is actually there.

---

[51] John Bargh et al., 'The Automated Will: Nonconscious Activation and Pursuit of Behavioral Goals', *Journal of Personality and Social Psychology*, 81(6) (2001), 1014.

[52] Wendy Wood and Dennis Rünger, 'Psychology of Habit', *Annual Review of Psychology*, 67 (2016), 289–314.

[53] Daniel Kahneman, *Thinking, Fast and Slow* (London: Penguin, 2011).

[54] David G. Myers, *Intuition: Its Powers and Perils* (New Haven, CT: Yale University Press, 2002).

[55] Hawkins and Blakeslee, *On Intelligence*.

[56] Andy Clark, 'Whatever Next? Predictive Brains, Situated Agents, and the Future of Cognitive Science'. *Behavioral and Brain Sciences*, 36(3) (2013), 181–204.

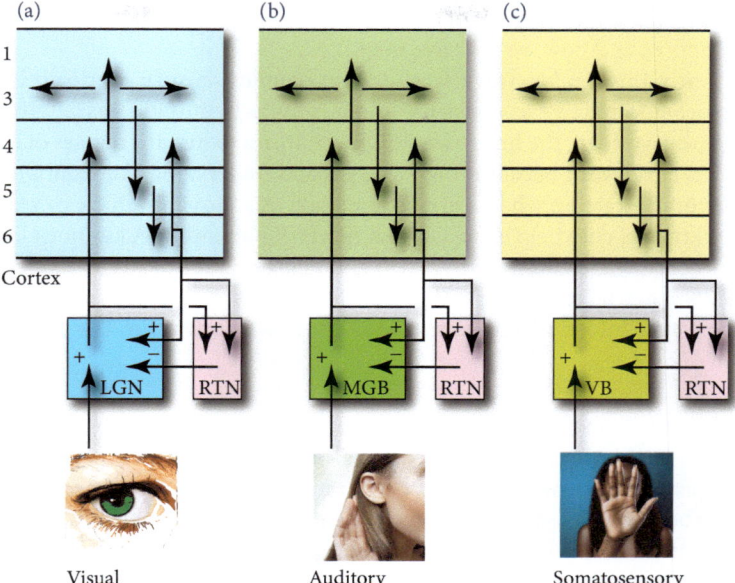

Figure 5.13 Corticothalamic feedback circuits: visual, auditor, and somatosensory circuits. The sensory organs do not have direct access to the relevant part of the cortex. Their input is to the relevant part of the thalamus. This links to cortex L4 only after being shaped by feedback from cortex L6 to the thalamus.

Figure adapted by Mauro Carfora from Henry J. Alitto and W. Martin Usrey, 'Corticothalamic Feedback and Sensory Processing', *Current Opinion in Neurobiology*, 13 (2003), 440–5.

In physiological terms, this occurs via corticothalamic feedback circuits in the brain (**Figure 5.13**).[57] Information from sensory organs (the eyes, ears, touch, balance and so on) is not routed directly to the relevant cortical area. Instead it is sent to the relevant part of a subcortical nucleus, the thalamus[58] (in the case of seeing, the lateral geniculate nucleus, or LGN). Here this incoming signal is compared with a downward signal from level **L6** in the visual cortex, which conveys information on what is expected to be the incoming information. The difference between what is actually seen and what is expected is fed up from the LGN to level **L4** in the cortex. From there it is fed up to higher levels **L3–L1** where it is interpreted (what is the actual situation?), and that interpretation is then fed back down to levels **L5** and **L6** and then to the LGN (with a complication being further feedback involving the thalamic reticular nucleus). This is the mechanism underlying sensory predictive processing[59] and so the various effects occurring in **Figure 5.12**.

---

[57] Henry J. Alitto and W. Martin Usrey, 'Corticothalamic Feedback and Sensory Processing', *Current Opinion in Neurobiology*, 13 (2003), 440–5. See also https://en.wikipedia.org/wiki/Recurrent_thalamo-cortical_resonance.

[58] See the Wikipedia article Thalamus.

[59] See the Wikipedia article 'Predictive Processing'.

Expectations are the outcome of this predictively processed data in the light of our past experiences and how we perceive them.

Importantly, this is a key way that the brain handles sensory overload: billions of bits of information arrive at our senses on a minute-by-minute basis.[60] In order to avoid sensory overload, the brain uses this mechanism to discard masses of irrelevant information (irrelevant because it is expected and understood, for example, a static wall) and concentrate on what is important (such as a moving vehicle or person).

Various kinds of cues help us select what needs our attention. A key question then is cue selection: how are cues recognized? Felin and Koenderink argue that organism-specific, top-down factors play a role in transforming 'raw' optical structure and latent or dormant cues into clues-for-something.[61] In functional terms, the process works like this:[62]

'The whole function of the brain is summed up in: error correction.' So wrote W. Ross Ashby, the British psychiatrist and cyberneticist, some half a century ago. Computational neuroscience has come a very long way since then. There is now increasing reason to believe that Ashby's (admittedly somewhat vague) statement is correct, and that it captures something crucial about the way that spending metabolic money to build complex brains pays dividends in the search for adaptive success. In particular, one of the brain's key tricks, it now seems, is to implement dumb processes that correct a certain kind of error: error in the multi-layered prediction of input. In mammalian brains, such errors look to be corrected within a cascade of cortical processing events in which higher-level systems attempt to predict the inputs to lower-level ones on the basis of their own emerging models of the causal structure of the world (i.e., the signal source). Errors in predicting lower level inputs cause the higher-level models to adapt so as to reduce the discrepancy. Such a process, operating over multiple linked higher-level models, yields a brain that encodes a rich body of information about the source of the signals that regularly perturb it.

Predictive processing is key to competent reading.[63] It underlies the way we perceive text (**Figure 5.12, right**). Crucially, it underlies the way that we simply do not see the blind spot resulting from the fact that there are no light-detecting photoreceptor cells on the part of the retina where the optic nerve enters it.[64] We receive no data at all from a region in front of us about the size of three fingers when your arm is held out straight in front of you. But the brain predicts what we ought to see there—so that is what we 'see'! They way this happens is that we constantly scan the text by small movements called saccade,[65] so we have the relevant information needed to fill in the

[60] See the Wikipedia article Sensory overload.

[61] Teppo Felin and Jan Koenderink, 'A Generative View of Rationality and Growing Awareness', *Frontiers in Psychology*, 13 (2022), 807261.

[62] Clark, 'Whatever Next?'. https://www.cambridge.org/core/journals/behavioral-and-brain-sciences/article/whatever-nextpredictive-brains-situated-agents-and-the-future-of-cognitivescience/33542C736E17E3D1D44E8D03BE5F4CD9

[63] Kenneth Goodman, Peter Fries and Steven L. Strauss, *Reading—The Grand Illusion: How and Why People Make Sense of Print* (London: Routledge, 2016).

[64] See the Wikipedia article Blind spot (vision).

[65] See the Wikipedia article Saccade.

text at each moment, even if we don't see it right then. What Goodman et al. call the 'grand illusion' is the way you can read this page as if no blind spot exists.[66]

5.3.4 Taste and Smell

These interrelated senses (smell affects how things taste) are the only ones that do not work in this predictive processing way. Instead, they are both due to molecules coming from the environment detected by specific taste buds in the tongue and olfactory sensory neurons in the nose.[67] The primary tastes detected by humans are sweet, sour, bitter, salty and savouriness (unami), while odour molecules are detected by about 350 olfactory receptor types. The various combination of these primary odours leads to us experiencing about 10,000 different odours. Animals such as mice and dogs are far more sensitive to smells than we are.

Expectations play a role in how we experience taste as well. The best wine in the world tastes different when it is drunk from an elegant wine glass or a plastic cup. The brain expects a different taste in these two contexts. Even the shape of a plate, or how food is placed on a plate, alters how we it tastes to us.[68]

5.3.5 Rationality, Action Choices and Abstract Causation

Underlying the way that rationality enables action choices is the fact that we are a symbolic species.[69] This is a key distinction between us and all other living beings, including the great apes. Spoken language plays a crucial role by representing opinions, emotions, facts, plans and so on, thereby enabling the functioning of societies. It is learnt by an informal social process of teaching.[70] Written language is a key feature of modern society, enabling the sharing of ideas and emotions through both time and space. This is beautifully stated by Carl Sagan:[71]

> What an astonishing thing a book is. It's a flat object made from a tree with flexible parts on which are imprinted lots of funny dark squiggles. But one glance at it and you're inside the mind of another person, maybe somebody dead for thousands of years. Across the millennia, an author is speaking clearly and silently inside your head, directly to you. Writing is perhaps the greatest of human inventions, binding together people who never knew each other, citizens of distant epochs. Books break the shackles of time. A book is proof that humans are capable of working magic.

---

[66] Goodman et al., *Reading—The Grand Illusion*.

[67] See the Wikipedia articles Sense of smell, Olfactory system, and Taste.

[68] Peter Stewart and Erica Goss, 'Plate Shape and Colour Interact to Influence Taste and Quality Judgements, *Flavour*, 2 (2013), 1–9.

[69] Terrence Deacon, *The Symbolic Species: The Co-evolution of Language and the Brain* (London: W. W. Norton, 1998).

[70] Michael Tomasello, *Constructing a Language: A Usage-Based Theory of Language Acquisition* (Cambridge, MA: Harvard University Press, 2005).

[71] Carl Sagan, *Cosmos*, episode 11: 'The Persistence of Memory' [video], produced by the Austrailian Broadcast Commission, televised by PBS 7 December 1980.

Learning to read and write is key in any modern society: literacy is the key to high-level social functioning. However, it is strongly disputed how literacy should be taught, with reductionist and holistic views competing. A paper I wrote with my wife Carole Bloch, reproduced in Appendix A, gives a holistic view.[72]

Rationality leads to outcomes: deductive causation takes place. Rational thought leads to goals that shape our actions. Thereby mental states have causal power[73]—the existence in the mind of plans for a building or road or digital computer can change physical outcomes (**Figure 3.10**: bottom left). Rational decision making involves choosing between a set of alternative possibilities, using explicit or implicit evaluative criteria with cognitive, psychological, emotional and normative aspects. The process is, however, threatened by information overload, disinformation and the problem of unknown unknowns. The understandings on which we base our decisions in society are based on mental models, perhaps the rational choice models proposed by economists. We need analytic models to plan outcomes, but they are limited in what they can represent and by reliability of outcomes. This, of course, applies particularly to standard economic models that have repeatedly failed to predict real-world events, influenced by emotions and social and political factors not recognized by many models.

Abstract causation. A remarkable feature of the world is the existence of abstract causation[74]—the way that non-physical entities can cause physical effects via the workings of the human mind. Consider the rules of chess (**Figure 5.14**). They control the physical movement of chess pieces on chess boards throughout the world. But they are not physical things. They are not made of wood or stone or steel, and they don't involve electrical or magnetic forces acting on the chess pieces. So how is this control possible?

They are socially agreed rules of play arrived at by a transgenerational process of negotiation (they emerged at the end of the fifteenth century). Once they are agreed, they instruct all players around the world what moves are and are not allowed; for example, the queen can move any number of vacant squares horizontally, vertically or diagonally, but cannot jump as a knight can. This is learnt by all chess players, and through brain plasticity, stored in the detailed patterns of neural connections in their brains. Action potential spike trains in these neural networks shape their thoughts as a player considers her next move, in the light of her game strategy shaped by a study of many games played by others in the past. Finally, a chess piece is moved by her muscles in accord with the rules of chess—abstract entities that shape these physical events. We do not know in detail how particular thoughts, such as the rules for moves by a queen, are stored in the brain's neural networks, or represented by these spike chains. Nevertheless we do know that this happens.

I claim that this is an example of abstract causation. The reductionist response is 'Hold on!—in the end the rules of chess are implemented physically, for example

---

[72] George Ellis and Carole Bloch, 'Neuroscience and Literacy: An Integrative View', *Transactions of the Royal Society of South Africa*, 76 (2021), 157–88.
[73] Peter Menzies, 'The Causal Efficacy of Mental States', in Sven Walter and Heinz-Dieter Heckmann, eds, *Physicalism and Mental Causation* (Exeter, UK: Imprint Academic, 2003), 195–224.
[74] The discussion of abstract causation given here is reproduced, with permission, from an article I wrote for the Institute of Art and Ideas (IAI).

Figure 5.14 The rules of chess as an example of abstract causation.

Figure by Mauro Carfora.

through electrons moving in the brain as just mentioned, so they are nothing but particular brain states at the physical level—configurations of electrons and protons that change over time—so just another example of how physics controls everything'.

However, this is not correct. That is how they are implemented in one particular case. But the rules of chess can be spoken about, represented by sound waves; printed on a page, represented in printed text; explained in a video; talked about in a chess class; represented in algorithms in a chess playing computer. The rules of chess can exist in these multiple forms, not just in individual minds. Furthermore, in response to the claim 'they are nothing but brain states', then the issue is whose brain? Gary Kasparov's? So the rules of chess will cease to exist when he dies? This is obviously not correct. They are represented in the brains and minds of millions of chess players around the world. They are not identical to their representation in any individual brain.

So what are they? They are abstract entities, defined as the equivalence class of all these representations.[75] That is, they are the set of all representations of how chess pieces can move that correctly represent the rules stated by the International Chess Federation.[76] This is not a physical entity; it is abstract. It gains its causal powers when local chess clubs and their members agree to abide by these rules. Thus it is a clear case of abstract causation: abstract entities having physical effects. Physics does not determine what happens: the brain does. The overall causal chain is that the players'

[75] See the Wikipedia article Equivalence class.
[76] See the Wikipedia article FIDE.

mind/brain considers the rules of chess and the state of play, and on this basis decides what move to make next. She then makes the move through her agency: causing specific muscle states in her hand to be activated as desired. This causes chess pieces to move on chess boards according to the rules of chess, or images of chess pieces on a computer screen to move similarly. Physics enables this to take place; thoughts about chess moves, and their implications for the way the game will develop, determine the specific outcomes that occur.

There are many other examples of abstract causation, apart from the rules of other games such as cricket and tennis and baseball. They include (see Figure P.8 in the Prologue) traffic lights that control traffic movements by changing colour from red to green; signs on a wall indicating if a restroom is for men or women or both; signs advertising what a shop sells; social media influencing attitudes to climate change, thus altering actions that either mitigate climate change or exacerbate it. These all influence physical outcomes through their social significance, based on the fact that we are a symbolic species.[77]

Digital computers. In all these examples of abstract causation, we can clearly establish its occurrence, but cannot understand in full detail the neural events that lead to its existence. However, there is one case where many levels of abstract causation occur, and we can understand precisely what is going on: namely the way that digital computers function. This is characterized by two key features.

The first is that their functioning is based in two related kinds of abstract entities: algorithms and data. **Algorithms** are abstract specifications of the logic the machine will carry out, for example Bubblesort. This is an abstract specification of the logic whereby data represented as a list can be sorted into alphabetic order—a key need in many applications. In pseudocode, it can be represented as follows:

```
procedure bubbleSort(A : list of sortable items)
    n := length(A)
    repeat
        swapped := false
        for i := 1 to n-1 inclusive do
            { if this pair is out of order }
            if A[i-1] > A[i] then
                { swap them and remember something changed }
                swap(A[i-1], A[i])
                swapped := true
            end if
        end for
    until not swapped
end procedure
```

It is not specified here in any particular computer language: it can be coded in any high-level language such as Fortran, Cobol, Lisp, Basic, C++ and Java. Thus it is multiply realizable.

---

[77] Deacon, *Symbolic Species.*

The algorithm acts on the data in the list. *Data* is an abstract representation of some physical situation, and must be formatted as required by the implementation language chosen. The data is abstract because it is also multiply realizable. It can be spoken about, written down on paper or on a blackboard, represented in a video or, as in this case, stored in a digital computer memory. It is also multiply realizable in terms of the units used (feet and inches or meters and centimetres) and in terms of computer representation (binary, hexadecimal or decimal).

The second key feature is the existence of a tower of virtual machines that enable computer function, as described by Andrew Tanenbaum and Todd Austin in their book *Structured Computer Organization*.[78] Each virtual machine (an abstract entity) is described by its own language and rules. The different levels are related to each other by compilers or interpreters that chain down instructions and data from higher to lower levels until they reach the binary code level, where they control whether transistors are on or off. This results in an outcome at the binary code level. This outcome then chains back up to the high-level language where it is displayed on a screen, printed on paper or used to activate a mechanical device that alters real-world outcomes. An example is an automatic landing system in an aircraft which receives data from many sources, uses that data to calculate a desired trajectory to result in safely landing on a runway, and activates motors to change control surfaces so as to achieve this desired outcome.

Returning to the game of chess, when a chess game is played by a digital computer, this adds another level of abstraction to what's going on: that of the algorithms that play the game, which are constrained by symbolic representation of the rules of chess. In the brilliant book *Computer Science: A Very Short Introduction*, Subrata Dasgupta claims that both physical and abstract causation occur in digital computers.[79] He states,

> As regards digital computers, in its most fundamental, the stuff of computers is systems of symbols, forming symbol structures—that is entities that stand for, denote, or represent other entities. Computing is symbol processing.

Symbols are by their very nature multiply representable. He goes on to state,

> Some computational artefacts are entirely abstract: they not only process symbol structures, they themselves are symbol structures and are devoid of any physicality, though they may be made visible by physical media such as marks on paper or computer screens. Physico-chemical laws do not apply to them. They neither occupy physical space nor do they consume physical time.

These abstract entities have causal powers via what Dasgupta refers to (p. 23) as liminal structures, where the abstract becomes physical and can result in images on a screen, marks on paper, three-dimensional printing of artefacts, automatic landing of an aircraft and so on. A detailed explanation of how this all takes place in physical

---

[78] Andrew Tanenbaum and Todd Austin, *Structured Computer Organisation* (Boston: Pearson, 2013).

[79] Subrata Dasgupta, *Computer Science: A Very Short Introduction* (Oxford: Oxford University Press, 2016), 22–3.

terms is given in a paper I wrote with my colleague Barbara Drossel.[80] The logic embodied in the algorithm chains down to turn transistors on and off in accord with the data by altering a potential energy term in the quantum equations of motion. This is the detailed physics that enables the liminal structures referred to by Dasgupta to change physical outcomes.

Downward causation is taking place **logically** in the tower of virtual machines, where higher logical levels determine what happens at lower logical levels, and **physically** in the way that data entered on a keyboard (a macro-level entity) shapes movement of electrons in transistors at the micro level. The existence of this downward causation means that physics laws can't by themselves determine the specifics of the computer's workings in any specific case.

Physics, of course, enables this by underlying how transistors turn on and off according to the gate voltage. **When** that happens **in what order** in which specific transistors is dictated by the computer program embodying the underlying algorithms, together with the data the program acts on. These are both initially provided by the user, but they may be modified on a minute-to-minute basis due to the computer being connected to the Internet. And the very existence of the both the computer and the data is a result of the fact that human minds have agency (§5.6), allowing them to design and create digital computers—and to use them for specific purposes.[81]

Mathematics. A further key form of abstract causation is through another abstract symbolic system: mathematics and mathematical equations.[82] Through enabling us to codify the laws of physics, they underlie all engineering projects; through enabling accurate tracking of resources, they underlie all commerce. In this way, a number of equations have changed the world.[83]

Sometimes general solutions can be found by using mathematical methods, but this is rare. In the case of realistic models of real-world situations (simulating airflow over an aircraft wing to predict its flight characteristics, weather forecasting, analysing correlations in data and so on), they are usually solved by using digital computers. The numerical methods used to do so are coded in algorithms, so there are multiple levels of abstract causation involved: the equations, the numerical methods, the algorithms.

Another case of abstract causation is using digital computers to simulate neural networks to create artificial neural networks (ANNs), as in the case of large language models such as ChatGPT. An extraordinary recent feat was the creation of the programme AlphaFold, enabling the prediction of the very complex folding patterns of proteins. Two of the team that created AlphaFold won the Nobel Prize in Chemistry in 2024 for their work on protein structure prediction.[84]

---

[80] George Ellis and Barbara Drossel, 'How Downwards Causation Occurs in Digital Computers', *Foundations of Physics*, 49(11) (2019), 1253–77.

[81] Albert. Bandura, 'Toward a psychology of human agency', *Perspectives on psychological science*, 1(2) (2006), 164–180.

[82] See the Wikipedia articles Mathematics and Equation.

[83] Robert Crease, *A Brief Guide to the Great Equations: The Hunt for Cosmic Beauty in Numbers* (London: Hachette UK, 2010). Ian Stewart, *In Pursuit of the Unknown: 17 Equations That Changed the World* (London: Hachette UK, 2012).

[84] See the Wikipedia article AlphaFold and the Nobel Prize websites Popular information and Advanced information.

However, the mathematics does not have to be complex in order to be useful. A key innovation was the introduction of spreadsheets such as Microsoft Excel, which just uses arithmetic.[85] This transformed how financial records were kept and how projections and budgets were made, because one could simply update relevant data on the screen and immediately see the effect on an organization's finances.

5.3.6 Possibilities and Imagination

What can be achieved by us individually or collectively is determined not just by what is possible—we can't achieve something if it is impossible, for whatever reason—but by our understanding of what is possible. To explore this requires not just rational exploration of the possibilities we are considering, but the creative use of imagination to generate possibilities we can explore.[86] Crucial progress results from 'thinking outside the box' in a sensible way.[87]

5.3.7 Stories and Narratives

Key influences on our choices are stories,[88] for example the (untrue) stories about immigrants eating dogs and cats in Ohio that helped shape the outcome of the recent presidential election in the USA. Stories convey claimed facts and events, whether true or untrue, and emotional responses to the events related. Narratives embody patterns of understanding arising out of the stories we hear,[89] and can provide a key input to making decisions in the face of radical uncertainty.[90] They are an alternative to extensive analytical thought because they concisely convey holistic patterns of causation in an understandable way.

5.3.8 Shortcuts: Emotions, Heuristics, Satisficing and Lock-in

In terms of real-world decision making, our problem is the pressure of time. We need to make key decisions, and if we spend too much time collecting data and analysing it, the opportunity may be gone forever. We need fast methods of making decisions in the current context, based on our overall goals and values.

---

[85] See the Wikipedia article Spreadsheet.
[86] Harold Rugg, *Imagination* (London: Harper and Row, 1963). Paul L. Harris, *The Work of the Imagination* (Oxford: Blackwell Publishing, 2000).
[87] Sabrina Meherally, 'What It Means to Truly Think Outside the Box', *Harvard Business Review*, 2 March 2022.
[88] Jonathan Gottschall, *The Storytelling Animal: How Stories Make Us Human* (New York: Houghton Mifflin Harcourt, 2012).
[89] See the Wikipedia article Narrative.
[90] Samuel Johnson, Avri Bilovich and David Tuckett, 'Conviction Narrative Theory: A Theory of Choice under Radical Uncertainty', *Behavioral and Brain Sciences*, 46 (2023), e82.

Kahneman's famous book *Thinking, Fast and Slow* presents a thoughtful analysis of how to do this, but his **System 1** is not homogeneous: it represents a variety of different kinds of effects.[91] From my viewpoint, there are four things to note.

First, emotions play a key role in our decision making: indeed that is why we have them (§5.1.3). They give a fast response to the situation, on the basis of either our evolutionary heritage or social context. I discuss this further in §5.4. The key point is that we must balance these emotional pressures with more rational assessments of the situation: we cannot afford to let our emotions run away with us.

Second, we use various rules of thumb—heuristics[92]—to decide what to do on the basis of limited data. These are based on past experience and enable choosing satisfactory outcomes on the basis of limited data.

Third, while we may have high hopes of what we can achieve, we must be prepared to accept a good enough solution by 'satisficing'[93]: that is, accepting a not quite ideal outcome but it will do.

Fourth, as pointed out by Ruth Chang,[94] lock-in takes place. That is, once we make a decision, the whole context of decision making changes, and it does so in such a way that it is hard to undo the decision that has been made. It's often an irreversible process.

5.3.9 The Social Brain and Mind Reading

Context is crucial, so we have a social brain:[95] the social context affects all these processes. Recent 'wide' perspectives on cognition (embodied, embedded, extended, enactive and distributed) should be seen as steps toward building integrated explanations of the mechanisms involved, including not only internal mechanisms but also interactions with others, groups, cognitive artefacts and their environment. In the context of irreducible uncertainties, *cultural mental models*—categories, concepts, social identities, narratives and worldviews—profoundly influence judgment and behaviour.[96]

In particular, perception extends to the social context where we are continually in effect reading the minds of those we encounter, as a basis of predicting what they will do.[97] This is an extension of predictive processing to social situations, and this process evolves in a culturally dependent way.[98] The extended mind includes emotional sharing and sharing norms.[99]

---

[91] See the Wikipedia article *Thinking, Fast and Slow*.
[92] See the Wikipedia articles Heuristic (psychology).
[93] See the Wikipedia article Satisficing.
[94] Ruth Change, 'Hard choices', *Journal of the American Philosophical Association*, 3(1) (2017), 1–21.
[95] Robin Dunbar, 'The Social Brain Hypothesis', *Evolutionary Anthropology*, 6 (1998), 178–90.
[96] Karla Hoff and Joseph E. Stiglitz, 'Striving for Balance in Economics: Towards a Theory of the Social Determination of Behavior', *Journal of Economic Behavior & Organization*, 126 (2016), 25–57.
[97] Frith, *Making up the Mind*. Chris Frith and Uta Frith, 'Theory of Mind', *Current Biology*, 15 (2005), R644–5.
[98] Cecelia Heyes and Chris D. Frith, 'The Cultural Evolution of Mind Reading', *Science*, 344 (2014), 1243091.
[99] Felipe León et al., 'Emotional Sharing and the Extended Mind', *Synthese*, 196 (2019), 4847–67. Cecilia Heyes, 'Rethinking Norm Psychology', *Perspectives on Psychological Science*, 7 June 2022.

5.3.10 Emergence and Agency

The confluence of these effects means that physics—the patterns of electrons and protons in the brain, and their interactions—can't by itself determine the mind's workings. Important as that is, physics is just one the mosaic of causal effects that shape brain outcomes. The mind transcends physics by shaping physical outcomes to achieve its goals. It has agency that allows it to do so by downwardly determining the context within which physical forces work. Our understanding of what is happening is central to the mind/brain's functioning and resultant physical outcomes. None of this is visible at the underlying physics level. In short, we have agency that is not determined by the underlying physics.[100]

I discuss this further in §5.6.

5.4 The nature and Function of Emotions

A key feature is the innate (genetically determined) emotional systems that guide the process of learning via neuromodulators such as dopamine and norepinephrine that are known to have emotional effects (see the discussion in §5.1.3 and **Figure 5.9**). This is disputed by the **theory of constructed emotion**,[101] which claims emotions are intellectual constructs in precisely the same way as the idea of money is. This theory is denying the status of emotions as qualia—experiences that we deeply feel. For a convincing response, see Karolina Westlund's comments, pointing out that the theory has no sound evolutionary basis, the meta-analysis supposed to support it mixes incompatible data sets and it ignores the neuromodulators known to play a key role in emotions.[102]

The plausible view is that primary emotional systems exist and are based on physiological systems ('ascending systems' or 'reticular formations') and associated neuromodulators such as dopamine and norepinephrine that have come into existence through natural selection because they are essential for our existence. The reasons I find this convincing, combining the work of Edelman and Panksepp. are set out in §5.1.3. Thus the view I take is that of Panksepp and Solms.[103] The primary emotions are genetically determined self-regulatory systems, and exist to help regulate our behaviour in a way that will enhance survival, which is why they exist. Hence they are not primarily aimed at conveying emotions via facial expressions and other bodily signalling, although that may be beneficial in group contexts. These systems are shared with all mammals, which is why we feel empathy with them and why so many people love their pets and grieve their deaths.

---

[100] Gazzaniga, *Who's in Charge?* Mitchell, *Free Agents.*

[101] Lisa Feldmann Barrett, 'The Theory of Constructed Emotion: An Active Inference Account of Interoception and Categorization', *Social Cognitive and Affective Neuroscience*, 12 (2017), 1–23. Lisa Feldmann Barrett, *How Emotions Are Made: The Secret Life of the Brain* (London: Pan Macmillan, (2017).

[102] Karolina Westlund, 'My Problems with the Constructed Theory of Emotions' (2021) https://illis.se/en/constructed-theory-of-emotions/.

[103] Panksepp, *Affective Neuroscience.* Ellis and Solms, *Beyond Evolutionary Psychology.*

Panksepp proposed seven primary (genetically determined) emotional systems. With Judith Toronchuk I have proposed including an additional two.[104] The resulting full set is shown in **Table 5.1**.

The SEEKING systems underlies exploration, and the fundamental human need to find meaning in what we experience.[105]

The DISGUST system is an anticipatory system, protecting us from harm before it has happened. In humans, it gets transformed to play a powerful role in social contexts.[106]

The PANIC/attachment system was given that name because of the distress infants experience when separated from caregivers, but gets extended in teenage and adult social contexts to attachment to social groups. This leads to the distress associated with being excluded from social groups, as emphasized by Stephens and Price.[107] It is the emotional driver underlying the social brain.[108]

The PLAY system is key to learning. It starts with rough and tumble play in animals as emphasized by Panksepp, but then extends to symbolic play in children, the power of stories and the existence of theatres with suspension of belief, and underlies the power of imagination and creativity.

The POWER/dominance system arose as an evolutionary solution to the problem of conflict in a group over access to resources. What evolution discovered is that a dominance hierarchy could solve the problem of conflict over resources once a group has formed. Members of the group compete for position in the social hierarchy, either by combat or by behaviour, but then accept their position in that hierarchy until the next struggle for position in the hierarchy takes place. It has ancient evolutionary roots, stretching back to lizards, the pecking order of chickens and the existence of alpha-male wolves. It is related to notions of selves because it is associated with the territorial imperative,[109] where 'territory' is not just physical territory, but also social, economic, political and academic territory that is deeply tied into human identity. Malfunctioning of this dominance system is identified by Stevens and Price as one of the major causes of psychiatric illness in humans.[110] It plays a strong role in social and academic conflict, and of course is central to competitive sport.

I have emphasized these systems because in my view they are a key feature of our nature. We are each born with all these systems, but with different relative strengths of the different systems in each individual. If any of them are genetically set either too high or too low, problems will ensue.

Secondary (social) emotions piggyback on the primary ones, and are both positive (e.g. pride) and negative (e.g. shame, guilt). They are, of course, strongly culturally dependent. What triggers these emotional systems is a complex psychological issue. It can occur due to external events, or internal musings.

---

[104] George Ellis and Judith Toronchuk, 'Affective Neuronal Selection: The Nature of the Primordial Emotion Systems', *Frontiers in Psychology*, 3 (2013), 589.

[105] Viktor Frankl, *Man's Search for Meaning* (Boston: Beacon Press, 2006).

[106] Martin Kavaliers, Klaus-Peter Ossenkopp, and Elena Choleris. 'Social neuroscience of disgust', *Genes, Brain and Behavior*, 18(1) (2019), e12508.

[107] Anthony Stevens and John Price, *Evolutionary Psychiatry: A New Beginning* (London: Routledge, 2015).

[108] Dunbar, The social mind hypothesis

[109] Robert Ardrey, *The Territorial Imperative: A Personal Inquiry into the Animal Origins of Property and Nations* (New York: Kodansha America, 1966).

[110] Stevens and Price, *Evolutionary Psychiatry*.

Table 5.1 Evolutionary Needs Met, and the Systems That Meet Them

Evolutionary needs met		Primary emotional system	Function
Individual needs			
Basic functioning	E1	SEEKING system	Situation evaluation, incentive salience, hedonic appraisal, facilitates learning
Basic survival	E2	DISGUST system*	Avoiding harmful foods, substances, environments
	E3	RAGE system	Defence: protection of organism, resources and conspecifics
	E4	FEAR system	Defence: flight, limiting of tissue damage
Social needs			
Reproduction	E5	LUST system (desire, satiation)	Ensuring procreation, enhancement of bonding
Group cohesion: Bonding, affiliation	E6	PANIC/attachment (affiliation, separation distress)	Protection of vulnerable individuals; creates bonding through need for others
	E7	CARE system	Caring for others, particularly offspring
Learning, cooperation	E8	PLAY system	Bonding with conspecifics, development of basic adaptive and social skills, creativity
Group function: regulating conflict	E9	POWER/dominance system* (rank, status, submission)	Limiting aggression in social groups: allocating resources, esp. sexual ones

Note: The two marked with an asterisk * are the extra ones proposed in Ellis and Toronchuk, 'Affective Neuronal Selection' (2013), in addition to those proposed by Panksepp.

5.5 What Is Innate in Our Brains and What Is Not

There are many claims of innate modules in the brain, such as Chomsky's proposal of existence of a language acquisition device, and other suggestions of 'folk physics' modules, 'folk psychology' modules and so on. Some evolutionary psychologists proclaim many 'just so' stories of innate modules such as these. However, they cannot, in fact, exist for evolutionary, genetic and developmental reasons. This is discussed in detail in a book by myself and Mark Solms.[111] The three key issues are this.

[111] George Ellis and Mark Solms, *Beyond Evolutionary Psychology: How and Why Neuropsychological Modules Arise* (Cambridge, UK: Cambridge University Press, 2018).

First, there is not enough detailed information in our DNA to specify specific neural connections in the brain. We have about 10^{11} neurons, each with about 10^3 synaptic connections, thus about 10^{14} synaptic connections in all, but we have only about 23,000 genes that code for proteins. In informational terms, there is not a fraction of the information available in DNA necessary to specify details of these connections, particularly when one remembers that genes have to specify existence of a skeleton, skin, heart, lungs, hands, feet, eyes, ears and so on. A lot of information is required to generate our bodies, including, for example, existence of the optic nerve connecting the eyes to the thalamus (which needs specific wiring).

Second, if that information was indeed there, there is no developmental process which could produce precise wiring diagrams in the cortex embodying specific concepts, ideas or rules such as the phrase structure rules that Chomsky alleges shape language. What actually happens is that initial neural connections are made randomly by a curious process inter alia involving neurons being created in one place and then migrating somewhere else. Then as we experience life and learning takes place, connections are strengthened, weakened or pruned in such a way as to represent what we have learned.[112] This process of neural plasticity at the micro level underlies brain plasticity at the macro level such as remembering events and ideas and learning muscle control.

Third, given that we have only about 23,000 coding genes, the issue is that evolutionary processes must ensure that through developmental processes, all the physiological systems that we need will be created; they all play key roles in our functioning and so survival (Chapter 4). While it is crucial for our survival that we develop a language-ready brain, capable of symbolic behaviour and so understanding language,[113] it is not remotely plausible that syntactic details are so important for survival that they would be genetically coded. Indeed it is precisely because we have brains capable of predictive processing of data (§5.3.3) that grammatical details are largely irrelevant in understanding spoken language and written text. What are, in fact, innate are the primary emotional systems discussed in §5.4. The ascending systems and associated neuromodulators such as dopamine play a crucial role in survival by guiding rationality, which is why they are innate systems. This is discussed in my book with Mark Solms mentioned earlier.[114]

5.6 The Existence of Agency and Free Will

The existence of free will is denied by many, including neuroscientists, for example Crick[115] and Sapolsky,[116] and physicists, for example Carroll[117] and Hossenfelder.[118]

---

[112] Lewis Wolpert et al., *Principles of Development* (Oxford: Oxford University Press, 2002).

[113] Goodman et al., *Reading—The Grand Illusion.*

[114] Ellis and Solms, *Beyond Evolutionary Psychology.*

[115] Francis Crick, *The Astonishing Hypothesis: The Scientific Search for the Soul* (New York: Scribner, 1994).

[116] Robert M. Sapolsky, *Determined: A Science of Life without Free Will* (London: Penguin Press, 2023).

[117] Sean Carroll, 'Consciousness and the Laws of Physics', *Journal of Consciousness Studies*, 28 (2021), 16–31.

[118] Sabine Hossenfelder, *Existential Physics: A Scientist's Guide to Life's Biggest Questions* (New York: Viking Press, 2022).

Physical determinists insist that if we could exactly reproduce every physical detail of our brain down to the smallest detail—the atomic structure of every neuron, the position and momentum of every electron and every proton, the details of their connections via synapses and the state of excitation of every synapse at an instant—one could then use the relevant laws of motion and the known forces acting on every particle to predict in detail every future brain state. Call all that data $D(t)$.

Then the claim is that data $D(t_0)$ at an initial time t_0 uniquely determines that data $D(t_1)$ at any later time t_1

$$D(t_0) \to D(t_1) \quad \text{for all } t_1 > t_0. \tag{5.1}$$

Because all the relevant forces are known,[119] and assuming (5.1) is true, agency and free will are illusions. Physics is doing all the work, and all higher states are epiphenomena.

We are not in charge of our fate; we cannot be held responsible for our actions in a court of law. Furthermore you may be proud of a book you wrote, a speech you made, a brilliant golf stroke, whatever, but actually it was not YOU that did any of this: it was just electrons and protons interacting with each other. Or it was neurons making you do it.

However, on the other side, many defend free will, including Donald,[120] Murphy and Brown,[121] Heisenberg,[122] Gazzaniga,[123] List,[124] Ismael,[125] and Mitchell.[126] Fischer[127] reviews the problems with Sapolsky's book arising inter alia because he does not even define free will.

This is an important issue, so I will consider it in depth. In this section I look at (§5.6.1) 'Multiple Competing Kinds of Deterministic Explanation'; (§5.6.2) 'The Nature of Causal Completeness: How Actions Are Determined'; (§5.6.3) 'The Influence of Molecular Stochasticity'; (§5.6.4) 'Irrelevance of Compatibilism and Incompatibilism: Our Existence as Open Systems'; (§5.6.5) 'The Cosmic Context'; and (§5.6.6) 'If Free Will Is Not True, Science and Philosophy Are Not Possible; Pragmatically, It's Apparent'.

---

[119] Carroll, 'Consciousness and the laws of physics'.

[120] Merlin Donald, *A Mind So Rare: The Evolution of Human Consciousness* (London: W. W. Norton, 2001).

[121] Nancey Murphy and Warren Brown, *Did My Neurons Make Me Do It?* (Oxford: Oxford University Press, 2007).

[122] Martin Heisenberg, 'Is Free Will an Illusion?', *Nature*, 459 (2009), 164–5. Martin Heisenberg, 'The Origin of Freedom in Animal Behaviour', in A. Suarez and P. Adams, eds, *Is Science Compatible with Free Will?* (New York: Springer, 2012), 95–103.

[123] Gazzaniga, *Who's in Charge?*

[124] Christian List, *Why Free Will Is Real* (Boston, MA: Harvard University Press, 2019).

[125] Jennan Ismael, *How Physics Makes Us Free* (Oxford: Oxford University Press, 2019).

[126] Mitchell, *Free Agents*.

[127] J. M. Fischer, 'Determined: A Science of Life without Freewill', *Notre Dame Philosophical Reviews*, 2023.11.3.

Table 5.2 Kinds of Determinism That Have Been Proposed[a]

Level	Type	Claim
L0	Metaphysical determinism	Fate (Karma) or God made me do it
L1	Social/cultural determinism	Society/upbringing made me do it
L2	Psychological determinism	Behaviourism
L3	Neuroscientific determinism	My neurons made me do it
L4	Biological determinism	My genes made me do it
L5	Physical/causal determinism	Physical interactions (natural law) made me do it

[a]See the Wikipedia articles Fatalism, Cultural determinism, Behaviourism and Biological determinism.

5.6.1 Multiple Competing Kinds of Deterministic Explanation

Over the years, many different kinds of determinism have been claimed to hold by a variety of philosophers and scientists. I indicate them in **Table 5.2**. Except for the highest level **L0**, this table broadly corresponds to the hierarchy of emergence that is central to the discussions of this book. The degree of certitude associated with this table varies greatly at the different levels indicated. There is no agreement at all on the nature of causality at the supposed highest level **L0**. At the social and psychological levels **L1, L2**, only broad themes emerge as genuine predictions. At the physiological levels **L3, L4** well-established, reliable (but very complex) laws emerge, but at the molecular level there is considerable stochasticity that I discuss in §5.6.3. At the classical physical level, determinism holds in principle, but at the underlying quantum physics level **L5** it is well established that predictability does not hold (Chapter 2). This is the experimental position, despite attempts by some to deny that this is the case.

Those claiming that physics determines everything often neglect to state which particular physical level is the one that determines everything. Is it the Standard Model of particle physics? Some assumed Grand Unified Theory? Some supposed theory of everything (TOE), which if it exists, underlies all the other physics levels and so is the most fundamental? But we have no well-established TOE: the claim that it exists is just a hope, and the supersymmetry that was supposed to underlie it seems to not be correct.

Carroll, as well as Laughlin and Pines, claims that the physics that matters for biology is the interactions of protons and electrons as described by the Schrödinger equation.[128] I agree with this, except that when spin matters, as it does for establishing the periodic table of the elements, it's the Dirac equation that is needed. But to get specific classical outcome you need wavefunction collapse in addition, as in quantum chemistry (Chapter 2). However, I accept the claim that quantum uncertainty does

---

[128] Carroll, 'Consciousness and the Laws of Physics'. Robert B. Laughlin and David Pines, 'The Theory of Everything', *Proceedings of the National Academy of Science*, 97 (2000), 28–31.

not per se play a role as regards the issue of free will. It takes place at the wrong scale to be relevant: it gets washed out.

Here is the way I view it. First, I deny that metaphysical determinism (**L0**) holds. Such proposals are in my view incompatible with science. Second, nowadays the two forms of determinism usually claimed to be true are physical determinism (**L5**) and neuroscientific determinism (**L3**).[129]

Now the key point is this: **they contradict each other.**

If **L5** is true, then no higher level uniquely determines outcomes: Sapolsky's discussion is simply irrelevant,[130] because physics **L5** determines everything with no remainder.[131] The stuff at **L3** is just epiphenomenal stuff that goes along for the ride. On the other hand, if determinism at level **L3** is true, then it is not the case that physics at level **L5** determines everything. It is **L3** that does that job, and it is **L5** that just goes along for the ride, via downward causal effects.[132] Similar comments apply to any claim that any other level is the only one that determines everything. It implies that, despite all the empirical evidence, all the others are epiphenomena. That is the core nature of strong reductionist viewpoints. Underlying both claims are reductionist philosophical attitudes (just one underlying causal theory is sufficient to explain everything) and psychological ones (reduction to the particular theory that I happen to understand is the one that matters).

So what is the resolution? It is to understand that **all of these levels are causally effective**, and replace the words 'determinism' in **Table 5.2** with the word 'efficacy'. Then every level is causally relevant, as claimed by Denis Noble,[133] but none by itself determines unique outcomes.

The problem is claiming that any single one of the levels suffices to determine outcomes uniquely. A network of causal influence shapes any specific outcome. For example, she died because the car was driving too fast; she died because it was raining, so the car skidded off the road; because she went out to buy cigarettes at the shop; because of Newton's Laws of Motion determining the car's trajectory; and so on. What we call **the Cause** is that part of this network we happen to be interested, in taking all the other contributing elements (the existence of the Universe, the Earth, Newton's Laws of Motion, and so on) for granted.

A holistic integrative view such as I propose takes into account all the emergent levels and their upward and downward interactions, the true nature of the cosmic context, the fact that physical causation is possible and that causal determinism can include our own choices.

---

[129] Car Hoefer, 'Causal Determinism', *The Stanford Encyclopedia of Philosophy* (summer 2024 edition), ed. by Edward N. Zalta and Uri Nodelman (2024), <https://plato.stanford.edu/archives/sum2024/entries/determinism-causal/>.

[130] Sapolsky, *Determined*.

[131] Carroll: Consciousness and the Laws of Physics.

[132] Samuel Lee, 'Building Low Level Causation out of High Level Causation', *Synthese*, 199 (2021), 9927–55.

[133] Denis Noble, 'A Theory of Biological Relativity: No Privileged Level of Causation', *Interface Focus*, 2 (2012), 55–64.

5.6.2 The Nature of Causal Completeness: How Are Actions Determined

An example of the interaction of causal effects is the way that social effects can cause genetic effects. A brilliant study of male zebra finches shows in detail how they display two song behaviours: directed and undirected singing.[134] Referred to collectively as 'the song system', they have two brain circuits underlying this: a posterior motor pathway necessary for song production, and an anterior pathway necessary for song acquisition.

'Directed' song is usually accompanied by a courtship dance and is addressed almost exclusively to females. 'Undirected' song is not accompanied by the dance and is produced in different circuits. This is all demonstrated in detail in the paper, showing the interlevel determination of action taking place between levels **L1** and **L2**. This is then related to brain physiology ('cortical–basal ganglia' subdivisions) with different gene activation at level **L4** depending on whether the bird sings female-directed or undirected song.

Obviously neurons are involved in the process, and when genes are controlled in this way it is enabled in the end by the underlying quantum physics causing the relevant macromolecular changes to occur.[135] Thus the paper describes the effects of social context on gene expression, and the way that causal completeness only occurs when the interactions between all the levels **L1** to **L5** in **Table 5.2** are taken into account.

5.6.3 The Influence of Molecular Stochasticity

Molecular unpredictability is rampant. The predictive equation (5.1) does not take into account the molecular randomness manifested in Brownian motion[136] and the molecular storm.[137] There is an immense stochasticity in molecular collisions in a cell; predictability (5.1) of outcomes at the molecular level is completely impracticable, not just in practice, but in principle, because with these huge numbers of collisions, the Heisenberg uncertainly principle prevents the initial data $D(t_0)$ being determinable at the molecular level at an accuracy that will allow precise enough predictions. Laplace's demon cannot even in principle know the initial data needed in (5.1) with sufficient accuracy to predict specific outcomes.[138] Chaotic dynamical systems occur in weather dynamics that affect our lives (should I take an umbrella to work today?), so arbitrarily small changes in microscopic initial conditions can lead to very different macro-level outcomes (which is the real butterfly effect[139]).

---

[134] Erich Jarvis et al., 'For Whom the Bird Sings: Context-Dependent Gene Expression', *Neuron*, 21 (1998), 775–88.

[135] Martin Karplus, 'Development of Multiscale Models for Complex Chemical Systems: From H+ H 2 to Biomolecules (Nobel Lecture).' *Angewandte Chemie International Edition*, 38 (2014), 9992–10005.

[136] See the Wikipedia article Brownian motion.

[137] Liam Graham, *Molecular Storms: The Physics, of Stars, Cells, and the Origin of Life* (Cham: Springer, 2023).

[138] See the Wikipedia article Laplace's demon.

[139] Tim Palmer, Andreas Döring, and G. Seregin, 'The Real Butterfly Effect', *Nonlinearity*, 27 (2014), R123.

Biological processes have evolved to take advantage of this randomness through the structure of molecular machines like kinesin and dynein that extract order from molecular chaos in order to achieve useful outcomes.[140] That dynamic opens up the way for higher levels to select lower-level desirable outcomes, freeing us from the tyranny of physical determinism and enabling agency to emerge,[141] since causal closure includes the logic of the emergent LIDA architecture where decisions are made,[142] and downward causation from that level can select lower-level states accordingly.

5.6.4   Irrelevance of Compatibilism and Incompatibilism: Our Existence as Open Systems

There is a sophisticated literature on compatibilism and incompatibilism:[143] is free will compatible with the kind of determinism represented by Eq. (5.1)? But that entire literature is irrelevant to the issue of free will, because it does not take into account the issue of what kind of system it applies to. Human beings are open systems (**Figure 5.15**).[144]

Given all that data $D(t_0)$ about the state of our body and mind at an initial time, it tells nothing about all the incoming data that will impinge on our senses or affect our bodies in the future. Rain and storms may occur, needing us to seek for shelter and warmth. An accident may occur, so that we must suddenly decide what to do (**Figure 5.15**): call an ambulance? call the police? try to apply first aid?

None of these events were known to our brain at an initial time t_0, so they are not encoded in the state of our brain $D(t_0)$, but they change our future thoughts and actions in critical ways. Furthermore we ingest new material at times after t_0 that may crucially affect outcomes: a fish that has gone off can causal violent illness and so alter our thoughts completely. Even the molecules that make up our brain alter over the course of time. Francis Crick puzzled over how memory could be preserved in the face of this molecular turnover.[145]

Thus in the real world, the deterministic relation (5.1) simply does not apply to open systems such as the human brain: new incoming data is not accounted for by that equation. That is why we have homeostatic systems to counter physical and chemical disturbances, adaptive immune systems to counter unpredictable viruses and predictive processing brains to predict what will happen, read the minds of those in the

---

[140] Paul M. Hoffmann, *Life's Ratchet: How Molecular Machines Extract Order from Chaos* (New York: Basic Books, 2012).

[141] Raymond Noble and Denis Noble, 'Harnessing Stochasticity: How Organisms Make Choices', *Chaos*, 28 (2018), 106309; Raymond Noble and Denis Noble, 'Can Reasons and Values Influence Action: How Might Intentional Agency Work Physiologically?', *Journal for General Philosophy of Science*, 52(2) (2021), 277–95.

[142] Franklin, et al . (2013). LIDA: A systems-level architecture for cognition, emotion, and learning.

[143] See the Wikipedia articles Free will and Determinism.

[144] George Ellis, 'Biological Emergence: A Key Exemplar of the Open Systems View', in Michael E. Cuffaro and Stephan Hartmann, eds, *Open Systems: Physics, Metaphysics, and Methodology* (Oxford: Oxford University Press, 2026). pp. 216–232.

[145] Francis Crick, 'Neurobiology: Memory and Molecular Turnover', *Nature*, 312 (1984), 101.

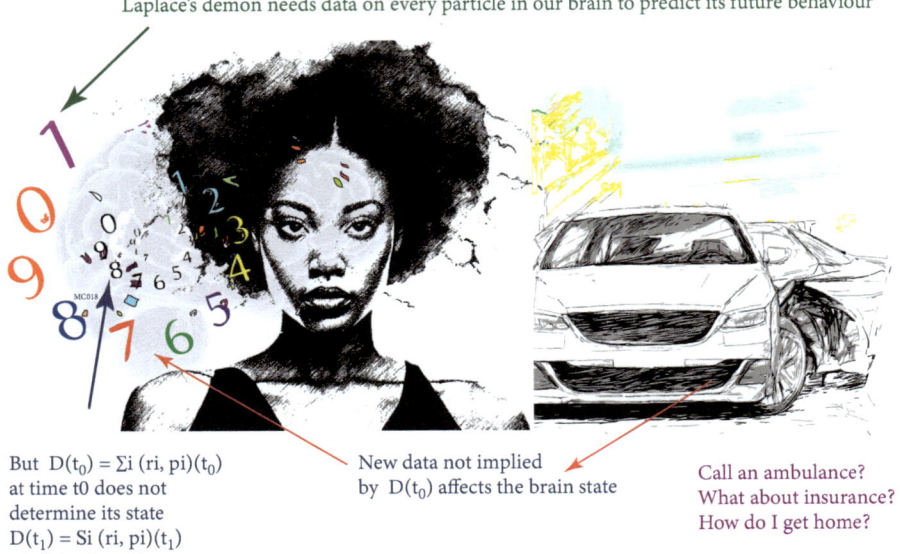

Laplace's demon needs data on every particle in our brain to predict its future behaviour

But $D(t_0) = \Sigma i\, (ri, pi)(t_0)$ at time t0 does not determine its state $D(t_1) = Si\, (ri, pi)(t_1)$ at a later time t1

New data not implied by $D(t_0)$ affects the brain state

Call an ambulance? What about insurance? How do I get home?

The Laplace's demon argument ignores the real-world context. Unpredictable data has entered not implied by her initial brain state

Figure 5.15 Unpredictability from initial data because we are open systems. The initial state of our brain at time t_0 does not have all the data needed to decide what to do at time t_1.

Figure by Mauro Carfora.

vicinity and plan what to do.[146] We have evolved to handle the fact that initial data $D(t_0)$ is not sufficient to determine what will happen in the future, because of external events. That is key to our survival.

5.6.5 The Cosmic Context

'Ah', the strong reductionists say, 'but if you know the initial state of every single atom in the Universe, then it is all determined after all! That incoming data to your mind is determined because the state of the entire Universe, considered as a whole, obeys the predictability relation (5.1). Your argument falls away.'

For the physicalist, this is a knock-down argument. But to apply this argument, we have to take that cosmic context seriously. Then this claim has many problems.

First, when is that data supposed to have been set? If this was to occur in terms of quantum fluctuations at the start of inflation (see Chapter 1), then irreducible uncertainty will occur as those fluctuations become classical. Then (5.1) does not apply. The argument for predictability fails.

---

[146] Frith, *Making up the Mind.*

Ignoring that issue, let's consider the initial data $D_{LSS}(t_0)$ that occurs on the last scattering surface (LSS) at time t_0, and determines later conditions in the Universe as follows:

$$D_{LSS}(t_0) \rightarrow D(t_1) \quad \text{for all } t_1 > t_0. \tag{5.2}$$

However, as galactic structures develop, the true butterfly effect occurs:[147] the existence of our own Galaxy is not uniquely predicted by that data, let alone the existence of our specific Solar System and the Earth.

Neglecting all those problems, what kind of data is there on the LSS that could possibly influence what is going on in our minds today? The standard view (Chapter 1), presented in Jim Peebles' recent book,[148] is that well-understood astrophysical processes lead to modulated random Gaussian fluctuations on the LSS (Figure 1.12(a)). This is the data $D_{LSS}(t_0)$. There is nothing in that data that can uniquely imply (via Eq. (5.2)) existence today of specific thoughts in our minds such as design principles for a digital computer.[149] If there was such data present in some coded form, there is no plausible process whereby it could get into our minds by developmental processes.[150]

But the crucial point is this: the astrophysical mechanisms described in Peebles' book cannot possibly have resulted in the initial data $D_{LSS}(t_0)$ including plans for an iPhone in some coded form (**Figure 5.16**).[151] Those processes simply don't lead to logical outcomes of this kind. Astrophysical processes don't know anything about digital computer design, or resultant iPhones.[152] But these both exist.

The only way that this can plausibly occur is this: $D_{LSS}(t_0)$ is of such a nature that galaxies, stars and planetary systems somewhat like our own come into existence, eventually leading to intelligent life arising. This life has sufficient agency that it can design and build objects like iPhones through developing mathematics, science and technology. This outcome is not uniquely written into $D_{LSS}(t_0)$, but is made possible by it.[153]

5.6.6 If Free Will Is Not True, Science and Philosophy Are Not Possible; Pragmatically, It's Apparent

The claim that free will does not exist is a self-refuting claim. If this were true, then the practice of science and philosophy would not be possible. You would not be able to make up a hypothesis, devise experiments to test it, carry out those experiments and decide on the basis of the outcome whether your hypothesis is supported by the

---

[147] Mark Neyrinck, Shy Genel and Jens Stücker, 'Boundaries of Chaos and Determinism in the Cosmos', *arXiv preprint arXiv:2206.10666* (2022). Dmytro Bandak et al., 'Spontaneous Stochasticity Amplifies Even Thermal Noise to the Largest Scales of Turbulence in a Few Eddy Turnover Times', *Physical Review Letters*, 132 (2024), 104002.

[148] P. James E. Peebles, *Cosmology's Century: An Insider History of our Modern Understanding of the Universe* (Princeton, NJ: Princeton University Press, 2020).

[149] David Harris and Sarah Harris, *Digital Design and Computer Architecture* (Amsterdam: Elsevier, 2013).

[150] Ellis and Solms, *Beyond Evolutionary Psychiatry*.

[151] See the Wikipedia article iPhone.

[152] Peebles, *Cosmology's Century*.

[153] Gazzaniga, *Who's in Charge?* Mitchell, *Free Agents*.

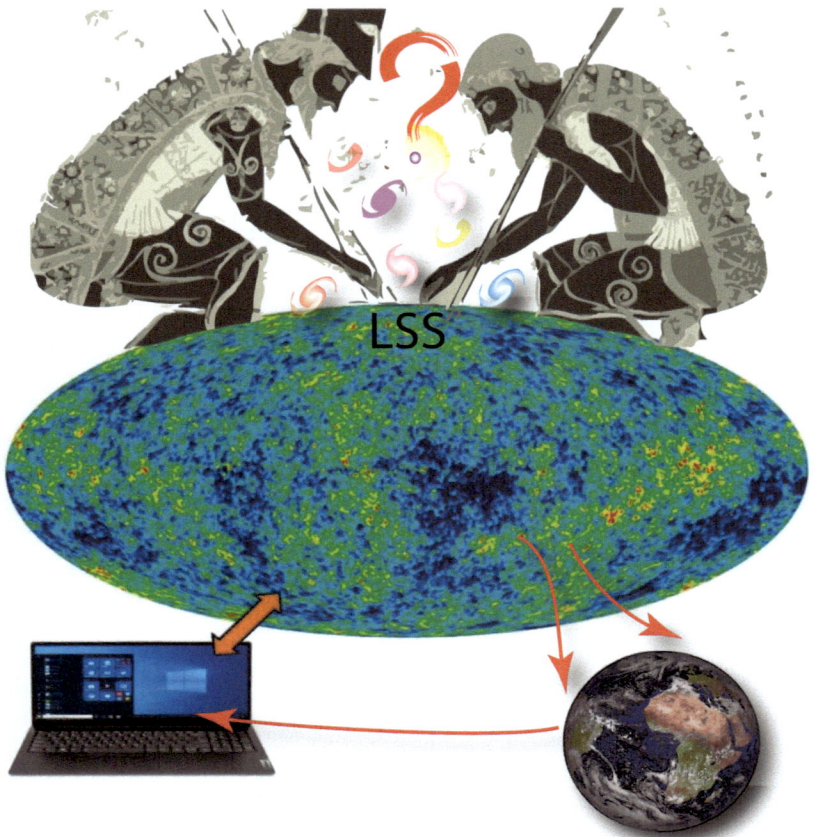

Figure 5.16 The cosmological context and outcomes today. If the data on the last scattering surface (Figure 1.12(a)) uniquely determines everything that exists on Earth, as implied by some physicists, then there is a coded data there for creation of every man-made object, including laptop computers. So the question is this: how did this data get there? What demiurge wrote it into the LSS in some coded form? Peebles' book *Cosmology's Century* contains no hint of a mechanism that could do this.

Figure by Mauro Carfora.

nature of reality or not. You would not have the freedom to carry out any of those action. The same would apply to philosophical analysis: you would not be able to propose a theory, and develop logical argumentation to either help support it or refute it. You require agency—the possibility of doing all this—to underlie the practice of philosophy, for example considering whether free will is possible. If your theory denies it, you need a better theory.

What about the famous action potential experiments by Benjamon Libet, where action potentials seem to initiate before a choice is made?[154] The point is that they are

---

[154] See the Wikipedia article Benjamon Libet.

not real choices:[155] they are not in fact about free will. The real choice involving free will occurred when the subject agreed to take part in the experiment.[156]

At a pragmatic level, the existence of free will is proven by the existence of technology such as aircraft, digital computers, buildings, books and iPhones. These are all products of the agency of the human mind, and indeed are based on the existence of the science that enables the technology to come into being.

How does this agency take place? In simple cases by deciding to move your arm or leg. In complex cases by considering options in the light of all choices and choosing between them on the basis of your choice criteria (§5.3.3). In all cases through the emergent causal power of the brain due to its neural network structure discussed earlier. We don't know the specifics of how thoughts are encoded in the brain, but as a matter of simple experience we have choice.

Ian McEwan has wonderful musings on this in his novel *Atonement* that are often repeated:[157]

> She raised one hand and flexed its fingers and wondered, as she had sometimes before, how this thing, this machine for gripping, this fleshy spider on the end of her arm, came to be hers, entirely at her command. Or did it have some little life of its own? She bent her finger and straightened it. The mystery was in the instant before it moved, the dividing moment between not moving and moving, when her intention took effect. It was like a wave breaking. If she could only find herself at the crest, she thought, she might find the secret of herself, that part of her that was really in charge. She brought her forefinger closer to her face and stared at it, urging it to move. It remained still because she was pretending, she was not entirely serious, and because willing it to move, or being about to move it, was not the same as actually moving it. And when she did crook it finally, the action seemed to start in the finger itself, not in some part of her mind. When did it know to move, when did she know to move it?

This affirms that we do indeed have the power to choose what to do, even if it is mysterious how it happens. But in the end, the point is this: the whole of the socially constructed world around us and its technological basis is the result of creative choice by the human mind, leading to resultant actions (§5.3.5). We have the ability to make this happen, and to do so in accord with our intentions.

5.7 Values, Purpose and Meaning Shape Outcomes

Absolutely fundamental to the human condition is the box 'Motives, Ethics, Meaning' **in Figure 5.11**. These are both motivators and constraints on all else that the brain does. In the end, they shape what we do. Our choice of purpose and understanding

---

[155] Uri Maoz, Gideon Yaffe, Christof Koch and Liad Mudrik, 'Neural Precursors of Decisions That Matter—An ERP Study of Deliberate and Arbitrary Choice', eLife Sciences Publications, Ltd. (23 October 2019).

[156] *"Merlin, A Mind so Rare".*

[157] Ian McEwan, *Atonement* (London: Jonathan Cape, 2001), 35–6. See also Frith, *Making up the Mind*; and Sarah-Jayne Blakemore and Chris Frith, 'Self-awareness and Action', *Current Opinion in Neurobiology*, 13 (2003), 219–24.

of meaning will shape what actions we take; for example, if I choose to be an artist, this will shape all I do. If I base my life in the meaning I get from some religious tradition, faithfully following its rituals and observances and guided by its dogma, that will shape a large part of my life. The same is true if I chose to be a politician, engineer, doctor, academic and so on. Each of these choices either explicitly or implicitly expresses the meaning underlying our lives and our values.[158]

Some of them—being a doctor, nurse, teacher—are at their core aimed at helping others, even if it does not always work out that way. Others may be aimed at generating profit for themselves with no regard to the damage caused to other people or the environment, perhaps doing all they can to get environmental protection laws removed. This is not, however, inevitable: making a profit can be compatible with both providing a needed service and accounting for and repairing environmental damage.[159]

The outcome is that values and meaning shape all outcomes at individual and social levels. In the case of an individual, this is represented in **Figure 5.17**. Values shape choices at the individual (psychological) level and this chains down to affect all the lower levels, including that of protons and electrons.[160]

If one believes it is acceptable to lie and cheat, some forms of action will occur that simply would not occur if one believes that this is not OK—one has a core of integrity that others can trust. And above all, one's ethical views may be totally self-centred, placing one's own needs and wants above those of all others; or one may be generous and considerate of others, even to the extent of sacrificing one's welfare on behalf of them; or one may lie between those extremes. In Chapter 6, I propose (§6.3) a

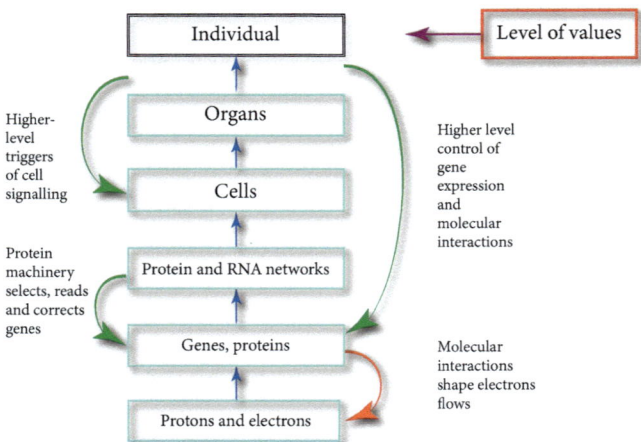

Figure 5.17 Values and meaning as the major level of control of what we do as individuals.

Figure by Mauro Carfora, adapted from Denis Noble.

---

[158] Denis Noble and George Ellis, 'Biological Relativity Revisited: The Pre-eminent Role of Values', *Theoretical Biology Forum*, 115 (2022), 45–70.

[159] Colin Mayer, *Capitalism and Crises: How to Fix Them* (Oxford: Oxford University Press, 2023).

[160] John M. Fischer, *The Freedom Required for Moral Responsibility* (Oxford: Oxford University Press, 2018).

related spectrum of possible values that applies to both individuals and organizations. I claim that given access to the relevant data. one can empirically determine where an individual or organization lies on this spectrum of values.

Moral action is possible. Taking all this into account, moral action is possible.[161] This requires use of one's agency to undertake meta-reflection on life, purpose, meaning and morality. Metacognition is key to this,[162] as discussed by Murphy and Brown:[163]

> The keys are sophisticated language and hierarchically ordered cognitive processes allowing humans to evaluate their own actions, motives, goals, and moral principles. This allows the processes of moral reflection and character building to influence outcome choices. Symbolism allows off-line prediction and logical branching of thought when assessing outcomes, thus enabling moral reasoning as a basis of action choice.

This is the **explicit normativity** referred to by Heyes,[164] which has a cultural basis.[165] A crucial issue then is how metacognition is possible. This is discussed in depth in a key series of papers by Stephen Fleming and colleagues.[166]

A key point then is this: through the lifelong process of the kind just outlined, we can change our nature until behaving morally and kindly becomes second nature. Character building takes place, and we have moulded ourselves to behave in a moral way. As emphasized by Alasdair MacIntyre, this is the process by which we turn ourselves into moral beings.[167] It is a specific case of self-determination:[168] through thoughts and actions shaping brain plasticity over the decades, and sufficient dedication, we can become close to what we want to be.

Two final comments are important.

Moral emotions are not the same as morality. Jonathan Haidt has written extensively on the moral emotions.[169] However, values/ethics are not the same as emotions, though they are informed by them. This is a crucial distinction because of the fact that just because something feels right does not mean that

---

[161] Matthew Talbert, 'Moral Responsibility', The Stanford Encyclopedia of Philosophy (Fall 2025 Edition), E N. Zalta & U Nodelman (eds.), https://plato.stanford.edu/archives/fall2025/entries/moral-responsibility/.

[162] Juliet Dunstone and Christine A. Caldwell, 'Cumulative Culture and Explicit Metacognition: A Review of Theories, Evidence and Key Predictions', *Palgrave Communications*, 4(1) (2018).

[163] Murphy and Warren Brown, *Did My Neurons Make Me Do It?*, 284–5.

[164] Heyes, 'Rethinking Norm Psychology'.

[165] Cecilia Heyes, Dan Bang, Nicholas Shea, Christopher D. Frith and Stephen M. Fleming, 'Knowing Ourselves Together: The Cultural Origins of Metacognition', *Trends in Cognitive Sciences*, 24 (2020), 349–62.

[166] Stephen Fleming and Raymond J. Dolan, 'The Neural Basis of Metacognitive Ability', *Philosophical Transactions of the Royal Society B: Biological Sciences*, 367 (2012), 1338–49. Stephen Fleming, Raymond J. Dolan and Christopher D. Frith, 'Metacognition: Computation, Biology and Function', *Philosophical Transactions of the Royal Society B: Biological Sciences*, 367 (2012), 1280–86.

[167] Alasdair MacIntyre, *After Virtue* (Notre Dame, IN: University of Notre Dame Press, 2007).

[168] See the Wikipedia article Self-determination theory.

[169] Jonathan Haidt, 'The Moral Emotions', in R. J. Davidson, K. R. Scherer and H. H. Goldsmith, eds, *Handbook of Affective Sciences* (Oxford: Oxford University Press, 2003), 852–70.

it is, in fact, right. To discern which is the case needs moral reflection ('explicit normativity'[170]) of the kind outlined earlier.

Moral realists propose that moral truths exist and are discovered by the mind, in analogy with how this is true for mathematical truths. I strongly support that position.[171] In my view it is crucial that we are able to say that both the Holocaust and Apartheid were evil as a matter of fact, not just as a matter of social convention. This view is compatible with Cecelia Heyes' claim that norms arise via cultural experiences overlaying implicit normativity.[172] I return to this issue in Chapter 7.

Further Reading

My top three are these, giving wonderful overviews of the nature of the mind/brain:

Merlin Donald, *A Mind So Rare: The Evolution of Human Consciousness* (London: W. W. Norton, 2001).

Eric R. Kandel, *The Age of Insight: The Quest to Understand the Unconscious in Art, Mind, and the Brain* (New York: Random House, 2012).

Chris Frith, *Making up the Mind: How the Brain Creates Our Mental World* (Malden, MA: Blackwell Publishing, 2013).

Also excellent are these two:

Philip Ball, *The Book of Minds: Understanding Ourselves and Other Beings, From Animals to Aliens* (London: Picador, 2022).

Kevin J. Mitchell, *Free Agents: How Evolution Gave Us Free Will* (Princeton, NJ: Princeton University Press, 2023).

More focused are these three, covering significant important aspects:

Esther Sternberg, *The Balance Within: The Science Connecting Health and Emotion* (San Francisco: W. H. Freeman, 2000).

George Ellis, *How Can Physics Underlie the Mind? Top Down Causation in the Human Context* (Berlin: Springer, 2016).

George Ellis and Mark Solms, *Beyond Evolutionary Psychology: How and Why Neuropsychological Modules Arise* (Cambridge, UK: Cambridge University Press, 2018).

Finally these two cover the key issues of the symbolic brain and language acquisition:

Terence Deacon, *The Symbolic Species: The Co-evolution of Language and the Brain* (London: W. W. Norton, 1998).

Michael Tomasello, *Constructing a Language: A Usage-Based Theory of Language Acquisition* (Cambridge, MA: Harvard University Press, 2003).

---

[170] Hayes, Rethinking norm psychology.

[171] George Ellis, 'On the Origin and Nature of Values, Tanner Lecture, Linacre College (2017).

[172] Heyes, 'Rethinking Norm Psychology'.

PART III
THE SOCIAL AND PERSONAL
DIMENSIONS

6

Society, the Environment, Technology and Values

We are social animals relying on our communities for existence, support and life opportunities. We create societies with varied cultures, and a variety of organizations that fulfil specific functions in those societies. We could not exist without these social structures. They support us, providing us with social, emotional and material needs, and shape the way we think and act.[1] We form social groups of various sizes and degrees of coherence; the organizations we create have agency in their own right.[2] This all takes place in an ecological context that is the basis of our physical existence. Our own development takes place in this complex context, which shapes our understandings, but which we help develop in a mutual interaction.[3]

Via the symbolic ability discussed in Chapter 5, humans have developed technology that has enabled us to conquer the Earth and travel to space.[4] This took place by a process of adaptive selection, These developments relied on harnessing new energy sources needed for agriculture, industry, transport, domestic use, computing and so on.

Societies involve complex interactions between individual people, each with their own agency and agendas. It is therefore much more difficult to characterize relevant emergent behaviours in a reliable way than in the physical sciences or physiology. Much effort has gone into developing sociological theories of how society works,[5] but I will not engage with these theories both because I do not have the expertise to do so and because my aim is to characterize social interactions in a way congruent with the view presented in the previous chapters about how life works. The key point is this:

Societies can be regarded as living systems, undergoing functional, developmental, and evolutionary processes just as in the case of all life. Because of this similarity, societies have also developed as modular hierarchical structures where homeostatic processes, predictive processing, and adaptive behaviour take place.

---

[1] Peter L. Berger and Thomas L. Luckmann, *The Social Construction of Reality: A Treatise in the Sociology of Knowledge* (New York: Anchor Books, 1966).

[2] David Elder-Vass, *The Causal Power of Social Structures: Emergence, Structure and Agency* (Cambridge, UK: Cambridge University Press, 2010).

[3] Peter L. Berger, *Invitation to Sociology: A Humanistic Perspective* (New York: Anchor Books, 1963).

[4] W. Brian Arthur, *The Nature of Technology: What It Is and How It Evolves* (New York: Simon and Schuster, 2009).

[5] Talcott Parsons, *Societies: Evolutionary and Comparatives Perspectives* (Englewood Cliffs, NJ: Prentice Hall, 1966). Anthony Giddens, *The Constitution of Society: Outline of the Theory of Structuration* (Berkeley: University of California Press, 1984).

How We Come to Be. George F. R. Ellis, Oxford University Press. © George Ellis (2026).
DOI: 10.1093/9780198950189.003.0008

I am, of course, not the first person to recognize this similarity. I am following in the tradition of the great pioneers of cybernetic theory: Ross Ashby, Kenneth Boulding, Russel Ackoff and Stafford Beer.[6] There is a more general line of thought developed by Ludwig Bertalanffy and many others that goes under the name of either systems theory or general systems theory.[7] This has similar concerns in many ways; in particular, it emphasizes the existence of emergence: 'A system is more than the sum of its parts', and the importance of general patterns of understanding that can be claimed to apply in many different areas. However, that theory has been criticized by some as overstating its claims, being too general in nature.

In this chapter I consider in turn, §6.1, 'Organizations and the Universal Needs of Societies'; §6.2, 'The Quality-of-Life Adaptive Feedback System'; §6.3, 'The Significance of Values for Social and Environmental Outcomes'; §6.4, 'The Environmental and Ecological Context'; §6.5, 'Technology as the Physical Basis of Society, and Its Change with Time'; §6.6, 'The Crises Facing Us Today'; and §6.7, 'The Great Struggle: Values and Views of Humanity'.

6.1 Organizations and the Universal Needs of Societies

The needs of a society are met by the people in them acting collectively through a large variety of organizations. These organizations are more than the sum of their parts: they have agency.[8]

In this section I look at, §6.1.1, 'The Functional Needs of a Society'; §6.1.2, 'Societies and Organizations Have Arisen by Cultural Evolutionary Processes'; §6.1.3, 'Society and Those in Them Interact in Upward and Downward Way'; §6.1.4, 'Meso-Level Organizations in Society, and Their Causal Powers'; §6.1.5, 'Modular Hierarchical Structures, Multiple Realizability and Decentralization'; §6.1.6, 'Emergence and Agency: The Dynamics of Organizations'; and §6.1.7, 'Organizational Practice'.

6.1.1 The Functional Needs of a Society

The functional needs of a society are discussed by Aberle et al.[9] They define a society as follows:

> A society is a group of human beings sharing a self-sufficient system of action which is capable of existing longer than the life span of an individual, the group being recruited at least in part by the sexual reproduction of the members.

---

[6] See the Wikipedia articles W. Ross Ashby, Kenneth E. Boulding, Russell L. Ackoff and Stafford Beer.
[7] See the Wikipedia article Systems theory.
[8] Elder-Vass, *The Causal Power of Social Structures*.
[9] D. F. Aberle, A. K. Cohen, A. K. Davis, M. J. Levy Jr and F. X. Sutton, 'The Functional Prerequisites of a Society', *Ethics* (1950), 100–11.

Material and sociocultural needs must be met in one way or another. The existential needs of a society are of a material nature, based in our physiological needs (Chapter 3): clean air, clean water, safe food, adequate waste disposal and toilets and sewage systems, and protection from the elements. Additionally there are the material needs of daily life: clothes, shoes, housing, furniture, refrigerators, stoves, ovens, other domestic appliances, shops and transport.

The sociocultural needs are just as real. These include educational, artistic, cultural, entertainment, sports, recreation and communication facilities. Religious organizations play a key role for many.

Organizational needs. In order that all these needs are met in an orderly and stable fashion, we rely on social institutions with suitable interactions. These include governments and administrative systems, a financial system, a legal system, an education system and health and welfare systems.

What about society itself? Aberle et al. identifies the following prerequisites for a society to exist:[10]

(a) Provision for adequate relation to the environment and for sexual recruitment,
(b) Role differentiation and role assignment,
(c) Communication,
(d) Shared cognitive orientations,
(e) A shared, articulated set of goals,
(f) The normative regulation of means,
(g) The regulation of affective expression,
(h) Socialization and
(i) The effective control of disruptive forms of behaviour.

Each of these needs is elaborated in the paper. They are each met, to a greater or lesser degree, by suitable social institutions, regulations or practices. Society has a variety of mechanisms to try to fix the problem if any of these needs are not being met. Taken together they comprise the **quality-of-life adaptive feedback system** that I discuss in §6.2.

Perhaps the most central of all is (c): no society, however simple, can exist without shared learned symbolic modes of communication. This is possible because we have a symbolic brain (Chapter 5). Key points are (f) and (i). Peter Berger emphasizes that, in the end, all societies rely on the use of force—or the threat of that use—to safeguard their integrity.[11] The related institutions are police, prisons, courts and so on. Crucial is (h), which is how values get shared. It takes place via families, on the one hand, and social institutions such as play groups and schools, on the other. Key to success is existence of good role models.[12]

It should be emphasized that these are prerequisites for a **well-functioning** society. However, items (d) and (e) are not features of many societies today; particularly,

---

[10] Aberle et al., 'The Functional Prerequisites of a Society.
[11] Berger, *Invitation to Sociology.*
[12] John F. Longres, *Human Behaviour in the Social Environment* (London: E. F. Peacock, 1990).

as I write, in the case of the USA. Part of the issue is that trust is central to a well-functioning society, which is lacking in this case.[13]

These needs are met by interacting health, social, economic, political and legal systems, as I discuss in §6.2. Organizations of all kinds exist to meet all our physical needs: these include agricultural, manufacturing, distribution, sales and transport by road, rail, water or flying. Infrastructure to support them are energy systems, water supply and disposal systems, sewerage and waste disposal systems. Social and cultural needs are met by a large variety of organizations, supported by news and telecommunication systems, computers including the Internet, and cell phones, which are ubiquitous nowadays.

6.1.2 Societies and Organizations Have Arisen by Cultural Evolutionary Processes

Analogously to the process of natural selection in biology, cultural evolutionary processes occur.[14] A process of adaptive selection by trial and error in society has developed and shaped social systems and organizations so as to meet all the needs just mentioned. This has resulted in key innovations such as control of fire, invention of the wheel, development of agriculture, harnessing of energy sources, the invention of money,[15] development of the concept of corporations[16] and so on.

It has also led to a huge increase in our understanding of the physical and biological worlds through science based in mathematics, leading to transformational technology based on this understanding (§6.5).[17]

6.1.3 Society and Those in Them Interact in Upward and Downward Ways

Societies emerge from the people making them up, particularly through organizations that are more than the sum of the parts: they have agency as organizations.[18] They act downward on the individuals making them up by the same mechanisms whereby this happens in general, as discussed in the Prologue to this book. This occurs through time-dependent constraints, on the one hand,[19] and through shaping the nature of lower-level elements, on the other.[20]

---

[13] Thomas W. Simpson, 'What is Trust?', *Pacific Philosophical Quarterly*, 93 (2012), 550–69.
[14] "Nicole Creanza, Oren Kolodny, and Marcus W. Feldman. "Cultural evolutionary theory: How culture evolves and why it matters." *Proceedings of the National Academy of Sciences* 114, no. 30 (2017): 7782–7789."
[15] See the Wikipedia article Money.
[16] Yuval N. Harari, *Sapiens: A Brief History of Humankind* (London: Random House, 2014), 31–3.
[17] Arthur, *The Nature of Technology*.
[18] Elder-Vass, *The Causal Power of Social Structures*.
[19] Shams-Ur Rahman, 'Theory of Constraints: A Review of the Philosophy and Its Applications', *International Journal of Operations & Production Management*, 18(4) (1998), 336–55. Alicia Juarrero, *Dynamics in Actions* (Cambridge, MA: MIT Press, 2009).
[20] George Ellis, 'Efficient, Formal, Material, and Final Causes in Biology and Technology', *Entropy*, 25 (2023), 1301.

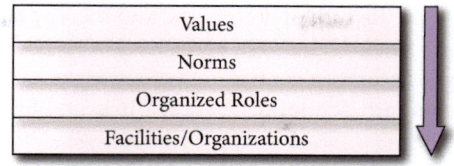

Figure 6.1 The relation between values, norms, roles and organizations ('situational facilities'). They each act downwards on the level below as indicated here.

Figure adapted from Neil Smelser, *Theory of Collective Behaviour* (London: Routledge, 2013), 32 by Mauro Carfora.

Economic and legal constraints delimit what resources individuals can have, and so what they can do. These are abstract constraints that shape outcomes, particularly laws defining ownership and use of resources. These constrain actions of individuals and thereby material outcomes, as do job descriptions, laws constraining what behaviour is allowed and rules of sports such as football. Relevant authorities decide on public holidays and celebrations (Freedom Day, Worker's Day, Christmas, Diwali, etc.). These then result in public holidays which alter physical outcomes. Symbolism shapes public discourse in this way.

Societies also shape the nature of their members in a downward way by processes that include informal education and formal education at schools and universities, apprenticeship programmes and training courses. These change novices into skilled pilots, engineers, scientists, doctors, lawyers and so on, thereby being made fit for roles that society needs to be filled. More generally, cultural influences and teaching act down to shape individual understandings and behaviour and values.[21] Through brain plasticity, these processes change our brains and result in 'cognitive gadgets'—special-purpose cognitive mechanisms built in during the course of development through social interaction.[22] These are products of cultural rather than genetic evolution. Crucially, these include values and norms that shape outcomes (**Figure 6.1**).[23] Through all these effects, our interpretation and indeed understanding of society and the world around us are shaped in a cultural way,[24] and then—through us—shape organizational structure and function.

6.1.4 Meso-Level Organizations in Society, and Their Causal Powers

David Elder-Vass in his key book *The Causal Power of Social Structures* (hereafter CPSS) explains in depth how meso-level organizations in society are emergent

---

[21] Berger, *Invitation to Sociology*.

[22] Cecelia Heyes, *Cognitive Gadgets: The Cultural Evolution of Thinking* (Cambridge, MA: Harvard University Press, 2018); Cecelia Heyes, 'Précis of Cognitive Gadgets: The Cultural Evolution of Thinking', *Behavioral and Brain Sciences*, 42 (2019), e169.

[23] Neil Smelser, *Theory of Collective Behaviour* (London: Routledge, 2013), 32–4. Cecelia Heyes, 'Rethinking Norm Psychology', *Perspectives on Psychological Science*, 7 June 2022.

[24] Berger and Luckmann, *The Social Construction of Reality*. Berger, *Invitation to Sociology*.

entities that are more than the sum of their parts, and thereby have agency that affects social and physical outcomes.[25] This is in agreement with Giddens, and Wilson and Snower.[26] Such emergence forms a central part of my argument in this chapter; however, it is denied in standard economic theory.[27] It is emphasized by Sawyer, who states[28]

> Social emergence is the central problem of the social sciences. The science of social emergence is the basic science underlying all of the social sciences, because social emergence is foundation to all of them. Political science, economics, education, history, and sociology study phenomena that socially emerge from complex systems of individuals in interaction.

The careful analysis in CPSS, taking both philosophy and mechanisms into account, explains in depth how meso-level social structures can have emergent causal powers.

The causal powers of meso-level structures. Apart from any arguments one may have, the evidence is in: firms produce goods, shops sell stuff, universities award degrees, banks lend money, power utilities provide power, football teams win or lose games, restaurants serve meals and so on. None of this could happen unless those structures had the relevant powers, enabled by the individuals who are appointed to specific roles in them.

Society as a whole influences individuals in key ways, but that happens mainly through specific meso-level structures and through associated expectations and cultures. Elder-Vass states,

> There are social structures that possess causal powers, but these entities are not whole societies. Instead there are many social entities, and indeed kinds of social entities, that possess social powers. . . . Most of the powers that usually have been attributed to society as a whole belong to somewhat smaller and more clearly definable social entities: structures at an intermediate level between individual and society that can have more specific effects.[29]

This is illustrated in **Figure 6.2**. I largely agree, but argue that some emergent powers of a society overall do indeed occur and affect individuals, as for example when one society declares war on another and invades.

A president makes the declaration, but it is the society as a whole that is at war, and individuals are affected. Importantly, national legislative assemblies have the capacity to pass laws on behalf of the society as a whole. They can pass laws relating to taxation or social benefits or reproductive rights, thus having causal powers that act down to

---

[25] Elder-Vass, The Causal Power of Social Structures, 23, 26–7.
[26] Giddens, *The Constitution of Society*. David Sloan Wilson and Dennis Snower, 'Rethinking the Theoretical Foundation of Economics I: The Multilevel Paradigm', *Economics*, 18(1) (2024), 20220070.
[27] Wilson and Snower, 'Rethinking the Theoretical Foundation of Economics I'. Steven Keen, *Debunking Economics: The Naked Emperor Dethroned?* (London: Zed Books, 2011).
[28] R. Keith Sawyer, *Social Emergence: Societies as Complex Systems* (Cambridge, UK: Cambridge University Press, 2005), 189.
[29] Elder-Vass, The Causal Power of Social Structures, 7, 82.

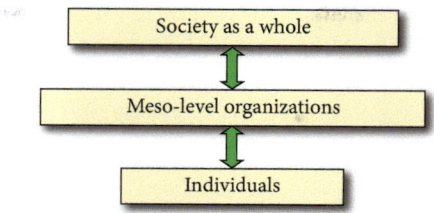

Figure 6.2 Individuals function in meso-level organizations, which taken together form society as a whole.

Figure by Mauro Carfora.

the level of individuals. Central to this chapter, some of the laws that governments pass underlie the existence of organizations by affording them legal status.

Nevertheless the majority of social structures that affect us on a daily basis are indeed meso-level organizations such as schools, shops, garages, hospitals, restaurants, universities, sports clubs, building contractors and so on. These ultimately derive their causal powers through laws and regulations that recognize their existence and regulate their powers, and some of those laws are passed at the national level.

So what are organizations? The key point is this:

> Organizations are at heart abstract structures, because constitutions define them and their powers. They are realized through physical entities: the people in them and the material stuff that they control. They are modular hierarchical structures that are multiply realizable in both organizational and physical terms. They attain their causal power because their abstract structure is legally and socially recognized. These abstract relations enable the organization as a whole to be more than the sum of its parts: it is not 'nothing but' a group of individuals. It is structured so as to act as entity in its own right, able to do things its members could not do on their own.

Thus an organization per se is an abstract structure that can own things, act in social contexts and be held responsible for those actions. These are all emergent properties,[30] made possible by the concept of a corporation.[31] These are constituted specifically so as to have causal powers. Developing the concept of a corporation was a key social innovation that transformed economic history. Yuval Harari discusses this in the case of the Peugeot SA company. He points out the company is not the same as the founder, the vehicles produced, the factories and machinery, or the mechanics, accountants and secretaries employed. Rather,[32]

> Peugeot is a figment of our collective imagination. Lawyers call this a 'legal fiction'. It can't be pointed to as it's not a physical object. But it exists as a legal entity. Just like

---

[30] Geofrey M. Hodgson, 'The Concept of Emergence in the Social Sciences: Its History and Importance', *Emergence, A Journal of Complexity Issues in Organizations and Management*, 2 (2000), 65–77.

[31] See the Wikipedia article Corporation.

[32] Harari, *Sapiens*, 31–3.

you and me, it is bound by the laws of the country in which it operates. It can open a bank account and own property, It pays taxes, and can be sued and even prosecuted separately from any of the people who own or work for it.[33]

Jensen and Meckling emphasize this point:

> It is important to recognize that most organizations are simply legal fictions which serve as a nexus for a set of contracting relationships among individuals. This includes firms, non-profit institutions such as universities, hospitals, and foundations, mutual organizations such as mutual savings banks and insurance companies and co-operatives, some private clubs, and even governmental bodies such as cities, states, and the federal government, government enterprises Such as TVA, the Post Office, transit systems, and so forth.[34]

Not all organizations will be corporations whose existence is recognized by law, but they will have a similar abstract structure recognized by the relevant interested parties and agreed by them. A charitable organization in the UK will be required to register as such by the Charity Commissioners, whereas a sports club or other recreational club can just have a constitution agreed to by the founding members.

In South Africa there is a set of regulations governing the creation of non-profit organizations and community-based organizations.[35] A constitution will inter alia state the aims of the organization, what officers will be appointed by what process, how new members can be admitted to the organization, how they can be ejected from the organization, in what circumstances it can be closed down and how its assets will be dealt with in that case. Members joining the organization will be required to agree to obey the constitution.

Organizations thus have both abstract and physical structures. The mechanisms through which their parts and the characteristic relations between them produce the emergent properties of the whole range from physical relationships whereby cement and bricks form buildings of a particular design to abstract relations curated by the officers of the organization.[36] These include the constitution that formally structures the organization, operating procedures that codify day-to-day operations, organograms and role descriptions that identify responsibilities of managers and employees, rights and duties set out in employment contracts and related human resources documents.[37]

The members of the organization will be assigned to committees and other roles (head of department, team captain, secretary, salesperson, manager, machinist, etc.) defined by these structures. In general these relations will be of a hierarchical nature, with the whole having parts (national office, regional offices, local branches), each of

---

[33] Harari, *Sapiens*, 31–3.

[34] Michael C. Jensen and William H. Meckling, 'Theory of the Firm: Managerial Behavior, Agency Costs, and Ownership Structure', *Journal of Financial Economics* 305 (1976).

[35] See *Education and Training Unit for Democracy and Development*: Constitutions for Non-Profit Organisations.

[36] Elder-Vass, *The Causal Power of Social Structures*.

[37] See Investopedia: Human Resources (HR): Meaning and Responsibilities.

which in turn will have components as needed: sections, departments, committees and so on, each with associated roles: head of finance, pay clerks, accountants, head of school, teachers, sports coaches, janitors, team members, etc.

These interact on a day-to-day basis at each emergent level of their modular hierarchical structuring, as well as between levels, to enable organizational functioning as required. Thus each organization will have a structure of control[38] relating boards of directors, CEOs, finance directors, local branch directors, committee members and so on, in the case of a commercial organization, or chairmen, committee, team captains, referees, linesmen, players and similar in social/sports organizations.

These will be formalized in documents laying out these structures, its purposes and the responsibilities required of occupants of any specific role in the organization. This is the abstract structure of the organization, which is realized by appointing specific individuals to each of these posts. These are multiply realizable in that they can be filled by different individuals: the role itself is logically independent of those who carry it out. Expectancies related to different roles—policeman, judges, doctors, nurses, teachers and so on—shape the way people understand the social context when interacting with those filling these roles.[39]

The people making them up are not just any set of people: they are people with specific abilities, linked by the abstract structures that are the heart of the organization.[40] Parallel to physiological structures in biology (see Chapter 3), units in any organization exist to accomplish something: they have a mission and objectives, and a manager accountable for the accomplishment of those objectives.

6.1.5 Modular Hierarchical Structures, Multiple Realizability and Decentralization

Like biological organisms, organizations are generically structured in a modular hierarchical way so as to be able to perform complex tasks.[41] This is a key principle for all complex emergence[42] and so is necessarily the case for large organizations[43]—those that employ more than say fifty to a hundred people.

However, their abstract structure is more flexible and variable than the physical structures in biology and engineering, as it is not restricted by the nature of physical possibilities. As in the case of the logical hierarchy in digital computers, it is a logical hierarchical structure that must obey the rules of logic in the context of

[38] Richard Scott and Gerald Davies, *Organizations and Organizing: Rational, Natural, and Open Systems Perspectives* (London: Pearson Education, 2007), 202–14.

[39] Longres, *Human Behaviour in the Social Environment.*

[40] Peter Vail, 'Notes on Running an Organization', *Journal of Management Inquiry*, 1 (1992), 130–8.

[41] Herbert A. Simon, 'The Architecture of Complexity', in P. Klaus and S. Müller, eds, *The Roots of Logistics* (Berlin: Springer, 2012), 335–61; Herbert A. Simon, *The Sciences of the Artificial* (Cambridge, MA: MIT Press, 2019).

[42] Grady Booch et al., *Object-Oriented Analysis and Design with Applications* (Reading, MA: Addison Wesley, 2007). George Ellis and Paolo Di Sia, 'Complexity Theory in Biology and Technology: Broken Symmetries and Emergence', *Symmetry*, 15 (2023), 1945.

[43] Christoph Schneeweiß, 'Hierarchical Structures in Organisations: A Conceptual Framework', *European Journal of Operational Research*, 86 (1995), 4–31.

the definitions involved. Upward and downward causation occurs by essentially the same mechanisms as in biology and engineering as just discussed,[44] right down to the physical levels whereby the abstract structures are instantiated in physical terms.

Why modular hierarchical structures? Organizations will generally be modular hierarchical structures,[45] because that is the only way to successfully carry out complex tasks in a reliable way.[46] One takes the complex task and splits it up into a set of simpler tasks. Then one takes that set of tasks and splits them up into a set of even simpler tasks. At each level one will need to devise structures and procedures needed to carry out those tasks. Continuing in this way, one reaches a level where the tasks required are easy to do. One carries them out, integrates the results to get the desired results at the next level up, and continues in this way until one has accomplished the complex task one was aiming for at the beginning.

Key features of abstract modular hierarchical structures are set out in the fundamental book by Grady Booch et al.[47] They are **abstraction** (pp. 44–50), **encapsulation** (pp. 50–4), **modularity** (pp. 54–8) and **hierarchy** (pp. 64–7). An abstraction denotes the externally visible essential characteristics of an object that distinguishes it from all other kinds of objects. Encapsulation hides the internal workings of the object from the external world. The two are related by an interface definition that clearly states in what format the external world sends requests to the module, and in what way it returns its outputs to the external world.

Because of this structuring, **multiple realizability** occurs: any internal structure and associated dynamics is allowed in any module, as long as it does the required job. This applies at each downward step to each lower level, including the physical levels involved in the material realization of the organization. There are many billions of different possible arrangements of atoms underlying an organization's physical aspects. However, some particular internal structures will be preferred if they do the job faster, use less energy or cost less. Multiple realizability is a key feature that will recur later.

The three kinds of organizational hierarchies. There is not just one kind of hierarchy involved. To be specific, I'll consider the organizational hierarchies involved in constructing an aircraft. They are of three kinds.

- **Physical hierarchy: material.** An aircraft is a modular hierarchical structure of physical stuff. The top level is the aircraft itself, the next level is the major components such as the fuselage, tails, wings, engines and so on. The level next down is the parts from which they are made such as metal sheets or moulded plastic components (**Figure 6.3**), and so on down to the screws and rivets and glue that hold them all together. There is a similar hierarchy for the factories that make the aircraft.
- **Abstract hierarchy: the design.** All the physical stuff—the entire modular hierarchy—has to be designed to the last detail. That design is abstract because it is multiply realizable: it can be talked about, represented visually (as in

---

[44] George Ellis, 'Efficient, formal, material, and final causes in biology and technology'. *Entropy*, 25 (2023), 1301.
[45] See the Wikipedia article Hierarchical Organization.
[46] Scott and Davies, *Organizations and Organizing*, 202–14.
[47] Booch et al., *Object-Oriented Analysis and Design with Applications*.

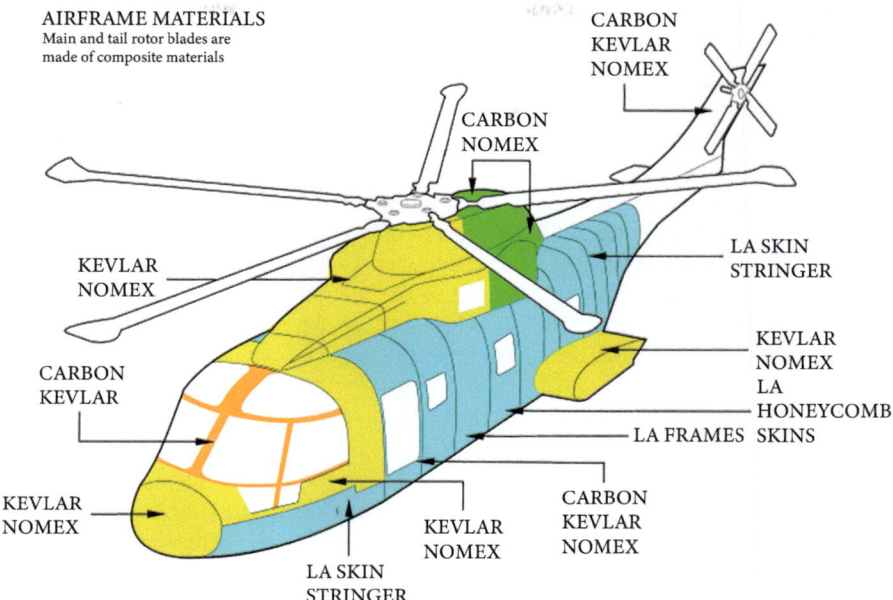

AIRFRAME MATERIALS
Main and tail rotor blades are
made of composite materials

CARBON
KEVLAR
NOMEX

CARBON
NOMEX

KEVLAR
NOMEX

LA SKIN
STRINGER

CARBON
KEVLAR

KEVLAR
NOMEX
LA
HONEYCOMB
LA FRAMES SKINS

KEVLAR
NOMEX

CARBON
KEVLAR
NOMEX

KEVLAR
NOMEX

LA SKIN
STRINGER

Figure 6.3 The physical hierarchical structure of an aircraft: the materials used in an
AgustaWestland AW101 helicopter.

From Wikimedia Commons.

Figure 6.3) at many levels of detail: there are diagrams for the aircraft as a
whole, from the structures making it up to the rivets and screws that hold it
together. Most importantly, the design can be represented in terms of data for a
computer-aided design computer program that can output instructions for automated assembly lines,[48] thereby making it happen. The abstract design will be a
modular hierarchy because it represents the physical hierarchy.

- **Organizational hierarchy: making it happen.** This is where the abstract design
becomes physical: a case of liminal causation (see §5.3.3)). One has to organize
for every single part to be manufactured, from the smallest (nuts, bolts, screws)
to the largest (wings, engines, the aircraft itself) in accordance with the abstract
design. This does not necessarily have to be all done by a single organization;
in fact, that is impractical: one will order many precreated parts (screw, rivets,
electric motors, aircraft engines and so on) according to a design specification
in each case. But there has to be a single organization that coordinates this,
arranging for all the requisite parts to be available at the right place and time as
required for assembly. The whole process is an illustration of why modular structuring is so important: it makes manufacture of very complex objects possible.[49]

---

[48] Ravio Rao, 'What are Robotic Assembly Lines? History, Components, Advantages, Limitations,
Applications, and Future', *Wevolver*, 3 March 2023.
[49] Scott and Davies, *Organizations and Organizing*, 202–14.

Membership of the organization will be hierarchically structured whenever it is a large organization.

Decentralization of agency. A fundamental issue, based in the encapsulation of internal functioning of modules, is this: to what degree is agency allowed for the structures at the periphery of an organization as against control from the centre?[50] Thus in the case of biology, modules such as the heart are self-regulating to a great degree: the brain does not give each muscle in the heart detailed instructions as to what to do; in effect the body as a whole tells the heart that more oxygen is needed and please just do it. The heart is constructed so as to respond as needed;[51] the cells in the heart ensure this happens. In effect agency is delegated to modules (the heart and cells), provided they do what they are required to do to meet the needs of the whole. How they do it is their affair.

The same issue of agency arises for the parts that make up the organization, as it does for the whole. Such decentralized agency exists if organizational design allows it. Consequently delegation of decision making is a key issue.[52] This is essentially the issues of abstraction and information hiding that are key to functioning of all modular hierarchical structures.[53] The issue is the tension between centralized planning, needed to provide large-scale supply of water and dams, electricity, drainage, sewerage systems, transport and other infrastructure and associated urban design, and a decentralized system that allows for local adaptive developmental processes to shape outcomes in a locally responsive way. Such a local adaptive process is described in Christopher Alexander's book *The Timeless way of Being*.[54] Agency of parts of an organization is a central issue, as much as is that of organizations themselves.[55]

Lack of decentralization of power in an organization results in micromanagement,[56] which is problematic for three reasons.

Firstly, for informational reasons: those at the centre (a head office in London) won't know that a roof is leaking or a truck needs repair in Colchester. Those at the local office are aware of these facts and should be able to take timely action to deal with them. The central office won't know whether procedures they insist on imposing, e.g. centralized control of ordering all office stationery, are having the desired effect.

Secondly, because of urgency. It does not affect those at head office if the repairs don't take place soon, so they will often wait to consider the issue till they have had a cup of tea and then discuss it with their colleagues. The procedures

---

[50] Stafford Beer, *Brain of the Firm* (London: Wiley, 1981). Stafford Beer, *Platform for Change* (London: Wiley, 1978), §6.2.

[51] Denis Noble, 'Modelling the Heart—From Genes to Cells to the Whole Organ', *Science*, 295 (2002), 1678–82.

[52] See the Wikipedia article Delegation. Beer, *Brain of the Firm*. Beer, *Platform for Change*, §6.2. Mark Carney, *Values: An Economist's Guide to Everything That Matters* (Glasgow: William Collins, 2021).

[53] Scott and Davies, *Organizations and Organizing*, 202–14.

[54] Christopher Alexander, *The Timeless Way of Building* (Oxford: Oxford University Press, 1979), See the Wikipedia articles 'The Timeless Way of Being' and 'Design Pattern'

[55] Elder-Vass, *The Causal Power of Social Structures*.

[56] See the Wikipedia article Micromanagement.

imposed centrally may cause immediate problems, and those at the centre won't be affected by them and so won't feel the need to act very soon.

And thirdly, it has a demotivating effect. If those at the periphery are denied agency, they will soon not care what the outcome is, because they were unable to affect the decisions that resulted in bad outcomes.

6.1.6 Emergence and Agency: The Dynamics of organizations

Through the processes and structures discussed here, just like life, organizations are examples of emergent entities with agency. Ultimately this organizational agency is based in the individual agency discussed in §5.6 of this book, and considered in detail in the carefully considered Chapter 5 of CPSS.

In analogy with the biological case, organizations undergo developmental and evolutionary processes. Developmental processes take place when individual organizations start up from small beginnings, and then enlarge the scope of their activities and their physical basis. The difference from biology is that there is no physical genetic material (DNA) shaping outcomes. Rather the underlying drivers are imagination, planning and agency of those creating and then developing the organization. The analogue of genetic material is the sharing of ideas via social interactions and institutions. A key issue is procedures for deciding which new people will join the organization, and training them to fulfil their functions.

Evolutionary processes take place in that competition takes place between organizations to fill social and economic niches, with some flourishing and some withering away and dying out.[57] There is no physical DNA underlying this process: ideas are the basis of the dynamics. Abstract causation takes place (see Chapter 5).

Organizational functioning depends on the analogue of metabolic processes. They take in the material required to create their product, transform the nature of this material so as to become the kind of stuff needed to make it[58] and then make it by manufacturing or building processes. Doing so creates heat and waste material that must be disposed of. Sometimes there is toxic waste,[59] requiring very careful handling.

Homeostasis occurs, in that feedback processes are nowadays commonplace.[60] Customer satisfaction is requested either by filling in forms or by Internet surveys requesting a variety of questions to be answered on a scale of 1 to 5. This is done either by the firm itself or, in the case of franchises, by the central organization owning the franchise in order to check up on the quality of service provided by the various branches. This is central to many Internet-based organizations such as AirBnB, TripAdvisor, Uber and Amazon Books.

Organizations also may exhibit predictive processing activities, as the brain does. This involves collecting information, sorting out what is relevant, comparing it with

---

[57] Francis Heylighen and Donald Campbell, 'Selection of Organization at the Social Level: Obstacles and Facilitators of Metasystem Transitions', *World Futures: Journal of General Evolution*, 45 (1995), 181–212.

[58] Mark Miodownik, *Stuff Matters* (London: Penguin Random House, (2013).

[59] See the Wikipedia article Toxic waste.

[60] Antonio Damasio and Hannah Damasio, 'Exploring the Concept of Homeostasis and Considering its Implications for Economics', *Journal of Economic Behavior & Organization*, 126 (2016), 125–9.

predictions and altering internal models of the external world on this basis. This can happen at various levels of sophistication. It may just involve budgeting processes,[61] which are essential for running an organization properly. However, it can be much more complex, as in the use of the system dynamics models developed by Jay Forrester and others,[62] taking into account the effects of delays in operational processes and how they can lead to cycles occurring.

Organizational success is based on imagination: seeing possible niches in the organizational ecology. Additionally, emotional aspects are often important: success may relate to ability to recruit positive emotions associated with the organization. Taken together, this all underlies the way that organizations are very much more than the sum of their parts: they are strongly emergent, as argued in CPSS.[63]

Values and meaning drive all of this.[64] The abstract causation that takes place embodies meaning and values that change physical outcomes. A great example **(Figure 6.4)** is given by David Deutsch in his book *The Fabric of Reality*:[65]

> Consider one particular copper atom at the tip of the nose of the statue of Sir Winston Churchill that stands in Parliament Square in London. Let me try to explain why that copper atom is there. It is because Churchill served as Prime Minister in the House of Commons nearby; and because his ideas and leadership contributed to the Allied victory in the Second World War; and because it is customary to honour such people by putting up statues of them; and because bronze is the traditional material for such statues, and so on. Thus we explain a low-level physical observation—the presence of a copper atom at a particular location—through extremely high level theories about emergent phenomena such as ideas, leadership, war and tradition.

Determining the appropriate goals to drive the whole leads to the need for metacognition considering the meaning of the whole, which has a cultural basis.[66]

Overall what this all means is that organizations are to a considerable degree characterized by the same diagram (Figure 5.11) that characterizes the operation of the mind. The key difference is that there are not genetically based inherited emotional systems, as opposed to socially based emotions. Rather there are internal emotional relations within the organization, as compared to emotional relations relative to the external world. With this modification, Figure 5.11 in essence also applies in the case of organizations.

---

[61] See the Wikipedia article Budget; and Investopedia article, What Is a Budget? Plus 11 Budgeting Myths Holding you Back.

[62] See the Wikipedia page System dynamics and the website of the Systems Dynamics Society.

[63] Elder-Vass, The Causal Power of Social Structures, 23, 26–7.

[64] George Ellis and Denis Noble, 'Economics, Society, and the Pre-eminent Role of Values', Global Solutions Initiative (2024).

[65] David Deutsch, The Fabric of Reality (London: Allen Lane, 1997), 22.

[66] Nancey Murphy and Warren Brown, Did My Neurons Make Me Do It? (Oxford: Oxford University Press, 2007). Nicholas Shea et al., 'Supra-personal Cognitive Control and Metacognition', Trends in Cognitive Sciences, 18 (2014), 186–193. Cecelia Heyes et al., 'Knowing Ourselves Together: The Cultural Origins of Metacognition', Trends in Cognitive Sciences, 24(5) (2020), 349–62.

Figure 6.4 A copper atom on the nose of a statue of Winston Churchill: an example of abstract causation embodying meaning and values.

Figure by Mauro Carfora.

6.1.7 Organizational Practice

How to actually run organizations is taught at business schools, which are looked down on by academic economists. But this is a mistake. It is in fact a very interesting topic, because management is a difficult thing to do well, with many interacting dimensions.[67] How it is done crucially shapes the social world around us.

Constraints play a key role, as in all emergent systems. Bottom-up and top-down effects take place with a key issue being, on the one hand, decentralizing as much agency to the periphery as possible for motivational and causal reasons, as discussed earlier (the people at the coal face actually know what is going on there and can respond locally), but on the other hand ensuring that organizational purposes and values shape outcomes at the periphery as well as the centre.[68] Organizational culture makes a key difference here, as do leadership styles.[69]

There are a series of pitfalls that institutions fall prey to in practice, that are difficult to avoid. Specifically,

---

[67] Mary Jo Hatch, *Organizations: A Very Short Introduction* (Oxford: Oxford University Press, 2011).
[68] Stafford Beer, *Brain of the Firm.*
[69] Jim MacQueen, *The Flow of Organizational Culture* (London: Palgrave Macmillan, 2020). James MacGregor Burns, *Leadership* (New York: Harper Perennial Political Classics, 2010).

- **Parkinson's law,**[70] which has two parts: 'work expands so as to fill the time available for its completion', and 'the number of workers in a public bureaucracy tends to grow, regardless of the amount of work to be done'.
- **The Peter principle,**[71] namely, 'in a hierarchy, employees tend to rise to their level of incompetence'. This leads to Peter's corollary: 'over time, every post tends to be occupied by an employee who is incompetent to carry out its duties'.
- **The institutional imperative:**[72] for any institution, no matter what its founding goal is, eventually that purpose will be supplanted by the imperative to survive, irrespective of any other purpose.
- At a deeper level, **Murphy's law**[73] applies to all institutions: 'anything that can go wrong will go wrong'. Ellis's corollary is that this will usually take place for completely unexpected reasons. One can't imagine all the ways things can go wrong, or how weirdly people may behave.

6.2 The Quality-of-Life Adaptive Feedback System

A society has welfare, social, economic, political and legal aspects that interact with each other and with the environment in ways controlled by individuals and institutions.

Following the line of thought of the pioneering cybernetics thinkers,[74] the interaction between these domains can be viewed as a multilevel feedback (homeostatic) control system whereby any coherent group in society tries, with greater or lesser degree of success, to adjust the social situation to meet their needs.[75] The quality-of-life (QOL) adaptive feedback system works to increase the welfare of individuals and groups via these interactions.

In this section I discuss, §6.2.1, 'The Basic Idea'; §6.2.2, 'Fleshing It Out'; §6.2.3, 'An Example'; §6.2.4, 'Values and Meaning, Images and Narratives'; §6.2.5, 'The Inverse System and Causal Closure'; §6.2.6, 'Dimensions of Well-being: Welfare and Poverty'; §6.2.7, 'Interacting Groups, and the Environmental and Ecological Context'; and §6.2.8, 'The Scales of Social Interactions'.

6.2.1 The Basic Idea

One envisages society as a whole as a system whose purpose is to create a good quality of life for its constituent populations via homeostatic feedback control processes.

---

[70] See the Wikipedia article Parkinson's law.
[71] See the Wikipedia article Peter principle.
[72] Robert N. Kharasch, *The Institutional Imperative: How to Understand the United States Government and Other Bulky Objects* (Devon: Charterhouse Books, 1973).
[73] See the Wikipedia article Murphy's law.
[74] See the Wikipedia articles W. Ross Ashby, Kenneth E. Boulding, Russell L. Ackoff and Stafford Beer.
[75] George Ellis, *The Quality of Life Concept: An Overall Framework for Assessment Schemes.* South African Labour and Development Research Unit, University of Cape Town (SALDRU) Working Paper 30 (1980). Ellis and Noble, 'Economics, Society, and the Pre-eminent Role of Values'.

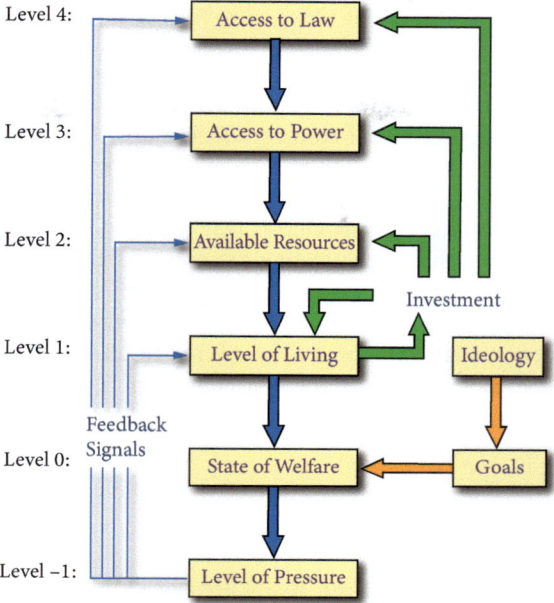

Figure 6.5 The societal quality-of-life feedback system. Figure 6.6 gives more details of the levels.

Figure by Mauro Carfora.

One can aim to characterize how well it succeeds in doing so. The mechanism (see §3.3 and Figures 3.7, 3.8) is that the difference $E(t) = S(t) - G$ between a set of system-state variables $S(t)$ and a chosen set of goals G drives the system to reduce that difference, as far as is allowed by the relevant constraints $C(t)$ on what is possible.

Thus in economic terms G is effectively the utility function for the system, shaping demand for resources and goods, but is a much broader multidimensional concept than in usual economic theory, as it covers physiological, psychological, social, political and legal issues as well as economic ones (**Figure 6.5**).

Because societies are changing all the time, goals will vary with time: $G = G(t)$. They depend on context: different societies have different goals which will, to some degree, be shaped by interactions with other societies and with the environment, which changes all the time, particularly due to population growth. Possible supply of goods is restricted by the constraints $C(t)$. All these variables are vectors of high dimension (they each have many components). Overall, this is a dynamic theory of how multidimensional utility is maximized, rather than an equilibrium theory of utility maximization. There is never equilibrium.[76]

Existence of this system does not necessarily imply that society will have been consciously constructed to act in this way, although for many aspects this will indeed be so. In other cases they will be the result of largely unmediated social actions,

---

[76] Keen, *Debunking Economics.*

such as the combinations of interests that lead to the 'invisible hand' of the economic market.[77]

However it happens, we can regard the emergent social system from the viewpoint of a particular population or group. What structures potentially enable them to succeed in attaining their desired goals? How capable is it to produce the desired results? It may be very efficient, so quality of life is good, or totally inefficient. In that case, quality of life will be bad; perhaps so bad that one characterizes the situation as being one of poverty, which can be of many kinds[78] (see §6.2.6).

Society can function either via activating existing feedback mechanisms that are there precisely in order to correct errors that may occur, as in any feedback control system, or by changing the system itself in an adaptive way, if the feedback process using existing structures fails to produce the desired result. That is, it has an evolutionary character. It can improve all these structures and processes if necessary to provide a better functioning society.

6.2.2 Fleshing It Out

More details of each level are shown in **Figure 6.6** (see also my article on 'The Dimensions of Poverty'[79]).

The basic level is **level 0,** the 'State of Welfare'. It is difficult to accurately assess this level, because of important social and psychological effects on well-being. It consists of two major components, one concerned with physical welfare (nutritional status, strength, disease, etc.), and one with sociocultural well-being (family life, community life, morale, etc.). Maslow's *Hierarchy of Needs* suggests how these aspects are related to each other.[80] These are measured by subjective indicators of contentment.

The first causal level is **level 1,** the 'Level of Living'. This is concerned with the deployment of the flow of available resources and amenities which make possible the maintenance and improvement of the State of Welfare; that is, with consumption patterns to enhance welfare, characterized as 'utility' in standard economic theory. This is where economic decisions are made for either consumption or investment. Thus one would find here the way available income is used in expenditure on food and medical services, on payments for housing and so on. Thus this is the level where market forces play a key role in shaping outcomes, decentralizing optimization decisions via the 'hidden hand' of economic theory, which is a way of decentralizing decision making.[81] However, the way that resources are used to increase welfare is not a narrow financial one. It encompasses all resource usage leading to increased welfare, using the term 'resource' in its widest sense to include, for example, clean air

---

[77] See the Wikipedia article Invisible hand.

[78] George Ellis, 'The Dimensions of Poverty', *Social Indicators Research*, 15 (1984), 229–53.

[79] Ellis, 'The Dimensions of Poverty'.

[80] Abraham Maslow, 'A Theory of Human Motivation', *Psychological Review*, 50 (1943), 370; Francis Heylighen, 'A Cognitive-Systemic Reconstruction of Maslow's Theory of Self-actualization, *Behavioral Science*, 37 (1992), 39–58.

[81] See the Wikipedia article Delegation and Robert Bellah et al., *The Good Society* (New York: Alfred Knopf, 1992).

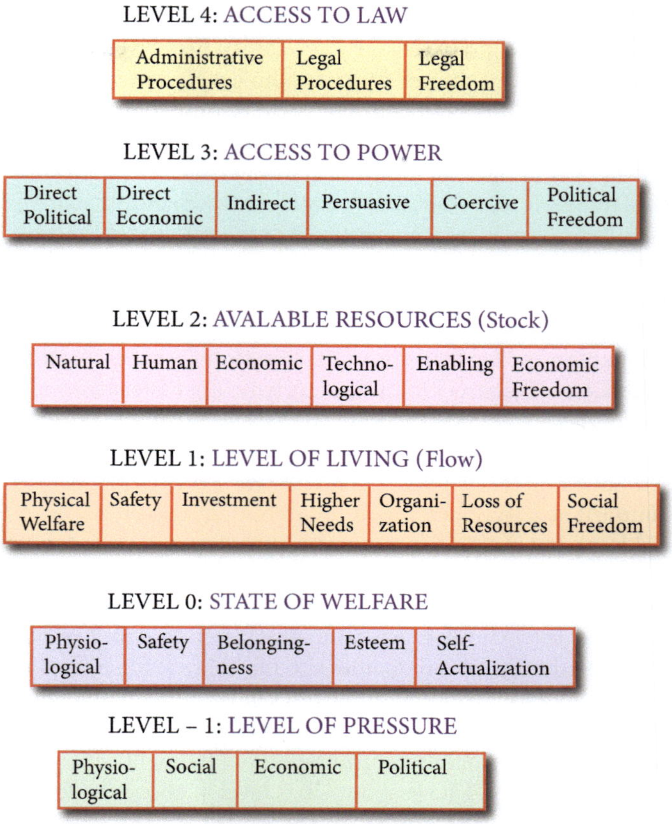

LEVEL 4: ACCESS TO LAW

Administrative Procedures	Legal Procedures	Legal Freedom

LEVEL 3: ACCESS TO POWER

Direct Political	Direct Economic	Indirect	Persuasive	Coercive	Political Freedom

LEVEL 2: AVALABLE RESOURCES (Stock)

Natural	Human	Economic	Techno-logical	Enabling	Economic Freedom

LEVEL 1: LEVEL OF LIVING (Flow)

Physical Welfare	Safety	Investment	Higher Needs	Organi-zation	Loss of Resources	Social Freedom

LEVEL 0: STATE OF WELFARE

Physio-logical	Safety	Belonging-ness	Esteem	Self-Actualization

LEVEL – 1: LEVEL OF PRESSURE

Physio-logical	Social	Economic	Political

Figure 6.6 Components of each level. A much more detailed breakdown is given in the Appendix to George Ellis, *The Quality of Life Concept: An Overall Framework for Assessment Schemes*, South African Labour and Development Research Unit, University of Cape Town (SALDRU) Working Paper 30 (1980).

Figure by Mauro Carfora

and quiet, unspoiled countryside, art galleries and symphony concerts. It is thus any feature whose use or deployment can be controlled to improve the group's welfare. The main ways the flow of available resources can be used to increase the group's welfare are for physical welfare and safety; for 'higher' needs; for organizational purposes; and for social, economic, political or legal investment on behalf of future welfare, rather than use for immediate consumption to increase present welfare. It is lessened by loss of resources for any reason, and constrained by the freedom available to direct this flow of resources as desired. Through this broad definition, it includes the kinds of multilevel social–economic interactions considered by David Sloan Wilson and Dennis Snower.[82]

---

[82] Wilson and Snower, 'Rethinking the Theoretical Foundation of Economics I'.

The second causal level (**level 2**) is that specifying the quantity and nature of 'Resources Available' to the group (level 1 was a **rate of flow** level; level 2 is a **stock** or quantity level). What can be achieved by group choices at level 1 is restricted by what they either own or control. We can consider these resources as divided into natural, human, economic, technological and enabling resources. They may be convertible to other resources by production/manufacturing processes or trading at some 'exchange rate' or 'terms of trade'. Human resources (people) are the engine that drive everything via their agency. Because of its easy convertibility, money is one of the most important of these resources.[83]

The resources available to the group at level 2 are in turn determined by the group's access to the power structure in society. Thus the third causal level (**level 3**) is 'Access to Power'. By definition, 'power' consists of all those features of the social system which can lead to allocation of resources to the group's benefit, or affect their allowed use by shaping laws at level 4. Such power can be exerted by virtue of coercion, trust, standing or authority in the community, or by political mechanisms, and by legislation. Access to power need not be direct, as long as it is effective; access to mass media or social media is an important feature here. Indeed controlling social media can be decisive at the political level. Institutional forms and controlling regulations are particularly important organizational resources in society, so the ability to control them is a key aspect of power.

The fourth causal level (**level 4**: 'Access to Law') is activated if relevant laws and regulations stipulate that specific outcomes should occur to the benefit of the group, but the relevant political or economic authorities nevertheless fail to do so. Thus one can make legal appeals to attempt to make them comply with the regulations. Because this kind of action can in principle control aspects of the political process as well as all other activities in society, I have shown it as level 4, above the level of access to power. In general we may expect groups to try activating these legal mechanisms if the others have failed to work as they are supposed to; for example, they may try to have the election or the activities of a politician declared invalid, or test proposed legislation against the country's constitution.

The four levels discussed so far (access to law, access to power, available resources and level of living) each control the level below in the sense of enabling or limiting what can be done at that level. The final level in the main causal set is **level -1**, which is the 'Level of Pressure'. This is the level manifesting the consequences of a lack of desired outcomes The first kind of pressure is 'consequential' pressure—automatic results of a poor state of welfare. This may be of a medical nature—a high death rate due to a poor health situation—or it may be of a social nature—a high crime rate or a high suicide rate. The second kind of pressure is purposeful pressure. This may occur through institutionalized means, with complaints laid through whatever official channels exist. If these do not work, it may occur in other ways. General discontent can manifest in publicized opinion polls, public protests and mass demonstrations, riots in the streets and so on. These can be aimed at changing any of the higher levels.

The final element in the main set of feedback loops, as in any feedback control system, is the **goals** of the group. This comprises factors acting as a reference signal for the

---

[83] See the Wikipedia article Money.

system, determining in which direction the group wishes to influence events. Comparing this desired state with the actual state results first in changes to alter outcomes of control signals (evidenced at the level of pressure) which flow through the major feedback loops to the four levels of control, and thereby attempt to influence the flow of resources and so alter the group's future state in the desired direction. The goals are a vision of a possible future.[84] They can potentially have components for every one of the aspects of welfare identified earlier, though many may be either absent or undetermined: the group is indifferent to them in these cases.

To complete the overall picture of the feedback system, two further features need to be considered. First, an important mode of resource use is the use of a resource flow to either build up future resources or increase future access to power, so that the group's position in the future will be improved; that is, **investment** may take place. In particular this covers **social investment**, such as organizing a major social event at considerable expense in the expectation of social returns, and **economic investment**, including investment in technology of various kinds, spending on research and financial investment. Particularly important is investment in education. **Political investment** takes place through expenditure on a political campaign, on mass media, or on lobbying politicians or trade union officials. **Legal investment** takes place by lobbying aimed at shaping the legal system, or through buying judges. Key choices are firstly how much to use for current needs and how much to invest to improve future prospects, and secondly what kind of investments are best for improving the future outlook.

Second, the goals that are consciously set by the group, or that are not explicit but are implied by their actions, are influenced by a variety of factors related to their aims and expectations. They partly depend on the group's perception of its situation, but also reflect their psychological, cultural, ethical and ideological or religious stances. For want of a better label, all such influences have been lumped together under the label **ideology**. This is where values and purpose and meaning enter, and shape outcomes as a whole.[85] Inclusion of this label reminds us that the responses of groups in society, and even their perceived levels of welfare, can be altered by changing these aspects while all else remains unchanged. The perceived welfare of the group can radically change as the result of a propaganda campaign or a religious conversion, while all the other elements represented here remain unaltered. A very rich company CEO whose sole purpose is to make as much money as possible might go to an Ashram, listen to a guru and change his aim in life to seeking enlightenment and helping others.

The key point I return to later is that goals depend on, and embody, the values of the group concerned, and the way they relate to meaning and purpose. This overall structure reflects the concerns of Bich et al., who consider different roles played by different elements in feedback control systems in the biological cases, separating those concerned with metabolism and those concerned more directly with control, in particular those shifting between different control regimes according to need through allosteric

---

[84] Kenneth E. Boulding, *The Image: Knowledge in Life and Society* (Ann Arbour, MI: Ann Arbour Paperbacks, 1961).
[85] Ellis and Noble, 'Economics, Society, and the Pre-eminent Role of Values'. Carney, *Values: An Economist's Guide to Everything that Matters.*

processes.[86] Metabolism in the biological case corresponds to economic interaction in the societal case. Because the whole process is the result of conscious actions involving predictive processing, this kind of switch of policy and corresponding planning and actions is a key part of the functioning of the social system I am describing.

This system can be represented at various levels of detail, in terms of both functions considered and timescales. How many components are there, in principle? The answer is a great many. Indeed each entry in the Yellow Pages of a telephone directory for a city represents needs that are being met by some organization or other, as do the ubiquitous advertisements on the Internet and social media.

An idea of these dimensions of welfare is given in the more detailed representation of these levels in the Appendix in Ellis (1980).[87] To be sure, there are equivalence classes of many kinds representing **classes of needs**: hairdressers, hotels, bookshops, cinemas, motor cars, holiday offers, travel options and so on, as categorized in the advertisements; but there are fine differences between them that can determine choices made, including geographic location, special times (summer, winter, Christmas, weddings, birthdays), and special events (Olympic Games, conventions, etc.).

Some outcomes are contextually determined and thereby have priority (the rent must be paid by Friday, the train leaves at 4pm and so on) and some are determined randomly (my friend told me about a holiday they had, the shop ran out of butter, it rained that day and so on).

6.2.3 An Example

Although this is not necessarily the way they will operate, it is convenient to think of the four main feedback loops as operating sequentially. An example will make this clear. Suppose the parents and headmistress of a school agree that smaller classes are desirable for the pupils (**Figure 6.7**)—an issue at level 0 due to conditions at level 1.

First, the pressures generated (meetings with the headmistress, parent–teacher association meetings, etc.) will result in feedback to level 1, where it is considered whether the need can be met by revised use of the existing flow of resources (a room can be converted from a storeroom to a classroom, teachers can be asked to teach extra classes instead of doing library duty and so on).

If this proves not to be a feasible way to solve the problem, the second feedback loop (to level 2) is activated. That is, it is considered if other resources from the stock of resources available to the school might be brought into play to provide the extra flow of resources needed, if necessary through conversion of the form of these resources (e.g. monies set aside in a development fund are used to build a classroom; school buildings are rented out at night, creating extra income which can be used to pay salaries).

Suppose it is not possible to meet the needs this way, because the school has already (within the possibilities allowed by the institutional framework) allocated all

---

[86] Leonardo Bich, Matteo Mossio, Kepa Ruiz-Mirazo and Alvaro Moreno, 'Biological Regulation: Controlling the System from within', *Biology & Philosophy*, 31 (2016), 237–65.

[87] Ellis, *The Quality of Life Concept*.

Figure 6.7 An overcrowded classroom in Malawi, implying there is a need for a reallocation of resources. There are different ways this can happen.
Picture by Carole Bloch (2005).

its resources; the level 3 feedback loop is then activated: the authority controlling the relevant power (the local education authority) is asked to make more resources available to the school. This department will, in turn, consider action through the main three feedback loops: it will consider whether it can reallocate its own existing flow of resources to deal with the problem (actions are considered that are 'without financial implications'); it may bring into play unallocated resources from its reserves (it may allocate funds in hand to pay new salaries); or it may apply in turn to its relevant resource-providing authority (perhaps a local authority executive committee) for further funds needed.

It may be that the relevant legislation governing the local authority stipulates that such funds should be made available to the school when needed but they refuse to do so. Then appeal can be made through courts or other statutory appeal mechanisms to try to get the situation rectified at level 4. This is the final available level.

6.2.4 Values and Meaning, Images, and Narratives

As mentioned in §6.2.2, the box 'Ideology' in **Figure 6.5** is where purpose, meaning and values occur. If they are changed, the outcome of the whole system changes. That is why values have a pre-eminent role in determining outcomes.[88] A change in values

---

[88] Carney, *Values: An Economist's Guide to Everything that Matters.* Ellis and Noble, 'Economics, Society, and the Pre-eminent Role of Values'.

corresponds to higher-level processes: adaptively changing goals in feedback control, or changing the criteria whereby adaptive selection takes place in evolutionary processes. This, in turn, depends on the selection criteria used in these higher-level processes.

While the highest level set of values—the understanding of values and meaning—may be arrived at by processes of metacognition,[89] this will often not be the case. Sophistication and patience is required to take that route. In practice, two alternatives are common.

First, various form of images of what is happening may be a key form of understanding what is going on:[90] 'a picture is worth a thousand words'.[91] They often represent and express specific values.

Second, a more profound form of understanding is via narratives, which can convey complex ideas and relations to values in situations of uncertainty.[92] Manipulating narratives is thus a key way of affecting outcomes, because it can change all the goals of the system and hence all its outcomes.

6.2.5 The Inverse System and Causal Closure

Individuals and groups affect the actions and values of society, through the institutions characterized by the societal feedback control system (**Figure 6.5**), but society, in turn, affects individual values through the inverse feedback system (**Figure 6.8**). In fact **the same** social mechanisms and institutions are involved (law courts, enacted laws, economic incentives, social sanctions, bad health outcomes), but in this case acting downward to influence individual behaviour and perceptions.[93]

Thus **Figures 6.5** and **6.6** represent how a coherent social group can attempt to influence structures and events so as to improve their welfare. **Figure 6.8** indicates how social institutions of the kinds indicated act down to influence the social groups and individuals that constitute them. The way this happens through various social institutions is detailed by Peter Berger,[94] showing that it is not only legal and political structures that act in this way but also economic and social institutions do so. Through such processes, the way members of a society understand the social situation is crucially affected, as detailed by Berger and Luckmann.[95] In particular, social institutions in a society will instil particular values in us. A key point indicated by the extra arrow in **Figure 6.8** is that political pressure may alter legal institutions so as to accomplish the aim of affecting individuals in such ways, for example by banning books (as in the USA today).

---

[89] Murphy and Brown, *Did My Neurons Make Me Do It?* Shea et al., 'Supra-personal Cognitive Control and Metacognition'.

[90] Boulding, *The Image.* John C. Farrell and Asa P. Smith, eds, *Image and Reality in World Politics* (New York: Columbia University Press, 1968).

[91] See the Wikipedia article A picture is worth a thousand words.

[92] S. Johnson, A. Bilovich and D. Tuckett, 'Conviction Narrative Theory: A Theory of Choice under Radical Uncertainty', *Behavioral and Brain Sciences* (2020), 1–47.

[93] Berger and Luckmann, *The Social Construction of Reality.* Berger, *Invitation to Sociology.*

[94] Berger, *Invitation to Sociology.*

[95] Berger and Luckmann, *The Social Construction of Reality.*

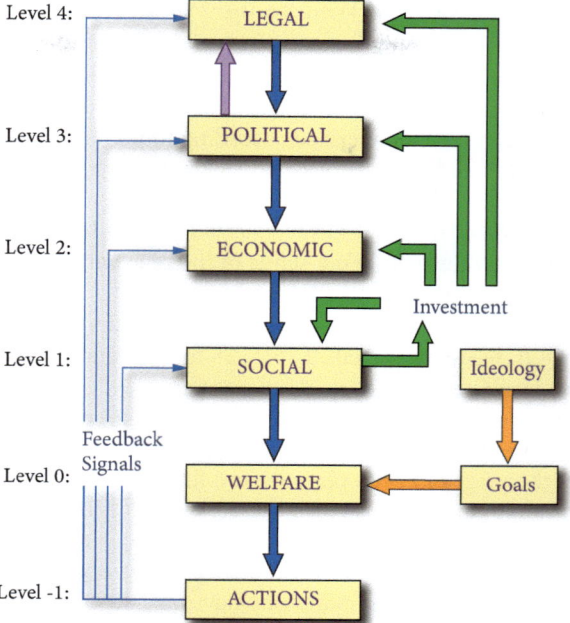

Figure 6.8 The inverse feedback system, whereby society affects individuals and groups.

Figure by Mauro Carfora.

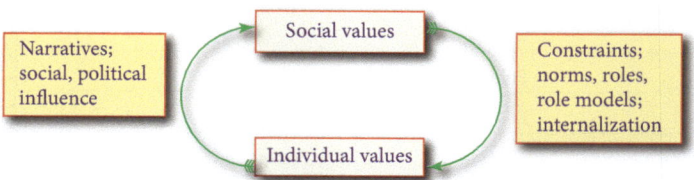

Figure 6.9 The feedback between social and individual values, leading to causal closure.

Figure by Mauro Carfora.

Hence we obtain the kind of causal closure that is required in all complex systems to determine unique outcomes:[96] we affect society, and society affects us. This applies in particular to values (**Figure 6.9**). This expresses the mutual dependence between people and society.

---

[96] Montevil Mossio, 'Causal Closure', in Werner Dubitzky et al., eds, *Encyclopedia of Systems Biology* (Berlin: Springer, 2013), 415–18.

6.2.6 Dimensions of Well-being: Welfare and Poverty

It is important to extend the concept of poverty beyond the purely economic one, as it may often be other aspects of the situation that are the true sources of welfare or lack of it. Welfare can be characterized according to the different aspects (health, social, economic, political, legal) represented by the QOL system, and therefore so can poverty.[97]

It is convenient to consider ten such aspects as in **Figure 6.10**. In each case (labelled **S1** to **S10**) one can rank the well-being status of the relevant group on a scale of +5 (high welfare) through 0 (neutral) to −5 (severe poverty).

Status labels S1 to S7 are the basic aspects of welfare deriving directly from the function of the QOL feedback system. Labels S8 to S10 refer to underlying aspects affecting perceived welfare, and so driving the system.

The fact one lives in a rich nation does not guarantee good welfare. The USA suffers severe legal poverty (S1 = −5) due to a corrupt supreme court, and as I write appears to be on the verge of a well-planned and financed anti-democratic takeover (S2 = −4.5), as is made clear by the Project 2025 Playbook.[98] Financial welfare is very varied in many countries, tending to be high in cities and low in the rust belts and rural areas. Education in my own country (South Africa) is also very varied, but in a state of severe poverty (S4 = −4) for a great many who have been let down by the educational system: they are barely literate. Social welfare (strong supportive communities) is very variable, and plausibly not strongly linked to economic status. Well-being in the USA is not good. There is health poverty due to strict anti-abortion laws which cause much

Nature	Category	Subcategory	Status Label
Main	Legal		S1
	Political		S2
	Economic	Financial	S3
		Educational	S4
	Social		S5
	Well-being	Health	S6
		Safety	S7
Foundations	Psychological		S8
	Perceptual/conceptual		S9
	Ideological/values		S10

Figure 6.10 Dimensions of welfare and poverty.

Figure by Mauro Carfora.

[97] Ellis, 'The Dimensions of Poverty'.
[98] See Project 2025 and The 180 day Playbook.

distress and even deaths, and health insurance policies designed to maximize profits for health insurers rather than providing the needed life-sustaining health care, leading to avoidable deaths (S6 = −4). There is very poor safety for school children due to extreme gun laws and resultant school massacres (S7 = −5).

Psychological welfare relates to psychological resilience. This is difficult to assess, but suicide rates are a measure showing that even some wealthy countries do not rank well.[99]

Conceptual poverty occurs in well-off countries due to poor education and the effects of social media. If an individual or group does not comprehend the concept of demonstrable cause-and-effect relations, they will be unable to understand that the introduction of vaccinations for a whole range of infectious diseases has been a key cause of improved health over the past hundred years.[100]

Ideological poverty relates closely to values. If your ideology classifies all foreigners as dangerous and evil, despite your own family having come as illegal immigrants, you are living an ungenerous life that demeans both you and those you interact with: S10 = −4. If your value system explicitly or implicitly is based on the idea that the only thing that matters is accumulating wealth and power for yourself, even to the extent of costing lives, then S10 = −4 to −5. If your central motivation is revenge backed up by death threats, then S10 = −5. If it is based on the idea that we are here on Earth to help each other, maybe sacrificing on behalf of others just because they are human, then S10 = +4 to +5. I return to the range of possible values in §6.3.

6.2.7 Interacting Groups, and the Environmental and Ecological Context

There is a QOL welfare feedback system (**Figures 6.5** and **6.6**) for each group in a society that is coherent enough that they can meaningfully be represented by one. These groups compete with each other for dominance in the context of society as a whole, through legal, political, economic and social influences.

The interaction between them is represented in **Figure 6.11**, showing the interaction of three systems for three different coherent groups. They could be political parties; social classes, insofar as they are organized; competing companies/corporations in an economy; groups identifying themselves as such due to language or national origin; and so on.

The ecological and environmental context. The whole set of such interacting social entities, in turn, interact with the surrounding environmental and ecological context (**Figure 6.11**). I will only briefly mention these crucial contexts, both because I am not expert in them and to do so in detail would require a whole book.

The natural environment comprises the land (continents, islands, mountains, valleys, plains and so on), water (rivers, lakes, seas, dams and so on) and atmosphere that are the physical context for our existence that has come into being by natural

---

[99] See the Wikipedia article List of countries by suicide rate.
[100] Vaclav Smil, *How the World Really Works: A Scientist's Guide to Our Past, Present, and Future* (London: Penguin, 2022).

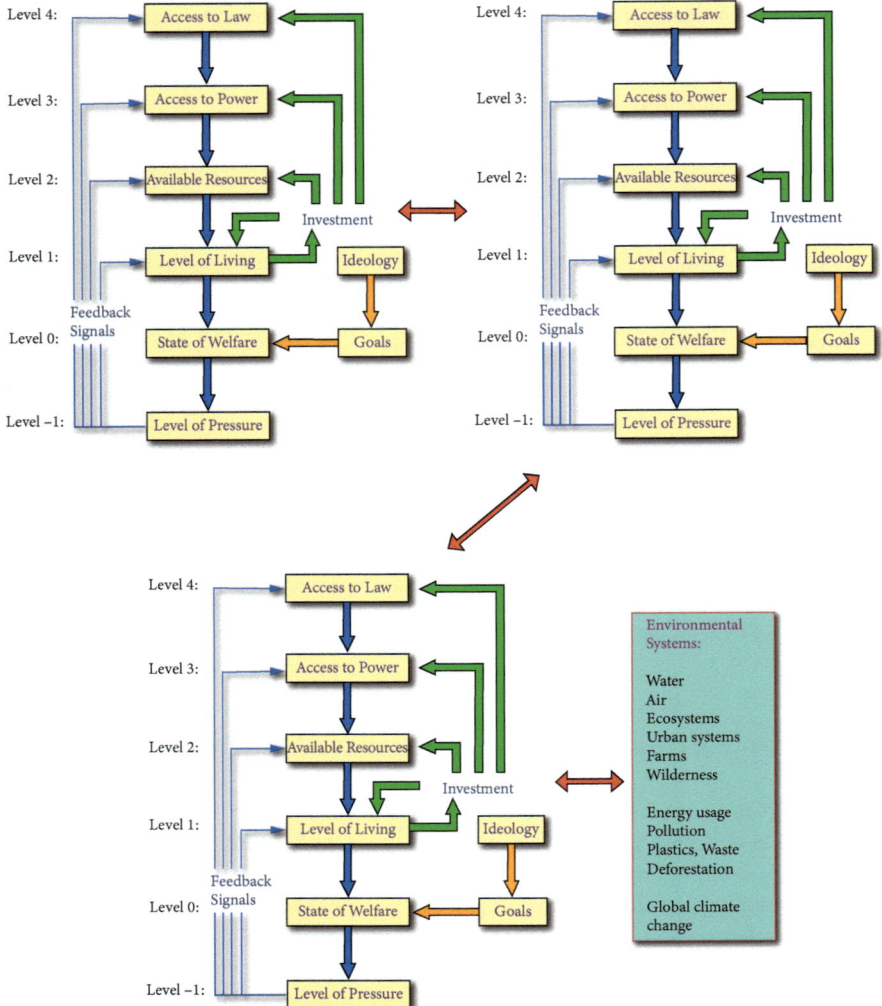

Figure 6.11 Three interacting groups and the environment.

Figure by Mauro Carfora.

processes.[101] Geological processes lead to the existence of mountain chains, plains, valleys and so on; continental drift occurs due to movement of tectonics plates; and global geochemical cycles turn over water, nitrogen, carbon and oxygen on very long timescales. This environment plays a key role in social and political outcome.[102]

Ecosystems consist of all life in some region, and ecology characterizes the interactions between them.[103] On land, plants and trees harvest energy from the Sun,

---

[101] See the Wikipedia article The natural environment. See also the Wikipedia articles Geology, Plate tectonics and Geochemical cycle.

[102] Tim Marshall, *The Power of Geography* (London: Elliott & Thompson, 2021).

[103] See the Wikipedia articles Ecosystems, Ecology, Food webs, Trophic levels and Ecological niche.

herbivores eat plants and predators eat herbivores, together forming food webs and trophic levels. Similar interactions take place in seas and lakes. Populations take advantage of ecological niches, and are capable of niche creation by altering the environment: ants make anthills, birds make nests, beavers make dams, many animals make burrows.

The biosphere is the sum of all ecosystems.[104] Our activities have affected the environment in two key ways. First, by replacing the natural environment with farms, towns and cities, thereby destroying forests and the habitats of many animals. Second, by causing all the emissions that are leading to global climate change, resulting in floods, extreme heat and fires. It is crucial to understand these interactions and take remedial action as a matter of urgency (§5.7).[105]

6.2.8 The Scales of Social Interactions

These interactions take place at all scales from the truly global scale **S1** to the family level **S8** (**Figure 6.12**), all based on individual people (scale **S9**) and their agency. Formal and informal organizations exist at each scale that help shape social and environmental outcomes.

Global scale S1: humanity as a whole. The United Nations (UN) has been set up to act politically (level 3) at the global level, and is only partially successful at helping to keep peace because of conflicts over goals. Nevertheless it sometimes acts in a way that matters, for example through the **Universal Declaration of Human Rights** and agreement on the **Sustainable Development Goals** (SDGs), which have made some difference to outcomes in many countries.[106] The **International Criminal Court**, endorsed by the United Nations, is able to prosecute individuals for crimes against humanity, genocide, and war crimes.[107] The **Intergovernmental Panel on Climate Change**, also endorsed by the United Nations, was set up to advance scientific knowledge about climate change caused by human activities and propose suitable remedies.[108] The **World Bank** provides loans to governments to assist economic development.[109]

International scale S2: groups of nations. Various organizations exist to promote the common interests of groups of nations: the G20, G7 and BRICS being examples.[110] While these represent the interests of countries and their inhabitants as a whole, the international capitalist system operates at this level, based in international trade. It promotes the interest of individuals and international corporations, with some of

---

[104] See the Wikipedia article Biosphere.
[105] See the Wikipedia article 'Climate Change' and the United Nations website 'What is climate change?'
[106] See the United Nations Foundation article The Universal Declaration of Human Rights is Turning 75: Here's What You Need to Know. See also the United Nations article The 17 Goals and the Wikipedia article Sustainable Development Goals.
[107] See the Wikipedia articles International Criminal Court.
[108] See the Wikipedia article 'Intergovernmental Panel on Climate Change'.
[109] See the Wikipedia article 'World Bank Group'.
[110] See the Wikipedia articles G7, G20, BRICS.

Scale	Nature	Entities	Organizations
S1	Global	Humanity as a whole	UN, IPCC, ICC, World Bank
S2	International	Groups of nations Capitalism as a system	G8, G20 and similar; multinational corporations
S3	National	Country national economic system	National government, national corporations
S4	Regional	State	State government, national corporations
S5	Local	City	Municipality
S6	Meso Level	City centres, suburbs	Firms, universities, banks, local corporations
S7	Community	Local areas, villages	Sports clubs, churches, Schools
S8	Family units	Related or voluntarily conjoined individuals	
S9	Individuals	People as such	

Figure 6.12 The scales of social interactions. Each scale affects outcomes on other scales via upward and downward interactions.

Figure by Mauro Carfora.

them becoming exceedingly rich through Internet-based organizations. I return to this issue in §6.6.3).

National scale S3: nations. This is where coherent dynamics occurs through national organizations specifically dedicated to carrying out tasks at the various levels of the QOL feedback system, particularly parliaments, houses of representatives, senates and so on, and supreme courts setting laws for the nation. National laws and national economic policy will control many outcomes across the nation.

Regional scale S4. This is where communities involved are large enough that the same kind of organizations as at the national scale occur, but at smaller scales. Local ordinances may be enacted and local budgets determine outcomes.

Local (city) scale S5. This will have similar organizations: city councils and mayors will determine local ordinances, plans and financing.

Meso-level S6 and community level S7. These are where local dynamics occur based in local organizations, and ultimately in the family units at level S8 where interpersonal relations rather than formal structures dominate.

A key form of structuring at this level is when a marriage takes place that involves legal and economic aspects as well as the personal ones.

6.3 The Significance of Values for Social and Environmental Outcomes

In the end values are the final arbiter of how this all works (see **Figure 6.1**). A paper by Denis Noble and myself [111] proposed this for the case of an individual (see §5.7); we then generalized this to the case of social institutions in a follow up paper.[112] In this section, I look at, §6.3.1, 'Values as the Highest Arbiter of Social and Physical Outcomes'; §6.3.2, 'The Spectrum of Implicit and Explicit Values'; §6.3.3, 'Utility of the Spectrum of Values'; and §6.3.4, 'Moral Motives and the Tragedy of the Commons'.

6.3.1 Values as the Highest Arbiter of Social and Physical Outcomes

Organizational purpose and goals. All organizations have a purpose; otherwise, they would not exist: it is the reason they were created. However, problem definition is a complex process,[113] and deciding what organizational purpose should be—specifically, how wide its scope should be—is a key part of leadership and management.[114] The outcome of considering this issue is creation of a hierarchy of goals that supply the value premises which underlie decisions.[115] In this hierarchical structure of ends,

> Each member of a set of behavior alternatives is weighted in terms of a comprehensive scale of values—the 'ultimate' ends.[116]

The function of goals is multidimensional: they have a cognitive function, a motivational function, a symbolic function, a justification function for actions taken and an evaluator function.[117] But additionally they have a normative nature: they implicitly or explicitly relate to what is desirable in a moral sense (**Figure 6.1**). They may do so in a positive way or negative way. As regards the QOL feedback system, values are the top level in the box labelled 'Ideology' in **Figure 6.5** (see §6.2.4). If you change values, you have to change all your goals so as to agree with them.

6.3.2 The Spectrum of Implicit and Explicit Values

The question then is, can we characterize values in a useful way so that we can assess them? The answer is yes. The basis is this:

---

[111] Denis Noble and George Ellis, 'Biological Relativity Revisited: The Pre-eminent Role of Values', *Theoretical Biology Forum*, 115 (2022), 45–70.
[112] To be published.
[113] George E. Dieter and Linda C. Schmidt, *Engineering Design* (New York: McGraw Hill, 2021).
[114] Peter Drucker, *The Practice of Management* (London: Routledge, 2012).
[115] Larence B. Mohr, 'The Concept of an Organisational Goal', *American Political Science Review*, 67 (1973), 470–81. Herbert Simon, *Administrative Behaviour: A Study of Decision-Making Processes in Administrative Organizations* (New York: Simon and Schuster, 1947).
[116] Scott and Davies, *Organizations and Organizing*, 54.
[117] Mohr, 'The Concept of an Organisational Goal'.

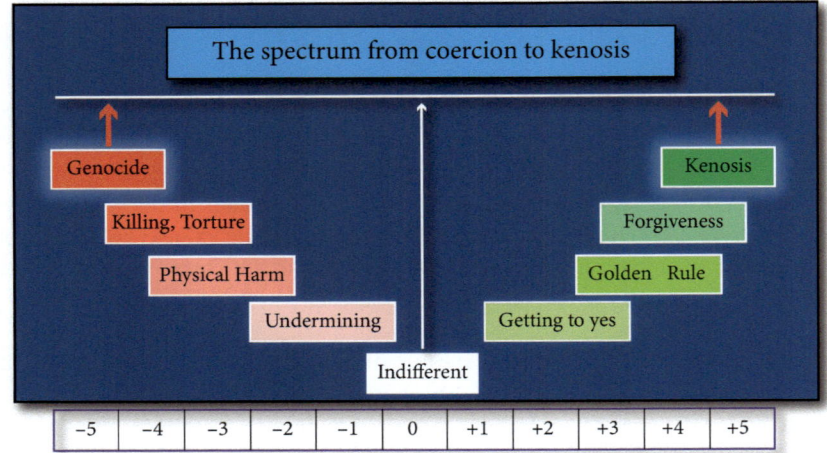

Figure 6.13 The spectrum of values, from extreme coercion to kenosis. All implicit or explicit personal and organizational values lie somewhere on this spectrum.

From George Ellis, 'On the Origin and Nature of Values', Tanner Lecture (Linacre College, Oxford, 2017). https://podcasts.ox.ac.uk/origin-and-nature-values

> Life's most persistent and urgent question is, 'What are you doing for others?'
> —Martin Luther King, Jr.

A spectrum of values exists expressing the possible set of ethical positions based on this understanding, ranging from the extreme negative value **V** = –5 to the extreme positive value **V** = +5.[118] It is sketched in **Figure 6.13**, with the very negative values on the far left, and the very positive on the far right. I'll class values into negative values, neutral values, positive values and transformational values.

Negative values: they have a destructive nature, stretching to the purely evil.

Value state V = –5, on the far left in **Figure 6.13**, is when the group concerned wishes to wipe out other groups and takes action to do so: 'Only my group's existence is tolerable, and I will try to exterminate yours by whatever means I can.' These are the values that underlay the Holocaust and all other cases of genocide,[119] the epitome of human evil. It is based in extreme hate and devaluation of the lives of other people. Organizations such as the Ku Klux Klan have the set same set of values, even if manifest on a smaller scale.

Value state V = –4 represents very similar values, falling short of total extinction of the victims, but the group involved are willing to kill and maybe torture other to attain their goals of domination and control. They are the only people who

---

[118] George Ellis, 'On the Origin and Nature of Values', Tanner Lecture (Linacre College, Oxford, 2017). https://podcasts.ox.ac.uk/origin-and-nature-values. Ellis and Noble, 'Economics, Society, and the Pre-eminent Role of Values'.

[119] See the Wikipedia article Genocide.

matter, all others are inferior and subject to them. This was the case for both the Catholic and the Anglican Churches some centuries ago, when dissidents were tortured and burnt at the stake. Many medieval castles in Europe featured torture chambers. The Apartheid regime in South Africa tortured and killed political opponents. There are still countries today where political opponents are tortured or assassinated. This behaviour expresses the values of the people in charge. Causing pain and death is the purpose of these actions.

Value state V = −3 involves being willing to take any action whatsoever in order to benefit one's own group, no matter what harm one causes to the other. This is the position of the MAGA wing of the Republican party in the USA today, and their supporting evangelical churches, as is made clear by Project 2025.[120] The attitude is not 'These are my values and I'm going to stick to them' but 'These are my values and I'm going to force you to live by them, no matter how much it harms you.' Stirring up hatred is central to Donald Trump's mode of action; now that he dominates the Republican Party, it is central to its nature and the nature of those 'churches' that support him so strongly. Again, in many cases causing distress and pain seems to be a feature, not a bug. Life is no longer sacrosanct, as is apparent in the many cases of US school shootings enabled by gun legislation permitting possession of assault rifles without background checks, and in the deaths of young women caused by anti-abortion legislation.[121]

It includes a willingness to cause death by removing all environmental safeguards in order to increase personal and corporate profit, as captured brilliantly in the film *Erin Brockovich*.[122] The attitude 'I'm prepared to cause you harm as long as I profit' occurs in cases of denial of health insurance claims in the USA,[123] which has caused many deaths and anguish.

Neutral values: they are neither positive nor negative.

Value state V = 0 is the position of live and let live: I'll work for my good and you work for yours, but we won't help each other. This is a basically unsupportive society; we are centrally selfish but not actively harming others. This is arguably the position underlying Game Theory, widely used in economic studies.[124]

Value state V = +1 is the more positive position of 'getting to yes':[125] we can work together to make positive things happen for both of us, because that will mean I'm not losing out myself. It is still motivated by self-interest; it is just a good way of getting what I want for myself.

These two stances may represent the majority of interactions in modern large-scale societies, outside of social interactions in small communities where an aim of actually helping each other may occur.

---

[120] See the Wikipedia article 'Project 2025'.
[121] Alison Gemmill et al., 'Infant Deaths after Texas' 2021 Ban on Abortion in Early Pregnancy', *JAMA Pediatrics*, 178, 784–91.
[122] See the Wikipedia article *Erin Brockowich* (film).
[123] See the ValuePenguin article Insurance Company Denials: Worst Companies and How to Appeal.
[124] See the Wikipedia article Game theory.
[125] Roger Fisher et al., *Getting to Yes: Negotiating Agreement without Giving In* (London: Penguin, 2011).

Positive values: they aim to produce constructive outcomes.

Value state V = +3 is on the positive side of the spectrum is known as the **Golden Rule**: do unto others as you would have them do unto you.[126] Here there is a concern for the welfare of others: a willingness to help them. This is a positive foundation for a worthy social order: it is more than caring just for oneself and one's family.

Value state V = +3 is to be contrasted with Colin Mayer's **reformulated Golden Rule**, namely 'Do unto others as they would have done to them.'[127] The key difference is that the first is self-referential and the second is not:

it requires us to act towards others in a form that recognizes and respects them for who and what they are and truly want to be.[128]

These are both attainable and a good basis for business and social life.

Transformational values: they have the potential to transform the situation entirely in a positive way.

Value state V = +4 sees forgiveness as a route to reconciliation of those who have been bitterly opposed to each other, a practical step toward creating common good. This is the position of Nelson Mandela and Desmond Tutu, and lay behind the remarkable, largely peaceful transition to a democratic government in South Africa that they enabled. It resulted in the creation of the Truth and Reconciliation Commission as a practical way to implement forgiveness in South African public life for the actions of the security police involving torture and killing.[129] It is transformational in terms of changing relationships.

This is based in the African concept of **Ubuntu**, reflecting the positive aspect of our common human nature: 'umuntu ngumuntu ngabantu'. That is, 'a person is a person through other persons', or 'I am who I am because of you.'[130] This is a crucial step further because it recognizes the mutual interaction and caring that is the true foundation of human well-being. It is reciprocal in a way that the previous ones are not, and so recognizes the deep foundations of the emergence of society: the whole (the group) is more than the parts (those individuals who make it up) through caring interactions. They each reinforce the other in constructive and generous way, resulting in a profoundly supportive social order. In economic contexts, it is possible to regard the firm as a supportive community rather than a collection of individuals,[131] and understand the purpose of management as promoting the common good.[132]

---

[126] Harry Gensler, *Ethics and the Golden Rule* (London: Routledge, 2013).
[127] Colin Mayer, *Capitalism and Crises: How to Fix Them* (Oxford: Oxford University Press, 2024).
[128] Mayer, *Capitalism and Crises*, xiii.
[129] See the Wikipedia article Truth and Reconciliation Commission (South Africa).
[130] Abeba Birhane, 'Descartes Was Wrong: "A Person is a Person through Other Persons" ', *Aeon* 7 April 2017.
[131] An example is the Scott-Bader Commonwealth
[132] David W. Lutz, 'African Ubuntu Philosophy and Global Management,' *Journal of Business Ethics*, 84(Suppl 3) (2009), 313.

Value state V = +5 is that of *kenosis*, that is, self-emptying.[133] It is a centrally generous and compassionate viewpoint recognizing the humanity and needs of other individuals and communities. It encompasses a theme of giving up: being prepared to voluntarily sacrifice your own interests on behalf of the greater good, even being prepared to do so if they are an enemy. This attitude of 'letting go' has a transformational nature, with the possibility of completely changing the quality and meaning of the situation facing us.

The word 'voluntary' is key: it must be a choice freely made in order to have its true quality. This is extraordinarily difficult to do, requiring great dedication and courage. It is the only truly transformational ethic because it has the power to turn an enemy into a friend. I am not here referring to the specifically Christian view of the concept, but to a wider one that generalizes it without any theological connotations.

It is a way of living that any religious or atheist person can adopt. Despite the fact that all the major religions have branches that are precisely the opposite, being based in terror and fear, Sir John Templeton argued that all the great world religions also have branches based in what he called 'Agape Love',[134] which is kenosis.

The transformational nature of kenosis. The key point is that in the end this is a truly transformational way of doing things. I first understood this when reading the Prologue to *Reading in St John's Gospel* by Archbishop William Temple.[135] He explains there that according to St John's Gospel, Jesus went into the desert to consider how he should carry out his perceived mission of transforming society to what it ought to be.[136] The temptations reflect three possible ways to do so that were each rejected because they are fundamentally mistaken: they cannot lead to the desired outcome.

The first is economic: to establish the 'good time coming' of everyone having wealth. This is a dominant view of 'the good society' today. The second is political: to create the desired society by establishing an earthly monarchy with a wise ruler. The third is intellectual: to provide irresistible proof of what a good life should be like and how society should be ordered.

Temple then says that each of these options contain some truth, but if any of them are taken as the whole truth they have a fatal defect: they each represent ways of securing outward obedience by bribery or coercion or providing incontrovertible evidence, but they cannot create inner loyalty. They are ways of controlling conduct, but not transforming hearts and wills. That requires a ministry not just teaching the precepts of perfect love but living such a life with all the sacrifice that entailed—a life of a kenotic nature. An example is the life of the bishop Monseigneur Bienvenu described in the great novel *Les Misérables* written by Victor Hugo.[137] The novel describes

---

[133] George Ellis, Tanner Lecture 2017: https://podcasts.ox.ac.uk/origin-and-nature-values.
[134] Sir John Templeton, *Agape Love: A Tradition Found in Eight World Religions* (Conshohocken, PA: Templeton Foundation Press, 1999).
[135] William Temple, *Readings in St John's Gospel* (London: Macmillan, 1947), xxvi–xxviii.
[136] In theological terms, to inaugurate the Kingdom of God. In this text, I am reinterpreting such theological understandings in lay terms that make sense in the context of the argument of this book.
[137] See the Wikipedia article *Les Misérables*.

the total transformation in the life of the former convict Jean Valjean through the generous and caring behaviour, plausibly kenotic, of Monseigneur Bienvenu.

Can such an approach work in a political context? It was the position of Mahatma Gandhi and Martin Luther King,[138] who thought deeply about it and demonstrated it could be used to radically change the situation in a political context.[139] Gandhi was an example of 'servant leadership', which is essentially the kenotic approach.[140] Richard Gush of Salem, South Africa, demonstrated this could be done in the context of a war between settlers and immigrants in South Africa.[141]

In the South African context of the struggle for freedom during the Apartheid years, two examples are this. Firstly, a horrific practice arose in the townships where police brutality against black people was rife. People who were suspected of being *impimpis* (informers) were burned alive by violent mobs in a process called *necklacing*: putting a tyre filled with petrol around the victims neck and then setting the petrol alight. Regrettably, Winnie Mandela supported this horrible practice. However, I am aware of two people who had the immense courage of striding into such a mob situation and demanding that it stop at once—and succeeding in shaming the crowd so that the victim was released. The two were Archbishop Desmond Tutu,[142] and a remarkable person called Rommel Roberts, employed by the Quakers as a peace worker in Cape Town.[143] This is a kenotic action because it involved huge bravery: the crowd could easily have turned against them, and necklaced them as well. But it transformed the situation. Secondly, Nelson Mandela refused to treat his jailors on Robin Island as enemies, despite the hostile nature of the relationship between prisoners and their warders. He treated them as human beings of value, and made friends with them.[144]

What about pacifism and wars? Should one not take part in wars to prevent the occurrence of unspeakable events such as the Holocaust? There has always been this tension arising around pacifism, and the Quakers developed a suitable response: they set up the Friends Ambulance Unit, which worked on the frontline in both the First and Second World Wars.[145] Thus they were doing their part to help fight against such dreadful events, putting their own lives at stake, but not actually killing other people. On the contrary, they were prepared to help the wounded from either side when they were in desperate need of help.

Can the approach be used in an economic context? That this may be so has been argued by Stucke and Ariel.[146] They argue for what they call **noble competition**: helping your rivals to reach their full potential. How to make this real is discussed in depth

---

[138] See the Wikipedia articles Mahatma Gandhi and Martin Luther King Jr.

[139] John J. Ansbro, *Martin Luther King Jr: The Making of a Mind* (Maryknoll, NY: Orbis Press, 1982).

[140] Annette Barnabas and Paul Clifford, 'Mahatma Gandhi—An Indian Model of Servant Leadership', *International Journal of Leadership Studies*, 7 (2012), 132–50.

[141] See the Wikipedia article Richard Gush and The Life of Richard Gush: African Emigrant (SA Yearly Meeting, 2018).

[142] See the Wikipedia article Desmond Tutu.

[143] Rommel Roberts, *Seeds of Peace: Stories of Silent Heroes and Heroines in South Africa* (Küsnacht, Switzerland: Digiboo Verlag, 2018).

[144] Mike Nicol, *Nelson Mandela's Warders* (Nelson Mandela Foundation, 2019).

[145] See the Wikipedia article Friends' Ambulance Unit and the article Friends Ambulance Unit (FAU) in WWI (Quakers in the World).

[146] Maurice Stucke and Ariel Ezrachi, *Competition Overdose: How Free Market Mythology Transformed Us from Citizen Kings to Market Servants* (New York: Harper Business, 2020), 256.

in their Chapter 10, entitled 'Competition: From Toxic to Noble'. Such an approach transcends the calculus of the market, and so the basis of standard economic theory and game theory. It lifts the whole basis of interaction to a higher level based in appreciating the fundamental worth of all the people you are interacting with, no matter who they are. When you create conditions where your purpose is to help each other, economic activity is no longer a zero-sum game.

Taking this stand does not have to involve being a doormat who can be taken advantage of by others because they know you have a generous character. It involves a nature of generosity and willingness to behave in a kenotic way but with an awareness that it can be taken advantage of, and a determination that this should not be the case. This is possible if you have internal strength sustained by your own integrity. It applies also in other contexts such as education and the arts, and is the basis of social life in a family context, particularly as regards the relation of a mother to her children. I discuss this in Chapter 7. I believe it is a profound basis of deep ethics and social action in all spheres of life.

6.3.3 Utility of the Spectrum of Values

By observing social interactions of any individual or group, one can establish where they lie on this spectrum of values. Thus this position is not just a matter of opinion, it is empirically observable. I make two relevant claims.

Claim 1: position on the spectrum of values is empirically determinable. Organizations and individuals can be placed on the spectrum of values (**Figure 6.13**) using proposals made by Mark Carney (now Canada's Prime Minister and leader of the Liberal Party) in his key book: *Values: An Economist's Guide to Everything That Matters.*[147] The indicators he mentions can be classified into two kinds: direct indicators of where values lie on this spectrum, and indicators of conditions that support their deployment effectively.[148]

Direct indicators of values are trust, integrity, transparency, fairness, solidarity, sustainability. These key values can be used to place organizations or individuals on the spectrum of values discussed in §6.3.2 because they are invariably inversely related to this position, and so are implicitly represented there. Each is measurable on a scale of −5 to +5. If it can be demonstrated that almost every statement you make is a lie, one knows both that you have no integrity and that no statement you make can be trusted. Fact checking can often be done in an impartial way; if you object to such fact checking, this is a clear statement regarding your willingness to lie. If you demonstrably value the welfare of your group over the welfare of all others and act accordingly, not hesitating to inflict harm, you are lacking in both fairness and solidarity with others.

---

[147] Carney, *Values: An Economist's Guide to Everything That Matters.*
[148] He does not make the distinction given here between two classes of indicators. That distinction is mine.

If your policies are such as to benefit yourself and your group by acts that directly cause or support environmental damage, your actions damage sustainability.

> These indicators can all be measured empirically and used as indicators of true values. Thus they establish empirical values. These may or may not correspond to stated values. If they do not agree, there is value discord. If so, this is a further indication of lack of trust and integrity.

Indicators of conditions that will support deployment of such values are dynamism, resilience, responsibility and humility. While these are important in implementing outcomes based in values, I regard them as means to turn values into value, rather than as values per se. Indeed the first three can as equally be deployed for negative purposes as they can for positive ones: they don't directly represent position on the spectrum of values. The last (humility) arguably does, but is not in my view as easily measurable as the others, and is more centrally a way of attaining desired outcomes rather than being one.

I will return to these indicators and their use as regards the economic system in §6.6.3.

> **Claim 2: these values apply to every level in the QOL feedback system.** They shape the goals in this system (**Figures 6.5** and **6.6**), and hence affect the feedback processes occurring at every level. Thus they apply to the welfare level, the social level, the economic level, the political level and the legal level, and can be measured empirically at each of these levels. The system as a whole expresses the group values.

Thus an organization's values can be classified via its actions at each of these levels: they will usually be congruent. What is needed at all levels is a value system that is aimed at enhancing the common good.[149] Transformational leadership is about managing values and so changing not just outcomes, but the character of the people comprising the organization so that they will have values on the positive side of the spectrum.[150] The key issue is, how can we help develop cooperative values that will lead to development of a caring society, with economic institutions that share wealth and care for the environment? Transformative leadership has the potential to do so.

Finally, one can ask, **are values nothing but the utility functions that are central to economic analysis?**

My answer is no, they are quite different:

> Values are a metric on utility functions, characterizing which are better or worse in moral terms.

Standard economics assumes the utility function is simply maximizing profit. A more enlightened view extends the utility function to take 'externalities'—taking costs such

---

[149] Michael Sandel, *The Tyranny of Merit: What's Become of the Common Good?* (London: Penguin, 2021).
[150] Burns, *Leadership*.

as environmental damage into account, and include them in accounting systems and in the way international capitalism runs.[151] The proposal here provides an overall framework for relating welfare, social, economic, political and legal activity to moral standards. Sure it is idealistic, but not impossible to approach to some degree in reality and is exemplified by firms such as the Scott Bader Commonwealth, which I return to later.[152]

Significance of values. Because it is so important and is the central theme not just of this chapter but of the book, I will re-emphasize here that, in the end, values are the final arbiter of how all social interactions work.[153] A fuller version of the statement by Martin Luther King, Jr, that I started with is this:[154]

> An individual has not started living until he or she can rise above the narrow confines of his or her individualistic concerns to the broader concerns of all humanity. Every person must decide whether he or she will walk in the light of creative altruism or in the darkness of destructive selfishness. Life's most persistent and urgent question is, 'What are you doing for others?'

This works out in relation to the nature of economic systems, actions regarding global climate change, the way health insurance companies deny healthcare and so on. It is reflected in the social system, and the degree to which societies are built on trust and caring. It is reflected in the political and legal systems that regulate the others.

A key example as I write (17 December 2024) is that yesterday, yet another school shooting took place in the USA. A 15-year-old girl opened fire in a Wisconsin school classroom, fatally shooting a fellow student and a teacher and wounding six other people before killing herself. This ironically took place at the Abundant Life Christian School. There have been 322 school shootings this year in the USA, according to the K–12 School Shooting Database website.[155]

How can this happen? President Joe Biden yet again called on Congress to enact gun-control legislation to prevent further massacres. Similar calls have gone unheeded after every school shooting in recent memory. Why? Because of the gun lobby funded by the National Rifle Association, which unbelievably states, 'The National Rifle Association is America's longest-standing civil rights organization'.[156] They have the support of Republican Party politicians, who pass laws to allow anyone to own assault rifles with no background checks, and to openly carry these weapons in public.[157] This has been ratified by the US Supreme Court. There could not be any clearer example of how values affect life. This empirically shows the values of all of these people lie at **value state V = −3** in the spectrum of values (**Figure 6.13**).

---

[151] Colin Mayer, *Prosperity: Better Business Makes the Greater Good* (Oxford: Oxford University Press, 2018).
[152] See https://www.scottbader.com/commonwealth/.
[153] Amitai Etzioni, *Moral Dimension: Toward a New Economics* (New York: Simon and Schuster, 2010).
[154] Martin Luther King, Jr, *Strength to Love* (Boston: Beacon Press, 2019).
[155] See the K–12 School Shooting database.
[156] See the National Rifle Association website https://home.nra.org/.
[157] See the Wikipedia articles Gun law in the United States and Gun Laws in the United States by state.

The underlying key question is this: what kind of family life, school education and religion result in a young girl behaving in this way? What kind of family gives her a gun and ammunition? There is something profoundly amiss when this happens.

6.3.4 Moral Motives and the Tragedy of the Commons

The kind of economic issue where values make a crucial difference is the **tragedy of the commons,** pointed out by Garrett Harden in 1968.[158] When some common good is shared by many people, each will tend to maximize their own use of the resource to the detriment of others. Eventually this can lead to the destruction of the resource. For example, this applies to a shared pasture that will be overgrazed, or to common fisheries. We also use the commons for waste disposal, resulting in pollution or poisonous conditions. Harden ends by commenting that the fundamental underlying cause of the tragedy of the commons globally is overpopulation.

I comment on this in §6.4. However, it also arises through overuse of resources by some very rich countries. Specific contexts where the issue arises at the present time are deep sea mining plans and plastics treaties.[159]

A key point is that it is possible to design social institutions that help combat the destruction of valuable natural resources. The lead in doing so has been taken by Nobel Prize winner Elinor Ostrom.[160] This requires an analysis of the sustainability of social-ecological systems and design of institutions that will help overcome the problem.[161] Very helpful in making such proposals is her grammar of institutions and her proposal for polycentric governance of complex economic systems.[162]

I cannot do justice to all her work, but it will not come as a surprise to readers of this book that self-organized resource governance systems should be designed as complex adaptive systems. A crucial role is played by ensuring that all interested groups can take part in designing and then operating the system: that is, they should be given full agency in the process so that they are in agreement with it and buy-in occurs.

The factors that need to be taken into account are (1) clear boundary and memberships; (2) congruent rules; (3) collective area choices; (4) monitoring; (5) graduated sanctions; (6) conflict-resolution mechanisms; (7) recognized rights to organize; (8) nested units; and (9) robust institutional performance. The dynamics of

---

[158] See the Wikipedia article Tragedy of the commons, and Garrett Harden, 'The Tragedy of the Commons', *Science*, 162 (1968), 1243–8.

[159] Vitoria Heath, 'Arctic Deep Sea Mining Plans Halted in Norway', *Geographical Magazine* (2 December 2024). Stefan Anderson, 'UN Plastic Pollution Treaty Derailed as Fossil Fuel Nations Block Production Limits', *Health Policy Watch: Health & Environment* (3 December 2024).

[160] Elinor Ostrom, *Governing the Commons: The Evolution of Institutions for Collective Action* (Cambridge, UK: Cambridge University Press, 1990); Elinor Ostrom, 'Coping with Tragedies of the Commons', *Annual Review of Political Science*, 2 (1999), 493–535.

[161] Elinor Ostrom (2009) "A General Framework for Analyzing Sustainability of Social-Ecological Systems." *Science* 325: 419–422.

[162] Sue Crawford and Elinor Ostrom. 'A grammar of institutions.' American pol sci review 89 (1995): 582–600; Elinor Ostrom 'Beyond markets and states: polycentric governance of complex economic systems' American economic review 100 (2010): 641–672.

institutional choice are summarized in Figures 6.1 to 6.5 in her book. Above all, solving the problem requires social norms that will support such institutions[163]—there is a need for moral motives in economic and social decisions.[164]

6.4 The Environmental and Ecological Context

A society exists in its environmental and ecological context (see §6.2.6), which provides the basis for our physical existence. How we interact with that environment is therefore a key issue.[165] Crucially, we turn large areas of the natural environment into farms, towns and cities, thereby destroying forests and natural habitats for animals. We interact with it by extracting resources from it (fishing, hunting, farming), which is essential for our physical existence, and harvesting trees, mining, extracting petroleum from oil wells and so on, key to our economy. We to some degree restore damage that may be caused by these processes. However, we use it for disposal of both toxic and nontoxic waste, including industrial and medical waste and sewage. We also interact with it by caring for it, cleaning up and planting trees and flowers, and enjoying it in many ways: walking, climbing, sailing, skiing, swimming, surfing, picnicking and so on.

Theory is needed that links research at multiple organizational, spatial and temporal scales, from micro to meter to landscapes up to the biosphere. The multidimensional and hierarchical multiscale nature of biodiversity requires solutions that can address cross-scale questions and identify cross-scale phenomena, as is made clear in a study by the National Academy of Sciences.[166] They identify four key themes: (1) biodiversity and ecosystem function; (2) resilience and vulnerability; (3) connectivity; and (4) sustainability of ecosystem services. Crucially, this study represents the kind of integrative analysis advocated by this book.

Pragmatic environmental policy recognizes that all relevant choices are tradeoffs.[167] Our mere existence together with the second law of thermodynamics guarantees that we will cause some damage to the environment; however, we can minimize such damage and engage in restorative policies. Industrial-scale energy generation of whatever form will always cause some kind of negative environmental impact. All we can do is choose what kind of negative impact that will be, and take steps to minimize it. Key are measures to use less energy, efficient use of the energy we do use and measures to reduce food and water wastage.[168]

---

[163] Elinor Ostrom, 'Collective Action and the Evolution of Social Norms', *Journal of Economic Perspectives*, 14 (2000), 137–58.

[164] Marc Fleurbaey, Ravi Kanbur and Dennis J. Snower, 'An Analysis of Moral Motives in Economic and Social Decisions' (2024). Available at SSRN: 4697702.

[165] Stefan Anderson, 'UN Plastic Pollution Treaty Derailed as Fossil Fuel Nations Block Production Limits', *Health Policy Watch: Health & Environment* (3 December 2024).

[166] National Academy of Sciences A Vision for Continental-Scale Biology: Research Across Multiple Scales (2024).

[167] William F. Baxter, *People or Penguins: The Case for Optimal Pollution* (New York: Columbia University Press, 1974).

[168] Steve Sorrell, 'Reducing Energy Demand: A Review of Issues, Challenges and Approaches'. *Renewable and Sustainable Energy Reviews*, 47 (2015), 74–82.

In this section I consider, §6.4.1, 'Global Climate Change Is Happening, §6.4.2, 'The Key Factor of Global Population Growth', and §6.4.3, 'Planetary Boundaries and Planetary Health Diet'.

6.4.1 Global Climate Change Is Happening

Global climate change is a key aspect of our impact on the environment and will cause massive disruption.[169] The underlying issue is the greenhouse effect (**Figure 6.14**) caused by greenhouse gases: carbon dioxide, water, methane, nitrous oxide, F-gases such as chlorofluorocarbons, and aerosols.[170] However, the key one that needs to be reduced is carbon dioxide, as it is having the greatest effect.

The physics is this. Incoming solar radiation gets reflected by clouds or is absorbed by the Earth's surface or is reflected by that surface. Infrared radiation gets emitted by the Earth's surface, some of which is absorbed by the atmosphere or clouds. Some of this is re-emitted to space by the atmosphere and clouds, but some is reflected back

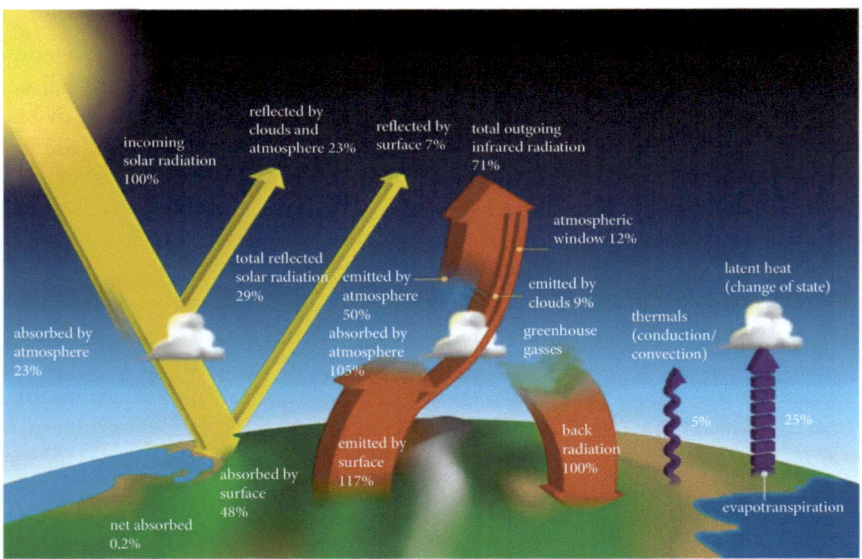

Figure 6.14 The global heat budget underlying climate change, with clouds (water vapour) and carbon dioxide both playing central roles. The greenhouse effect is the downward red arrow.

Image: NASA.

[169] United States Environmental Protection Agency, Basics of Climate Change. Nicholas Stern, *A Blueprint for a Safer Planet: How to Manage Climate Change and Create a New Era of Progress and Prosperity* (London: Penguin, 2009). World Economic Forum, *The State of Climate Action: Major Course Correction Needed from +1.5% to −7% Annual Emissions*, White Paper (November 2023). Carney, *Values: An Economist's Guide to Everything That Matters*.

[170] See the Wikipedia articles Greenhouse effect, Greenhouse gas and Greenhouse gas emissions.

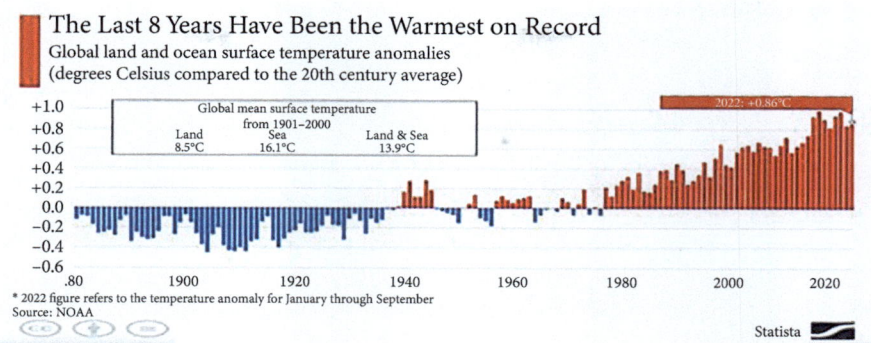

Figure 6.15 Evidence of global warming.

Image: Statista.

to the Earth by greenhouse gases. This is the greenhouse effect. Thermals (conduction/convection) and evaporation/transpiration complicate the picture because they also result in clouds.

Mark Carney's important book analyses the climate crisis in depth, summarizing the evidence of temperature anomalies (**Figure 6.15**), global see level increase and Arctic Sea ice volume.[171] Here it is important to note that global climate change involves both cooling and warming; see **Figure 6.16**. It involves rising seas and melting ice and extreme floods in some places, extreme heat, droughts and fires in others, but also cooler temperatures in some contexts. Global climate change can be directly linked to global carbon dioxide (CO_2) emissions,[172] so it is crucial to reduce the emissions due to industrial processes, buildings, transport, energy generation and food production/agriculture across the world. However, there is an important technical point.

Water (H_2O) is the primary regulator of climate change, as is evident from the way that clouds (water vapour) play a central role in the dynamics shown in **Figure 6.14**. Certainly CO_2 is a also a greenhouse gas,[173] but the key point is that there is a feedback mechanism whereby H_2O (in all three phases) can amplify the warming effect of increased CO_2, as discussed clearly by Tim Palmer.[174] This is why CO_2 emission is a key driver of climate change, and why it is crucial that carbon emissions are reduced. Complex physics underlies how clouds at different levels in the atmosphere play a vital role.[175]

---

[171] Carney, *Values: An Economist's Guide to Everything That Matters*, Chapter 11.

[172] See *Our World in Data*: https://ourworldindata.org/co2-emissions.

[173] See the Wikipedia article 'Greenhouse gas'.

[174] Tim N. Palmer, *The Primacy of Doubt: From Climate Change to Quantum Physics, How the Science of Uncertainty Can Help Predict and Understand Our Chaotic World* (Oxford: Oxford University Press, 2022), 112–17.

[175] Michael Allen, 'Cloudy with a Chance of Warming: How Physicists Are Studying the Dynamical Impact of Clouds on Climate Change', *Physics World* (26 November 2024). Laura Landrum and Alice DuVivier, 'Sea Ice is Shrinking during Antarctic Winter: Here's What It Means for Earth's Ocean and Atmosphere'. *Nature News and Views* (18 December 2024).

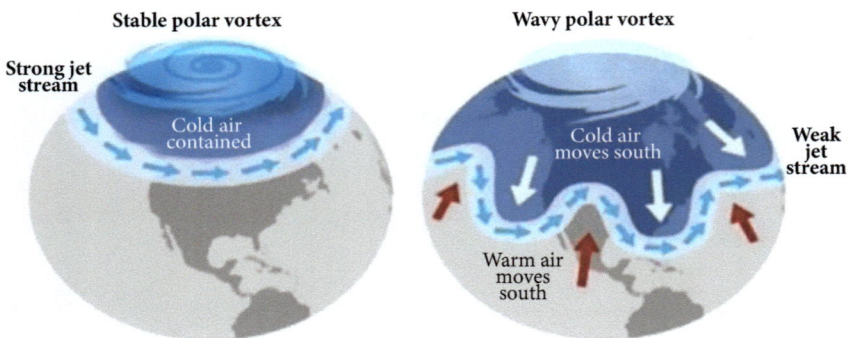

Figure 6.16 Global climate change has various effects. It causes warming overall, but also causes cooling in some areas.[a]

[a] For more details, see Robert McSweeney, 'Explainer: the Polar Vortex, Climate Change, and the "Beast from the East" ', *Skeptical Science* (7 March 2018).

Source: NOAA.

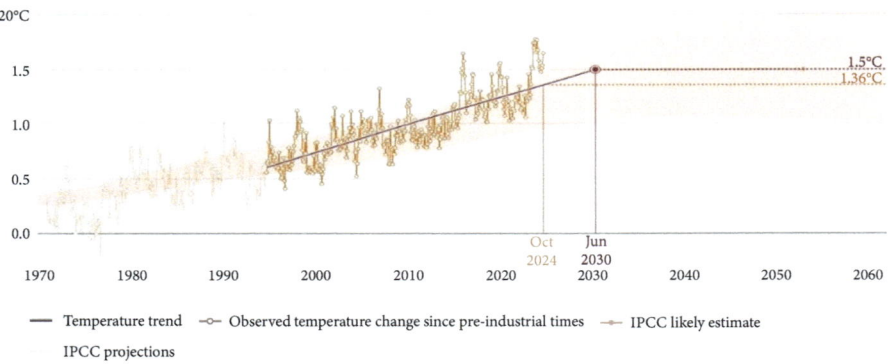

Figure 6.17 Predicted global temperature rise.
Image: IPCC.

The consequence of current rates of emission of greenhouse gases are shown in the historical record (**Figure 6.15**) and current predictions in **Figure 6.17**. If continued, this will lead to disastrous consequences of many kinds—but certainly not extinction of the human race, as some have claimed. The panic associated with that claim is not warranted. The problems arising from global climate change are documented by Mark Carney.[176] A key example is water stress due to climate change (**Figure 6.18**), because clean water access is a key human need.

Tipping points can happen as global warming occurs.[177] Energy generation and usage is a key issue; inter alia, nuclear power is needed as a component of

---

[176] Carney, *Values: An Economist's Guide to Everything That Matters*, 109–32.
[177] See Global Tipping Points.org. T. Lenton et al., 'Tipping Elements in the Earth's Climate System'. *Proceedings of the National Academy of Science*, 105 (2008), 1786–93. M. Armstrong et al. 'Exceeding 1.5°C Global Warming Could Trigger Multiple Climate Tipping Points', *Science*, 377 (2022), 6611.

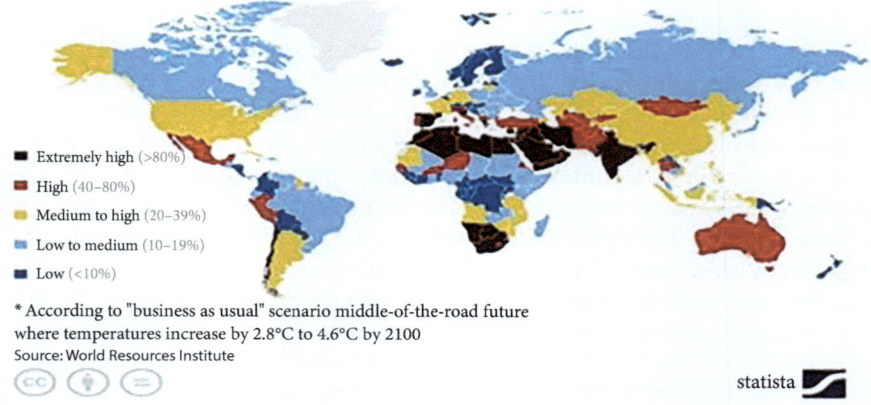

Figure 6.18 **Water stress predictions.**

Image: Statista.

electricity generation. Vaclav Smil states, 'Modern nuclear reactors, if properly built and carefully run, offer safe, long lasting and highly reliable ways of energy generation',[178] and I agree with that. However, fossil fuels will continue to be seen as indispensable in the functioning of the global economy for the time being. Decisions on these crucial issues need to be based in solid evidence;[179] see **Figures 6.15** and **6.17** as examples.

The problems arising from global climate change, and the result urgency to take action, are catalogued by Carney as follows.[180] Firstly, physical risks such as hurricanes, wild fires, flash flooding, damaging property and natural systems. Solid evidence of the results is the increase in underwriting risk for insurers. Secondly, climate change will affect the value of assets and the gross domestic product of countries. Thirdly, and crucially, it will cause forced migration, leading to climate refugees and can be expected to increase the incidence of disease and the risks of conflict. Fourthly, it is causing the destruction of natural habitats such as coral reefs, which results in the loss of species whose habitat is threated or lost.[181]

The main sources of anthropogenic greenhouse gases are burning fossils fuels, agriculture, deforestation and other changes in land use, cement manufacture and

---

[178] Smil, *How the World Really Works*, 40.

[179] Vaclav Smil, *Numbers Don't Lie: 71 Things You Need to Know about the World* (London: Penguin Random House, 2021).

[180] Carney, *Values: An Economist's Guide to Everything That Matters*, Chapter 11.

[181] Kunio Kaiho, 'Extinction Magnitude of Animals in the Near Future', *Scientific Reports*, 12 (2022), 19593, and the Wikipedia article Endangered Species.

aerosols.[182] To solve the problem, Carney advocates, firstly, **engineering technology,** driving scale and innovation.[183] This involves electrifying everything and developing green generation of electricity, increasing efficiency of power usage, reducing industrial emissions and so on. A key point where I disagree with him and agree with Smil is that, in my view, we should expand the use of nuclear energy, particularly pebble bed modular nuclear reactors.

Secondly, Carney advocates **political technology:** setting the right goals. He states that most fundamentally, this requires society to set out clearly its values and goals, and then for governments, companies, and all of us to work toward achieving them.[184] A step toward this has been agreement on the SDGs mentioned in §6.2.8.

Thirdly, **financial technology** is crucial and should ensure that every financial decision takes climate change into account. Reporting and effective risk management are key aspects in developing this (see §6.6.3). In Chapters 13–16, Carney emphasizes that values-based leadership is essential in achieving these goals. But of course this requires international agreement because the problem is a global issue.[185] Progress is being made through international climate negotiations, but tensions need to be resolved.[186] But there also needs to be a commitment from the well-off nations to reduce their inequitable use of energy to more sustainable levels.

Although international action is needed, it is also helpful when local action is taken by large numbers of people to cut down energy use. In the case of industrialized societies, this can involve sealing leaks and gaps and ducts; adding insulation to the house, pipes and ducts; turning down the thermostat and water heater; unplugging electrical devices when not in use; washing clothes in cold water and not overdrying them; getting rid of extra fridges and freezers; and using modern dishwaters rather than handwashing dishes. In any society one can use solar power for cooking (with solar reflectors focusing sunlight on a pot) or cooking partially on a stove, and then letting the cooking complete by putting it in a well-insulated **Wonder Box** with an insulating lid.[187] Any house can use solar energy to heat water with suitable collectors on the roof, and to generate electricity via solar panels. Wind generation of electricity is also possible. A key issue is energy storage, and an amazing new technology is to do so via sand batteries.[188]

A success story that should be more celebrated is the international action taken to deal with the ozone hole that appeared in the upper atmosphere, caused by ozone-depleting substances such as halocarbon refrigerants, chlorofluorocarbons (CFCs) and halons.[189] This is problematic because the ozone layer prevents harmful ultraviolet radiation (it causes skin cancer, sunburn, blindness and cataracts) from reaching the Earth's surface. A Montreal Protocol was agreed in 1987 that banned use of CFCs,

---

[182] British Geographical Survey (BGS), The greenhouse effect.

[183] Carney, *Values: An Economist's Guide to Everything That Matters*, Chapter 12.

[184] Carney, *Values: An Economist's Guide to Everything That Matters*.

[185] Carney, *Values: An Economist's Guide to Everything That Matters*.

[186] Ehsan Masood, 'Is the COP29 Climate Deal a Historic Breakthrough or Letdown? Researchers react', *Nature*, 636 (2024), 17–18. Editorial, 'Good COPs, Bad COPs: Science Struggles in a Year of Environmental Summits', *Nature*, 636 (2024), 521–2.

[187] See the Autodesk Instructables article How to Make a Wonder Box Cooker/Cooler.

[188] What Is a Sand Battery? *Polar Night Energy* (24/10/2024).

[189] See the Wikipedia article Ozone depletion.

halons and other ozone-depleting chemical, and as a result the ozone layer has largely recovered.[190]

Finally, a key point is that we actually need the greenhouse effect because it is crucial to keeping our planet at a suitable temperature for life.[191] Without the natural greenhouse effect, the heat emitted by the Earth would pass straight out from the Earth's surface into space, and the Earth would have an average temperature of about −20°C. No life as we know it could then exist.

6.4.2 The Key Factor of Global Population Growth

One should not forget that in the end all our major environmental problems are due to global population growth. They simply would not occur if the world's population were much smaller. Estimates of the Earth's sustainable population suggest it is about 8 billion.[192] We are already at that level, causing huge environmental stress. However, UN estimates are that the global population will grow to 11.2 billion by 2100 (**Figure 6.19**), leading to much greater stress.

This is a key driver of our climate change problems. Firstly, a larger population leads to higher food production dependent on irrigation with associated evaporation, and greater use of fertilizers and energy. Secondly, a larger population needs more heating, transport, construction and so on, all accompanied by greenhouse gas emission.

Achievement of the agreed-upon SDGs may succeed in reducing this growth by improving the standard of living globally. Sociocultural viewpoints on how many

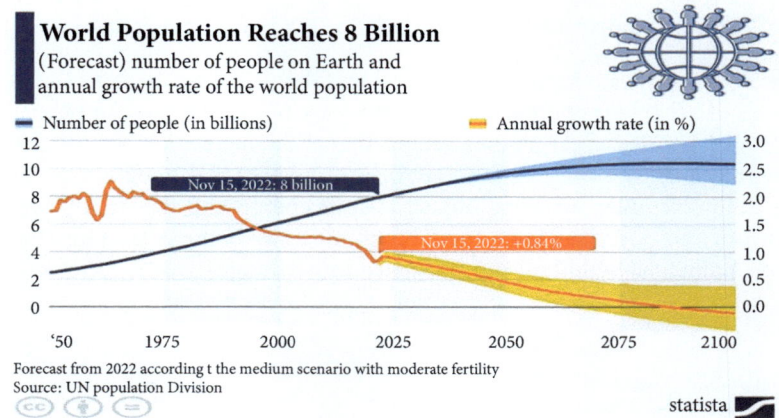

Figure 6.19 World population growth scenarios.

Image: Statista.

---

[190] See the European Environment Agency article What is the Current State of the Ozone Layer? (5 December 2024).

[191] Palmer, *The Primacy of Doubt*, 112–17.

[192] See the Wikipedia article Sustainable population.

children are desirable play a key role in this dynamic. What is needed is sexual education for girls and boys before they reach puberty, so that they understand their bodies and the changes that will occur to them, and family planning access to help reduce population growth. Also access to education is needed in general for women so they can have other roles in life in addition to child rearing.

Garrett Harden comments that the population problem has no technical solution: it requires a fundamental extension in morality, and associated commitment.[193]

6.4.3   Planetary Boundaries and Planetary Health Diet

To plan safe ways of navigating the future given all the relevant interlocking factors, one needs multiscale integrative views of the relevant factors as well as integrative views to deal with them.[194]

Planetary boundaries is a concept highlighting human-caused perturbations of Earth systems.[195] They relate biosphere integrity, climate change, novel entities, stratospheric ozone depletion, atmospheric aerosol loading, biogeochemical flows, freshwater use and land system change (**Figure 6.20**).

Planetary health diet A commission has reported on how to obtain healthy diets from sustainable food systems.[196] Complex feedback loops occur: for example, the connection between soil microbiomes and gut microbiomes means that using pesticides decreases biodiversity in both microbiomes. The complex interactions of social, political and economic factors between food and the environment are indicated in **Figures 6.21(a)** and **6.21(b)**, showing the formers' relation to the UN SDGs.[197] This integrative analysis is precisely the kind of approach that I advocate in this book.

How do we advocate for it? One method was discussed in Chapter 5 concerning the issues of behaviour and narratives, in this case occurring in a complex of interacting social, political and economic factors that affect the relation between food and the environment. The same issue arises for each activity that impinges on the environment, for example, providing industrial-scale energy, and specifically energy for agriculture; transport of all kinds, particularly air travel in the stratosphere; residential and commercial buildings, and associated urban design. Each requires key policy

[193] Harden, 'The Tragedy of the Commons'.

[194] Fleurbaey et al., 'An Analysis of Moral Motives in Economic and Social Decisions'. Anna Taylor et al., 'Climate-Resilient Development Pathways in South Africa', *South Africa Journal of Science*, 120(11/12) (2024). Anna Taylor et al., 'Operationalising Climate-Resilient Development Pathways in the Global South', *Current Opinion in Environmental Sustainability* 64 (2023), 101328.

[195] Johan Rockström et al., 'Planetary Boundaries: Exploring the Safe Operating Space for Humanity', *Ecology and Society* 14 (2009), 32. Johan Rockström et al., 'A Safe Operating Space for Humanity', *Nature*, 461 (2009), 472–5.

[196] Walter Willett et al., 'Food in the Anthropocene: The EAT–Lancet Commission on Healthy Diets from Sustainable Food Systems', *The Lancet*, 393 (2019), 447–92.

[197] Jessica Fanzo et al., 'Nutrients, Foods, Diets, People: Promoting Healthy Eating', *Current Developments in Nutrition*, 4 (2020), nzaa069. See the Wikipedia article Sustainable Development Goals.

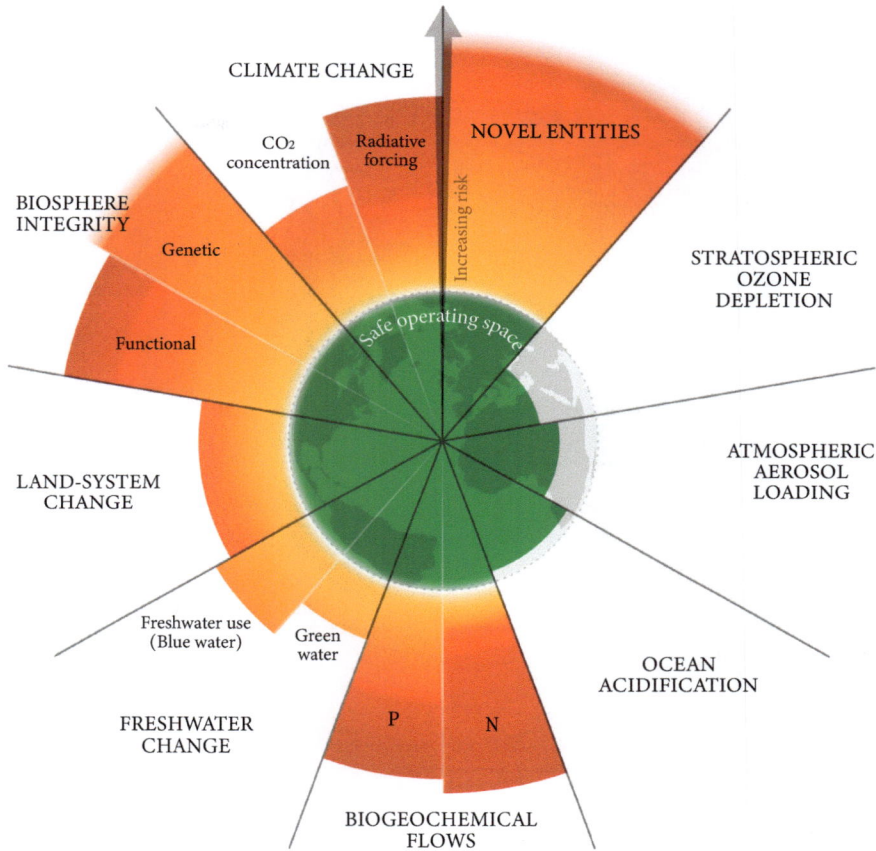

Figure 6.20 Planetary boundaries, state 2023 image from Stockholm Resilience Centre.

Source: Azote for Stockholm Resilience Centre, based on analysis in Katherine Richardson et al., 'Earth beyond Six of Nine Planetary Boundaries', *Science Advances* 9(37) (2023), eadh2458.

decisions. An overall needs-based policy that aims at providing a good life for all by targeted production and careful planning should be the goal.[198]

6.5 Technology as the Physical Basis of Society, and Its Change with Time

The rise of modern society is based in technology and the way it has changed over time.[199] We depend on technology for suitable clothing and shelter, health

---

[198] Jason Hickel and Dylan Sullivan, 'How Much Growth Is Required to Achieve Good Lives for All? Insights from Needs-Based Analysis', *World Development Perspectives*, 35 (2024), 100612.

[199] Arthur, *The Nature of Technology*. Harari, *Sapiens*, 31–3. Jacob Bronowski, *The Ascent of Man* [video] televised by BBC 5 May 1973.

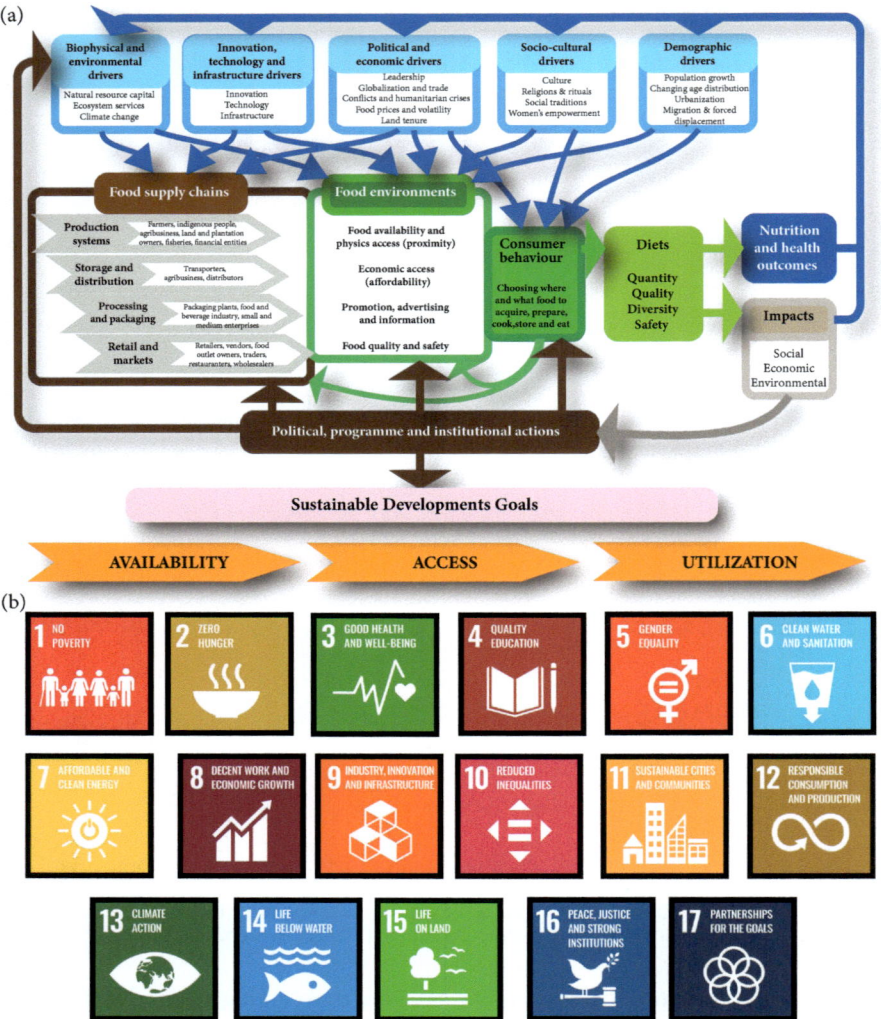

Figure 6.21 Environment and food: the interacting factors. (a) The complex interaction of factors affecting food supply chains and the environment. (b) The SDGs referenced in (a).

Image redrawn by Mauro Carfora from the version given in Jessica Fanzo et al., (2020). 'Nutrients, Foods, Diets, People: Promoting Healthy Eating', *Current Developments in 'Nutrition*, 4 (2020), nzaa069, where it is reproduced from the comprehensive HLPE Report #12—Nutrition and Food Systems

care, reliable food and water supplies, getting rid of solid and liquid waste, particularly sewage, and controllable energy. Technology also underlies manufacture of all the material things that meet the needs of daily life: furniture, stoves, refrigerators, kettles, dishwashers, washing machines, vacuum cleaners. It makes available transport of all types: bicycles, cars, lorries, trains, ships, aircraft, and communication systems, including radio and TV.

In recent times we have had the development of integrated circuits, digital computers, the Internet, cell phones and social media that taken together have transformed daily life out of recognition.

Technology is enabled by three key features. First, we are a cooperative species, enabled by our symbolic capacity. We have developed language that enables us to jointly set goals and plan how to achieve them. This is the social dimension that underlies technology. Second, we have developed mathematics and discovered how this enables us to determine the laws of nature governing physical outcomes. This is the abstract basis of technology. Third, we have discovered how to organize society in economic and engineering terms in order to achieve the physical outcomes we desire: flying overseas, eradicating smallpox, creating apps for cell phones, etc. This is the organizational aspect of technology.

In this section I look at, §6.5.1, 'The Historical Importance of Technology'; §6.5.2, 'Simple Technology'; §6.5.3, 'The Nature of Complex Technology'; §6.5.4, 'The Domains of Technology'; and §6.5.5, 'The Rise of Artificial Intelligence (AI) and Cryptocurrency'.

6.5.1 The Historical Importance of Technology

It is common for history to be taught in terms of the rise and fall of empires, the succession of kings, the rise and fall of democracy or communism and so on. But these are temporary changes that only last a century or so. The deep transformations in history are the technological ones, from the discovery of the wheel and agriculture and how to control fire, to the principles of thermodynamics, to the concept of money, to the way to control electricity to generate heat and light, the way to generate electromagnetic waves of all wavelengths to produce radios and radar, to the principles of digital computers and lasers. Similarly discovery of vaccinations, medical imaging such as CAT scans and fMRI have transformed medical care forever. Their influence has been profound across societies. But the key point is this. Once these have been discovered they are able to transform human life for ever after, for all people, no matter what the political system is, because that knowledge is publicly available and every nation can use them.

Our life today is deeply influenced by individual scientists and engineers who made these key discoveries. The lives of Napoleon, Stalin, Churchill and Mao Tse Tung were very important for a while, but they were not transformative in the same way that the lives of Isaac Newton, James Clerk Maxwell, Alan Turing, Alan Cormack, John Bardeen and Charles Townes have been.[200]

---

[200] Discoverers, respectively, of Newton's laws of motion, the laws of electromagnetism, the idea of a computer program, the idea of a CAT scanner, the idea of the transistor and the idea of the laser.

6.5.2 Simple Technology

Simple technology arises by understanding that one can use natural stuff, available in the environment, to attain outcomes one wants. Ancient technologies were making clothes, shoes and hats; constructing shelters against the weather; making fires for warmth and cooking; and making tools for cutting and grinding. The discovery of agriculture—collecting and planting seeds, watering them and protecting growing crops—led to a transformation of life, with the establishment of settlements instead of being nomads.[201] Transporting stuff led to the first important mechanical engineering discovery—wheels and axles, later greatly improved by use of ball bearings and lubrication. Further simple mechanical engineering discoveries were the lever (Figure 2.28), wedge and screw mechanism.[202] More complex were clocks and watches. Quite different were the discoveries of weaving, sewing, knitting and crocheting. Transformative were the discoveries of how to use glass to make windows, and how to make eyeglass lenses to improve eyesight (Figure 2.29). This is probably the simplest technology that makes a profound difference to millions.

6.5.3 The Nature of Complex Technology

Brian Arthur gives three definitions of technology,[203] which I summarize as follows:

A technology is a means to fulfil a human purpose. It is a combination of devices, methods and processes, the resulting entire collection of engineering practices being available to a culture to meet their needs.

He states that a technology is always organized round a central concept or principle: a phenomenon that is captured and put to use for some purpose.[204] Examples of such principles are gravity, heat, electromagnetism, chemistry, molecular biology and information.[205]

Arthur emphasizes that a complex technology involves modular hierarchical structures with a recursive nature, and is subject to evolutionary processes: it continually changes over time as ever better ways for meeting society's material needs are discovered.[206] I claim that such a structuring is inevitable, as it is the only way to solve the problem of creating purposeful complexity.[207] Technology, therefore, has a necessary structure that is similar in many ways to that of biology. At the start of this chapter I advocated that this is how to view societies. In parallel with what I said there, I now propose the following:

---

[201] See the Wikipedia article Agriculture.
[202] See the Wikipedia articles Lever, Wedge, Screw mechanism.
[203] Arthur, *The Nature of Technology*, 28.
[204] Arthur, *The Nature of Technology*, 33, 51, Chapter 4.
[205] See David Blockley, *Engineering: A Very Brief Introduction* (Oxford: Oxford University Press, 2012).
[206] Arthur, *The Nature of Technology*, 32–43.
[207] Vail, 'Notes on Running an Organization'. Booch et al., *Object-Oriented Analysis and Design with Applications*.

Technologies are structured broadly in the same way as living systems are. They are modular hierarchical structures that undergo functional, developmental and evolutionary processes on different timescales. Their functioning involves homeostatic processes, predictive processing and adaptative behaviour.

As in the case of biology, there is an essential function associated with each emergent level,[208] the whole fitting together to carry out the highest level function that is the reason the technology exists.

Functional processes shape minute-by-minute functioning, involving multilevel interactions—both upward and downward effects occur in both cases.[209] A key feature is *homeostasis*,[210] which maintains the stability of living systems facing inevitable stresses that occur due to both internal fluctuations and interaction with the external environment. Its central feature is feedback, leading to circular causation. The same effect in engineering contexts is labelled cybernetics.[211] Examples are controlling the speed of a steam engine or aircraft propellor via a governor or servomechanism, engineering process control and controlling the temperature of a room by means of a thermostat. That the same principle applies in both biology and engineering was noted by Norbert Wiener, and is made explicit in the title of his book.[212]

Predictive processing is a key way the brain functions.[213] It constantly generates predictions based on a mental model of the world, compares that prediction to incoming sensory input and updates that model accordingly. This is a key step further than homeostasis because it predicts the future and then uses this to decide action outcomes. It is common in advanced technology, for example aircraft automatic landing systems where one has to control the aircraft on the basis of predictions of where it will be in the future in the light of the current position, heading, speed, height, wind and control settings.[214] The system keeps on correcting the predictions as new data comes in regarding all these factors as wind changes and perturbs the plane's path. This process also underlies autonomous ship and automobile control systems,[215] and automation in factories and warehouses.[216]

Adaptive behaviour responds to the physical, biological, technological, social, financial, political and legal environment influencing what kind of technology can achieve desired goals.

Developmental processes take place when individual technological items—an aircraft, motor car, engine, whatever—are created out of the components that make them up. These manufacturing processes take place in a hierarchical way, making the parts first and then assembling them to create the whole. This all takes place according to

---

[208] Denis Noble, 'A Theory of Biological Relativity: No Privileged Level of Causation', *Interface Focus*, 2 (2012), 55–64.

[209] Ellis, 'Efficient, Formal, Material, and Final Causes in Biology and Technology'.

[210] See the Wikipedia article Homeostasis.

[211] See the Wikipedia article Cybernetics and W. Ross Ashby, *An Introduction to Cybernetics* (Boca Raton, FL: Chapman & Hall, 1956).

[212] Norbert Wiener, *Cybernetics: Or Control and Communication in the Animal and the Machine* (Cambridge, MA: MIT Press, 1961).

[213] See the Wikipedia article Predictive coding.

[214] See the Wikipedia article Autoland.

[215] For a specific examples see Sea machines and the Wikipedia article Tesla Autopilot.

[216] See the Wikipedia articles Automation and Industrial control system.

an abstract plan (see **Figure 6.3** and the text there). This abstract plan is in effect the DNA that contains the information for the manufacturing process. Just like DNA it can't carry out any actions by itself, but guides the construction process in the relevant context: a living cell in the case of DNA, a factory in the case of an automobile or aircraft or electric motor.

Evolutionary processes take place in the way new technologies or variants of a technology coming into being over time.[217] Variants are tried and accepted or discarded, depending on how well they work in relation to the competition. Examples are iron axes replacing stone axes, electric lights replacing gas lights, jet aircraft with pressurized cabins enabling flight in the stratosphere replacing piston engine aircraft and so on. Most of these evolutionary processes represent relatively small improvements that taken together make a large difference in outcomes. Examples in the case of an aircraft are small improvements in wing shape and its sweepback angle, improvements in turbine engine thrust, in aircraft instrumentation and so on, leading over time to a large improvement in performance. This is in close analogy to how small changes in DNA lead to relatively better survival rates of a population and so preferential selection of that DNA. However, some innovations represent the start of a completely new kind of technology based on some new scientific or technological insight. Examples are the possibility of sending and receiving radio signals, the use of nuclear energy to generate power, the use of biotechnology based on directly manipulating genes, using integrated circuits containing billions of transistors in programmable digital computers to enable the existence of laptop computers and cell phones.

This is where the analogy between biology and technology breaks down: in biology there is no mechanism whereby completely different kinds of DNA can suddenly come into existence. The mechanism that enables this to occur in technology is imagination:[218] the ability to think of totally new possibilities that have not been thought of before. To turn this into something new requires creativity:[219] the ability to see problems in new ways and escape the bounds of conventional thinking, even when this meets with resistance from others. This requires a confluence of six distinct but interrelated resources: intellectual abilities, knowledge, styles of thinking, personality, motivation and a supportive environment.[220]

Values and purpose. Finally, in accord with the view of brain function presented in Chapter 5, complex technologies also implicitly involve a world view based on purpose and values. The key question shaping our thoughts and actions in general is, what kind of problem is worth solving? What values do our solutions represent? This applies to technologies in particular: What kind of technology is worth investing in? What values do our technologies embody?

---

[217] Arthur, *The Nature of Technology*.
[218] Harold Rugg, *Imagination* (New York: Harper & Row, 1963). L. P. Harris, *The Work of the Imagination* (Oxford: Blackwell, 2000). Shen-yi Liao and Tamar Gendler, 'Imagination', *Stanford Encyclopedia of Philosophy* (Stanford, CA: Stanford University, 2019).
[219] See the Wikipedia article Creativity
[220] Robert J. Sternberg, 'The Nature of Creativity', *Creativity Research Journal*, 18 (2006), 87.

The way this all works out depends on the problems a particular technology or branch of engineering is designed to solve. These are the various domains of technology that I consider next.

6.5.4 The Domains of Technology

Following the line of thought just set out, it is not surprising that the different kinds of technology and associated branches of engineering correspond in broad terms to the different aspects of the physiology of living systems, and particularly the physiology of the human body.[221] I will distinguish four kinds of such systems. Two sets of physiological and technological systems are basic to structure and the ability to act, two are basic to ongoing functioning, two are basic to existence and two are the basis of higher-level function and action. They interact with each other, so each kind of system can play multiple roles.

6.5.4.1 Two Kinds of Systems Are Basic to Structure and the Ability to Act
- These are structural systems, and those that provide for action and movement.

6.5.4.1.1 Structural
The function required is to provide the basic structure of the system, define its boundaries and provide the ability to control what enters and exits across those boundaries.

In the case of biology, confining the discussion to vertebrates, the basic structural element is the skeletal system: a set of interlocking bones joined together by cartilage to form a backbone, skull, arms or wings, hands, fingers, legs, feet, toes and so on.[222] The boundary of the body is the integumentary system made of skin of some kind, covered by fur or hair or feather or scales, depending on the kind of vertebrate, and pores that allow perspiration.[223] The whole has various apertures through which material can enter and exit: the mouth, nose, urethra and anus.

In the case of technology, analogous elements are present to carry out similar functions. Buildings have foundations, a roof, floors, walls, doors, stairs, windows, chimneys or air conditioning systems, the whole supported by beams and columns. Aircraft have a fuselage, wings, landing gear, control surfaces, entry doors, emergency exits and an air conditioning system that brings in air from the exterior and heats it up. Automobiles have a chassis, doors and windows, wheels and suspension, maybe air conditioning. Computers have a case and chassis, input/output devices and cooling systems.

---

[221] Fanzo et al., 'Nutrients, Foods, Diets, People'. See also the Wikipedia articles Human body and Organ system.

[222] See the Wikipedia articles Vertebrate.

[223] See the Wikipedia article Integumentary System.

6.5.4.1.2 Action and Movement

The function required is to enable the system to act on the world to attain desired outcomes, and to be able to move if that is needed to do so.

In the case of biology, this ability is provided by the muscular system. It comprises skeletal muscle moving the arms, hands and fingers, legs and toes and head; cardiac muscle enabling heart function; and smooth muscle enabling digestion.

In the technological case, the ability to act is provided in three ways.

Firstly, there is using human muscle power. We use picks, spades, shovels, wheelbarrows and so on to move earth, dig trenches, make cement, etc. As regards movement, the bicycle enables great mobility and even can be used to transport stuff by using a small trailer,[224] and rowing enables getting a boat from place to place. Humans have also used horses, donkeys and oxen for mobility and the transport of goods.

Secondly, using wind as a source of power has a long tradition, mainly windmills for grinding grain but also for pumping water. Yachts of many kinds have provided transport for centuries.

Thirdly, power is provided by steam engines, internal combustion engines, jet engines, rockets, electric magnets and most importantly, electric motors.[225] These can be harnessed to provide transport by rail, road (cars, buses, lorries), water (boats) and flying. Furthermore they enable the construction of roads, railway, dams and so on by powering bulldozers, road graders and borers (tunnelling machines).

Importantly, as pointed out by Arthur, action is possible in a passive sense.[226] A dam acts when it keeps water in place (if the dam was not there, the water would flow down the valley). A bridge acts when it supports vehicles travelling across it (if it was not there, they would fall into the ravine). Such passive action takes place via physical deformation. The bridge carries the load by deforming a bit under pressure caused by a vehicle crossing over. The dam holds the water pressure in a similar way. This kind of action is clearly displayed by the way passenger aircraft wings deform as they take off. The wings then start to bear the weight of the aircraft, which on the ground was supported by the wheels. The wing shape is different in the two cases.

6.5.4.2 Two Kinds of Systems Are Basic to Ongoing Functioning

- These are metabolism and homeostasis/cybernetics.

6.5.4.2.1 Metabolism

The first function needed to enable ongoing functioning of the system is to provide for its energetic and material needs.

In the case of biology, metabolic systems are essential to our ongoing functioning. If we do not breathe, drink or eat, we die. Furthermore we need to excrete urine and faeces; otherwise, we would poison ourselves. The relevant physiological systems are

---

[224] See the Wikipedia article Bicycle
[225] Palmer, *The Primacy of Doubt*, 112–17.
[226] Arthur, *The Nature of Technology*.

first the respiratory and circulatory systems, involving the heart and lungs, and second the digestive and excretory systems, involving the stomach.

The latter takes in food of many kinds and transforms it into usable forms such as glucose (sugars to provide energy), amino acids (to make up protein) and fatty acids (to make up fats).

In the case of technology, in parallel to the biological case, we need to provide water, materials of many kinds and usable energy, and dispose of waste water, waste material and heat.

Water is provided by wells and dams, transported by pipelines and pumps to water treatment plants and reservoirs and distributed through pipes by pumps. Waste water, including sewage, is disposed of by pumping it to the sea, or by sending it to wastewater treatment plants and sewage treatment plants.[227]

Food is provided by farms of many kinds.[228] They include arable farms growing crops such as wheat or barley or corn, orchards, vineyards, coffee and rubber plantations, vegetable farms, fruit farms, paddy fields for rice, livestock farms for cattle, dairy farms, and pig, sheep, poultry and ostrich farms. Wood is provided by commercial forests. Many of these farms involve using farm machinery: tractors, trailers, threshing machines, seed planters, combine harvesters and so on. The whole enterprise depends on the availability of fertilizers based in ammonia. Food is also provided by fisheries of all kinds.[229]

Materials are provided by quarries and mines, using drills, explosives and jackhammers to extract the materials, and then lorries or trains to get it to factories or building sites. Mineral processing is the mechanical means of crushing, grinding and washing that enable the separation of valuable metals or minerals.[230] Crucial discoveries were cement, steel, plastics, paper and fabrics of many kinds.

The controllable use of **energy** has proved transformative in our history. It is provided by fire (burning wood or coal), water (hydropower), wind, natural gas, fuels obtained from crude oil or nuclear power. Electricity distribution includes the key component of transformers that change voltage.[231]

This is all standard metabolism: locating a resource, transforming it to a useful form and disposing of heat and waste. However, there is a specific example that demonstrates the core of how metabolism works: oil refineries (**Figure 6.22, left**) and fractionating columns (**Figure 6.22, right**). A key feature of modern technology is the way that crude oil from wells is separated into fractions by fractional distillation. The fractions at the top of the column have lower boiling points than those at the bottom. This enables separating the fractions out: the heavy bottom fractions are cracked into lighter, more useful products. This is an example of downward causation in at least two ways.

---

[227] See the Wikipedia articles Wastewater treatment and Sewage treatment.
[228] See the Wikipedia article Farm.
[229] See the Wikipedia article Fisheries.
[230] See the Wikipedia articles Mining and Mineral processing.
[231] See the Wikipedia article Transformer.

Figure 6.22 An oil refinery. Left: the complex as a whole. Right: the fractional distillation process that separates out oil into various products with different properties. The fractions at the top of the fractionating column have lower boiling points than those at the bottom. [232]

First, the incoming crude oil is sent to the first fractionating column and then to others in the complex; thus the structure as a whole and its control system determine the conditions for the dynamics in each column.

Second, the conditions in the column are controlled so as to shape the chemical dynamics at each level and thereby control the products at that level. Thus this is the dynamic I mentioned earlier: larger-scale structures reach down to the molecular level in order to attain desirable outcomes there.

Furthermore, plastics of many types (made of organic polymers) are also created from petrochemicals.[233] However, a crucial issue is plastic pollution,[234] and concerns about this (an issue of values) are beginning to change the dynamics of plastic production and use, for example through plastic recycling efforts.

Personal and public metabolism meet crucially in the need to provide public and private flushing toilets. That technology is a key aspect of civilization,[235] and plays an important role in well-functioning schools.[236]

6.5.4.2.2 Homeostasis/Cybernetics

- The second function needed is to counteract external or internal influences that undermine its stability. Homeostatic/cybernetic systems meet these needs. This is a key function in both cases that was discussed in §6.5.3.[237]

---

[232] See the Wikipedia article Oil refinery. Left: W. Siegmund, https://commons.wikimedia.org/ wiki/ File:Anacortes_Refinery_31911.JPGT). Right: https://en.wikipedia.org/wiki/Oil_refinery#/media/ File:Crude_Oil_Distillation-en.svg.

[233] See the Wikipedia articles Plastic and Plastic pollution.

[234] Sternberg, 'The Nature of Creativity'.

[235] Guido Corradi, 'Public Toilets are Vanishing and That's a Civic Catastrophe', *Aeon* (18 July 2024).

[236] Bill and Melinda Gates Foundation, 'How Do Better Toilets Lead to Better Attendance?' (25 March 2024).

[237] See the Wikipedia article Cybernetics and Ashby, *An Introduction to Cybernetics*.

6.5.4.3 Two Kinds of Systems Are Basic to Existence

- These are reproduction, and defence against existential threats.

6.5.4.3.1 Reproduction

- The first function required is to bring the system into existence by creating it out of its parts.

In the case of biology, this is the function of the reproductive system.[238] Confining the discussion to human beings, this is of course what the sexual organs are about, but it is a far wider issue than that. It has emotional, social, economic and legal aspects and consequences. Sometimes politics is involved as well.

Social customs and conventions usually play a big part. Similar issues arise for all social animals. The needed information to create a new human being lies in the genes—segments of DNA—which are read contextually by developmental systems according to position in the developing embryo. The biological agent that actually carries out reproduction is the living cell. All the other growth that takes place at higher levels is a consequence of cellular reproduction at the cellular level.

An important link between technology and biological reproduction is the contribution of technology to the many forms of birth control.[239] This shows how technological possibilities and imagination can completely alter the biological and social context.

In the case of technology, new stuff is created at building sites, dam sites, tunnel sites and so on. It is also created in workshops or in factories, often with assembly lines. In this case the 'DNA' is ideas and concepts, which shape both the product and the production process. In automated factories, that information will be encoded in algorithms and data. This was briefly discussed in §6.5.3, where it was pointed out that unlike the biological case, totally new ideas could arise that were not related to previous ones, and could completely alter both the production process and the product.

6.5.4.3.2 Defence against Existential Threats

The further function required is to counteract the two kinds of existential threats that face both animals and technologies: external threats and internal threats. Systems are needed to counteract both.

6.5.4.3.2.1 External Threats

In biology in general, existential threats are always there: one faces the reality of 'Nature red in tooth and claw'.[240] Animals and birds and sharks have teeth, fangs, claws and beaks which can cause severe wounds or death. This is one of the reasons life is a struggle for existence.

---

[238] See the Wikipedia article Human reproductive system.
[239] See the Wikipedia article Birth control.
[240] See the Wikipedia article In Memoriam A.H.H.

Considered as biological animals, humans are very weak: they have no chance by themselves against a lion, hyena, shark, rhino or elephant. However, through their technological capacity, they have developed weapons of many kinds (clubs, axes, knives, spears, bows and arrows, guns) that have made up for this deficiency and enabled them to gain the upper hand in the animal kingdom.

The downside is that in the technological case, humans are threatened by technologies specifically designed to cause existential threats to other people and their technologies. These include cannons, machine guns, rockets, tanks, bomber and fighter aircraft, battleships, cruisers, aircraft carriers, submarines and nuclear weapons. An ongoing arms race occurs, with each side vying for superiority. Military technology is therefore both a defence and a profound threat to our existence. Part of that technology is abstract: strategies, tactical plans, battle plans and supply plans—which is where this links to metabolism.

6.5.4.3.2.2 Internal Threats

In biology internal threats occur at both the systems and micro levels. Systems-level existential threats due to failure of physiological systems are heart attacks and strokes, suffocation or drowning and food poisoning. Cancer starts as a disease at the micro level but spreads to become a systems-level threat to life; in effect the body turns against itself as developmental systems fail. At the micro level, infectious diseases due to bacteria or viruses can be deadly, particularly flu, polio, TB, HIV/AIDS and COVID-19.[241] The adaptive and innate immune systems are there to provide protection against such diseases.[242] An important role is played by the wound healing system, because wounds allow entry of pathogens into the body.[243] It is a system at the cellular level rather than a large-scale physiological system.

A whole set of technologies exists to defend society from internal existential threats.

Firstly, regarding epidemics and medical emergencies as a threat to society, a whole variety of technologies has been developed to counter these problems. Crucially they include vaccinations for smallpox, flu, COVID-19, and other diseases that have been deadly in the past.[244] Medical technology includes a variety of small-scale protections: plasters, disinfectants, antiseptics and antibiotics, supported by thermometers and blood tests. It includes large-scale technologies: ambulances, clinics, hospitals, including intensive care units with ventilators that keep the patient breathing. This enables surgery involving anaesthesia, and crucially is supported by medical imaging: X-rays, CAT scans and fMRI which have had a transformational effect on medicine because you no longer have to open the patient up to get a good idea of what the problem is.

Secondly, there is a variety of technologies to counteract physical threats such as fires and flooding. On a small scale, protection against electric fires is provided by fuses and trip switches. Fire sprinkler systems are activated by fusible links. Helicopters and fixed wing aircraft are used to damp fires by dropping water on them.

---

[241] See the Wikipedia article Infection.
[242] See the Wikipedia articles Immune system.
[243] See the Wikipedia articles Wound healing.
[244] In 1974 the WHO adopted the goal of universal vaccination by 1990 to protect children against six preventable infectious diseases: measles, poliomyelitis, diphtheria, whooping cough, tetanus and tuberculosis.

A variety of barriers protect from flooding. On a larger scales, emergency services include fire engines with hoses, extensible ladders and water tanks.

Thirdly, there is a whole set of technologies supporting the institutions that protect against criminal activity such as robbery and murder. Police are provided with handcuffs, truncheons, pepper sprays, tasers and strong police wagons. There are surveillance systems with video cameras and often facial identification systems. On a larger scale there are secure prisons and courts.

6.5.4.4 Two Kinds of Systems Are the Basis of Higher-Level Functioning

These are systems associated with information, prediction, abstract causation and meaning; and those associated with emotional and recreational functions.

6.5.4.4.1 Information, Prediction, Abstract Causation, Meaning

The functional need is to take advantage of our intelligence and understanding as the basis of higher-level function and action.

In the biological case, this is provided by the nervous system and brain.[245] I discussed this in depth in Chapter 5. In the case of technology, the analogue of the brain is universities and research institutes.

The core of modern technology is measurement, collection and analysis of information and prediction on this basis. Its societal applications use information related to the problem the technology is designed to solve. This has already been implicit in much of what I have discussed already. Sensory systems of many kinds measure weight, temperature, pressure and speed, while radar and sonar detect vehicles and their motion. Crucial to society is the technology enabling communication: radios, TV, telephones, cell phones and digital cameras, making it possible to share images. Digital computers linked by the Internet are central to technology, allowing abstract causation and predictive processing to occur.[246] The Internet has led to the rise of global organizations that have transformed much of the economy, social interactions and politics, in ways I discuss in §6.6.3.

Abstract causation occurs via **abstract technologies** such as social agreements about the value of money supported by cash registers, contracts leading to the existence of corporations, constitutions underlying the existence of countries, laws determining what is legal and so on. It also occurs via instruction manuals explaining how specific processes should be carried out. The search for meaning is central to our higher-level functioning, and is supported by the availability of books, made possible by the printing press, and a variety of arts such as painting and sculpture, photography, films and plays. Each has an associated technology.

6.5.4.4.2 Emotional and Recreational Functions

The functional requirement is to help meet our emotional needs, including the need for recreation and relaxation. In the biological case, this is related to the innate primary emotional systems discussed in Chapter 5, particularly the play system and the

---

[245] See the Wikipedia article Nervous system.

[246] Subrata Dasgupta, *Computer Science: A Very Short Introduction* (Oxford: Oxford University Press, 2016).

ranking/hierarchy system. This results in two major branches of technology: those related to entertainment and those to sport.

Entertainment technology involves the creation of musical instruments of many kinds: violins, pianos, organs, trombones, trumpets, drums, synthesizers and so on. It also includes microphones, amplifiers, loudspeaker systems, TV cameras and screens, films, TV, YouTube and so on. Sports technology is very varied. It involves the construction of stadiums with lights and loudspeakers, sports field with field markings, goals and scoreboards. It involves making balls, bats, racquets, nets for competitive games. Recreational technology involves making surf boards, skis, ice skates and yachts; climbing ropes, pitons and bolts; gliders, paragliders and parachutes. In each case there has been enormous progress in recent decades. It also involves the construction of Big Wheels, roller coasters, carousels, Big Dippers and so on.

6.5.5 The Rise of Artificial Intelligence (AI) and Cryptocurrency

AI and its influence on society are now key concerns. They now handle such tasks as assessing visa requests for entry to a country, and whether a mortgage will be granted. However, their data sets can easily include biases such as discriminating against particular ethnic groups, so issues arise about how ethics relates to such AI programs.[247] Quite extraordinary has been the advent of publicly available programs such as ChatGPT with their abilities to write text and to generate images and videos to specification.[248] However, the programs are not always reliable. If they don't have the data needed—which depends on their training—they sometimes simply make things up. In brief, they are amazing but you can't trust them.

There are, however, areas where AI systems have made a real difference. They have an astonishing capacity to enhance discoveries, as for example in finding new materials that could be important for technology.[249] The prediction of protein structures by Alphafold has been revolutionary: indeed, it was behind the Nobel Prize in Chemistry in 2024.[250] Furthermore it seems that such systems can make highly reliable weather forecasts.[251]

Are they truly intelligent? This has been the subject of much discussion.[252] I believe that at the moment the consensus is that AI systems imitate intelligence rather than actually are intelligent. To do the latter would require the capacity of meta-analysis, which I briefly discussed in Chapter 5. They would have to make decisions on that basis. Are they conscious? Despite much hype about this, I simply don't believe they

---

[247] Bernd C. Stahl and Damian Eke, 'The Ethics of ChatGPT—Exploring the Ethical Issues of an Emerging Technology', *International Journal of Information Management*, 74 (2024), 102700.

[248] See the Wikipedia article ChatGPT.

[249] Amil Merchant et al., 'Scaling Deep Learning for Materials Discovery', *Nature*, 624 (2023), 80–5.

[250] See the Wikipedia article Alphafold. Janani Durairaj et al., 'Uncovering New Families and Folds in the Natural Protein Universe', *Nature*, 622 (2023), 646–53; Josh Abramson et al., 'Accurate Structure Prediction of Biomolecular Interactions with AlphaFold 3', *Nature*, 630 (2024), 493–500.

[251] Alix Soliman, 'DeepMind AI Weather Forecaster Beats World-Class System', *Nature*, 636 2024, 282–3.

[252] Nicolas Rouleau and Michael Levin, 'Discussions of Machine versus Living Intelligence Need More Clarity', *Nature Machine Intelligence*, 6 (2024), 1424–6; Graham Findlay et al., 'Dissociating Artificial Intelligence from Artificial Consciousness' (2024). arXiv preprint arXiv:2412.04571.

are, because that would require that they experience qualia and emotions. I can see no way that machines could have such experiences. But I could be wrong.

There are, however, two downsides to the development of AI at the current rate. The first is the associated problem of electronic waste.[253] Second, these systems use such a vast amount of energy that they are being seen as leading to a revival of the demand for nuclear power stations.[254] Thus in both cases they are a cause for environmental concern.

Cryptocurrency is, at its core, an unprecedented new way of owning and transferring assets.[255] Bitcoin, the best known, is just one tiny corner of a vast array of crypto-fuelled applications that change the game in favour of our right to hold and move value at our discretion, securely and without interference or threat. It is an extraordinary form of abstract technology, when value is derived from the difficulty of solving a mathematical problem by using a computer. There is no tangible asset associated with it. But it is playing an increasing role in economies around the world, and played a key role in the outcome of the last election in the USA because one party embraced it, and the other did not. It also uses vast amounts of energy, and a 2024 IMF working paper found that crypto mining could generate 450 million tons of CO_2 emissions by 2027, accounting for 0.7% of global emissions, or 1.2% of the world total.[256]

6.6 The Crises Facing Us Today

Part of our present-day condition is that we face a series of interlocking crises that threaten global society and local communities. I have already looked at global climate change and global population growth in §6.5. In this section I look at, §6.6.1, 'Global Pandemics:Flu and COVID-19'; §6.6.2, 'Global Organizations and Global Inequality'; §6.6.3, 'Anti-democratic Activity Powered by Social Media and Misinformation'; §6.6.4, 'Capitalism and Its Discontents', §6.6|.5, 'Other Options: Socialism and Marxism'; and §6.6.6, 'Decentralization of Agency: Local Power'.

6.6.1 Global Pandemics: Flu and COVID-19

These are expected, but unpredictable, and they have devastating global effects in both personal and economic terms.[257] They will happen again; indeed, some (like flu[258]) occur annually, but in that case the effects can be protected against by annual flu

---

[253] Isabelle Dumé, 'Generative AI Has an Electronic Waste Problem', *Physics World* (13 Dec 2024).

[254] Davide Castelvecchi, 'Will AI's Huge Energy Demands Spur a Nuclear Renaissance?', *Nature*, 635 (2024), 19–20.

[255] See the Wikipedia articles Cryptocurrency and Bitcoin.

[256] Shafik Hebous and Nate Vernon-Lin, 'Carbon Emissions from AI and Crypto Are Surging and Tax Policy Can Help', *IMF* (15 August 2024).

[257] See Project 2025 and The 180 day Playbook, 15.

[258] See the Wikipedia article Pandemic.

vaccinations: annual, because the virus mutates annually. Vaccinations demonstrably save lives, as in the case of measles,[259] polio and COVID-19 (**Figure 6.23**).

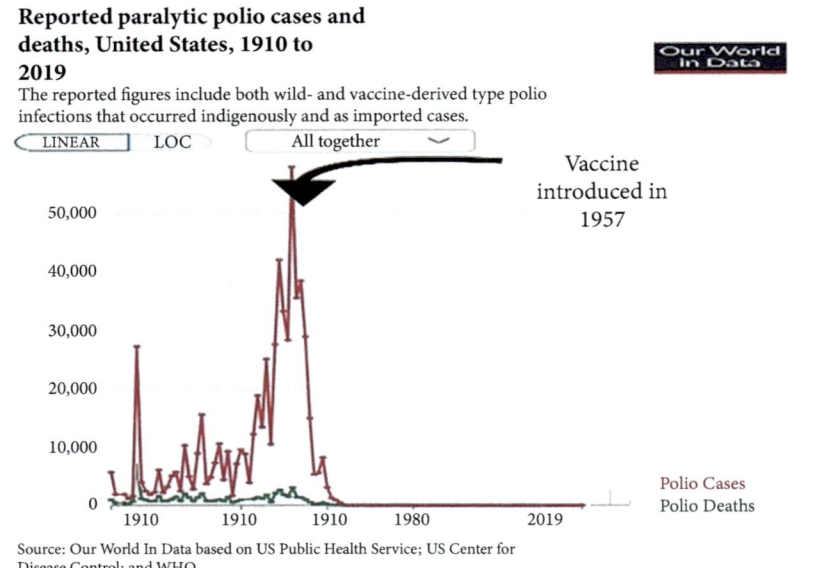

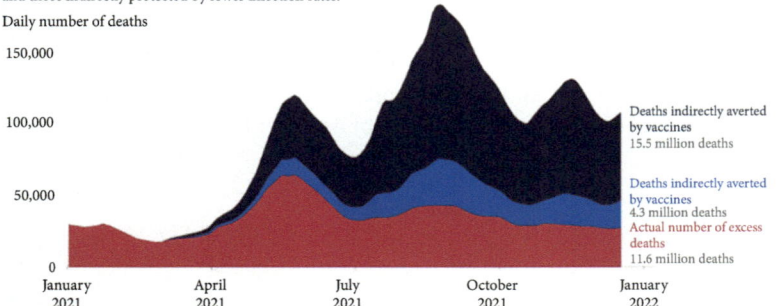

Figure 6.23 Vaccinations save lives. Left: Polio. Right: COVID-19. [260]

From Our World in Data.

---

[259] Our World in Data Measles vaccines have saved over 90 million lives in the last 50 years (2 December 2024).

[260] Our World in Data: Covid 19 Pandemic.

The use of vaccinations has been clouded by a key study claiming that they can cause autism, often referred to by anti-vaxers in their fight against vaccination. But this was a faulty study, and has at last been fully retracted.[261] The human nature of the issue is captured in this story by Roald Dahl in 1986:

Olivia, my eldest daughter, caught measles when she was seven years old. As the illness took its usual course I can remember reading to her often in bed and not feeling particularly alarmed about it. Then one morning, when she was well on the road to recovery, I was sitting on her bed showing her how to fashion little animals out of coloured pipe-cleaners, and when it came to her turn to make one herself, I noticed that her fingers and her mind were not working together and she couldn't do anything.

'Are you feeling all right?' I asked her.

'I feel all sleepy,' she said.

In an hour, she was unconscious. In twelve hours she was dead.

The measles had turned into a terrible thing called measles encephalitis and there was nothing the doctors could do to save her. That was in 1962, but even now, if a child with measles happens to develop the same deadly reaction from measles as Olivia did, there would still be nothing the doctors could do to help her. On the other hand, there is today something that parents can do to make sure that this sort of tragedy does not happen to a child of theirs. They can insist that their child is immunised against measles.

I dedicated two of my books to Olivia, the first was 'James and the Giant Peach'. That was when she was still alive. The second was 'The BFG', dedicated to her memory after she had died from measles. You will see her name at the beginning of each of these books. And I know how happy she would be if only she could know that her death had helped to save a good deal of illness and death among other children.[262]

6.6.2 Global Organizations and Global Inequality

The Internet has enabled the rise of a series of global Internet-based organizations that have been extraordinarily successful and allowed unparalleled amounts of wealth and power to be accrued by a few ultra-rich people (**Figure 6.24**). Examples include Amazon owned by Jeff Bezos; Alphabet (formerly Google) owned by Larry Page and Sergey Brin; Apple founded by Steve Jobs and Steve Wozniak; Meta (formerly Facebook) founded by Mark Zuckerberg; and Microsoft founded by Bill Gates and Paul

[261] Philippe Gautret et al. 'Hydroxychloroquine and Azithromycin as a Treatment of COVID-19: Results of an Open-Label Non-randomized Clinical Trial', *International Journal of Antimicrobial Agents*, 56 (2020), 105949. Richard Van Noorden, 'Controversial COVID Study That Promoted Unproven Treatment Retracted after Four-Year Saga', *Nature*, News (18 December 2024).

[262] 'Read Roald Dahl's Powerful Letter to Parents about Vaccination from 1988', *ScienceAlert*, 27 August 2018, available at https://www.sciencealert.com/read-roald-dahl-s-powerful-letter-to-parents-about-vaccination-from-1988.

Allen. Somewhat different is Tesla founded by Martin Eberhard and Marc Tarpenning but now with Elon Musk as CEO.[263] Other ultra-rich people have gained their wealth through chemicals, oil and so on,[264] but the new thing is the rise of these Internet organizations and the associated wealth.

Elon Musk is the extreme example. He is the richest man in the world. He has directly used his wealth politically to influence the outcome of the recent election in the USA by buying the social media platform Twitter, relabelling it X[265] and then using it to put out many posts supporting Donald Trump. He has been amply rewarded: his net worth climbed by more than $200 billion in 2024, a massive increase after he spent at least $277 million backing Trump and other Republican candidates. He is at the time of writing worth about $442 billion, and has literally bought himself a front role in running the USA by being appointed by Donald Trump to lead the Department of Government Efficiency (DOGE), which has caused immense damage both in the USA by firing thousands of people and internationally by shutting down USAID, leading to thousands of deaths. But he has not been elected to do so.

Other ultra-rich people such as Peter Thiel, co-founder of Paypal, also heavily financed the Republican Party campaign, so that Thiel's protégé J. D. Vance became vice president of the USA. Most of the billionaires mentioned above paid court to Donald Trump since his election success, and contributed large sums of money to him. Their reward came when he carried out his promise to reduce tax rates for billionaires and corporations, and reduce environmental constraints on their activities. It is estimated his proposed tax cuts will add $4.6 trillion to the national debt over the next ten years since Trump signed into law the Big Beautiful Bill, which gave the ultra-rich these promised tax breaks, on 4 July 2025.

An important point is that these organizations have become enormously wealthy because they have provided services that have improved the lives of billions of people. Google is so useful in providing information that 'Googling' has become a verb; Facebook and Twitter/X have become a major feature in billions of people's lives; and Amazon delivers a huge range of products to your front door in hundreds of countries. But there are also a range of smaller Internet-based organizations that have been transformative. Examples are AirBnB and Uber, which have made possible completely new models of providing services via apps on cell phones: renting out properties in the one case, and a taxi service available within minutes in the other.[266] Financial services available similarly include Paypal and Snapscan.[267] Entertainment services include YouTube and Netflix.[268] These are very useful services used by many.

By contrast, another way IT-based systems has led to a huge accumulation of wealth by a small number of individuals is via the trading of stocks and derivatives by quants.[269] Computers are used to track stock prices and predict futures, so that they

[263] See the Wikipedia articles Amazon (company), Alphabet, Inc, Apple Inc., Meta Platforms, Microsoft and Tesla, Inc.
[264] See the Wikipedia article High-net-worth individuals.
[265] See the Wikipedia article X Corp.
[266] See the Wikipedia articles Uber and Airbnb.
[267] See the Wikipedia article PayPal, and see SnapScan.
[268] See the Wikipedia articles YouTube and Netflix.
[269] See the Wikipedia article Quantitative analysis (finance).

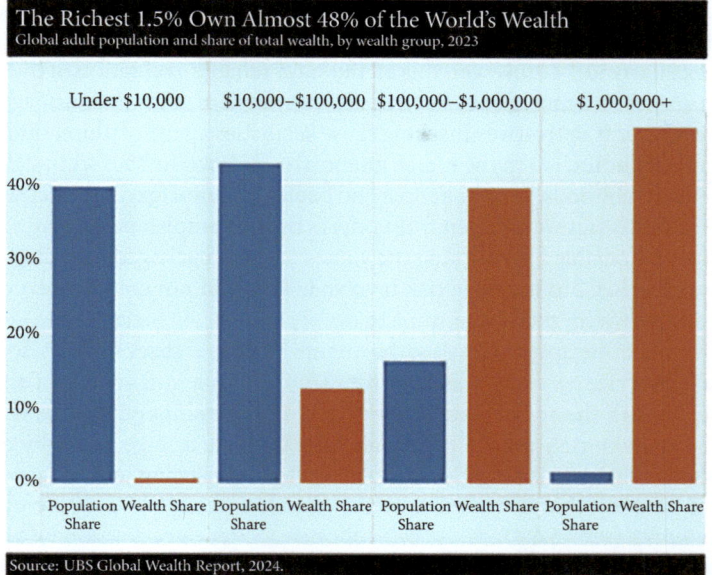

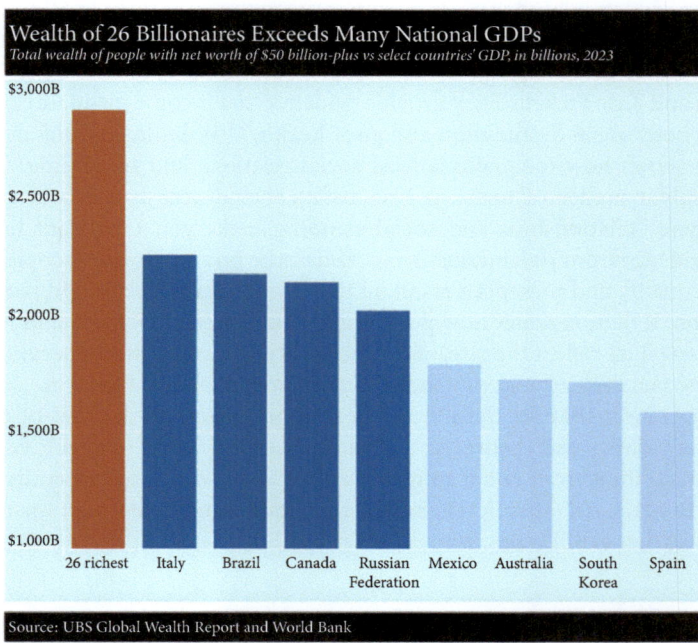

Figure 6.24   The wealth of the ultra-rich.

From https://inequality.org.

could be bought and sold in a way virtually guaranteed to make a profit. This was often done using the Black–Scholes equation: a clear example of abstract causation in the financial world.[270] However, this all relies on reliable evaluation of the worth of stocks and shares and mortgages. But a complex system of creating mortgage-backed securities and credit derivatives insuring these securities against failure, then repackaging them in bundles, led to the major financial meltdown in 2007–8 that destroyed many peoples livelihoods.[271] Poor people had been given mortgages they clearly could not afford, but this fact was hidden from buyers by the complex packaging. Crucially, the agencies responsible for verifying the values of the repackaged securities failed to do so correctly: they hid the huge risk involved. They did not competently carry out the key work they were paid large sums to do.

Furthermore, quite apart from this, the quants and other stock brokers do not provide a useful service to society as a whole. Their buying and selling of shares has nothing to do with the social value provided by the relevant companies, if any. The sole aim is to make rich people get richer, and to make a huge profit by doing so. If this results in factories being closed down and people losing jobs in some remote third world country, that is irrelevant to the financial traders. This is the way global capitalism works (see §6.6.4).

Inequality within countries is a major problem, for example in the UK where wealth is concentrated in London, and in the USA where the Rust Belt and many rural areas have been left behind. This is documented in the important book *Strangers in Their Own Land* by Arlie Hochschild.[272]

The effects of inequality have been studied by social epidemiologists Richard Wilkinson and Kate Pickett. They provide evidence that income inequality in a country lead to both social dysfunction and poor health.[273] Wilkinson's book developing the theme shows how inequality affects social relations and well-being.[274] Health is not simply a matter of material circumstances and access to health care; it is also how your relationships and social standing make you feel about life. Using detailed evidence from rich market democracies, the book addresses people's experience of inequality and presents a resultant theory of the psychosocial impact of class stratification. It demonstrates how poor health, high rates of violence and low levels of social capital all reflect the stresses of inequality and explains the pervasive sense that, despite material success, our societies are sometimes social failures.

A major issue is that for the ultra-rich, it appears that no amount of money is enough: they always want more. And in most cases they don't use their vast wealth to help others: they use it either to increase their own wealth or for vanity projects like flying to Mars. An exception to such self-aggrandisement amongst the ultra-rich mentioned earlier is Bill Gates, whose Bill and Melinda Gate's Foundation carries out

---

[270] See the Wikipedia article Black–Scholes model.

[271] See the Wikipedia article 2008 financial crisis.

[272] Arlie Hochschild, *Strangers in Their Own Land: Anger and Mourning on the American Right* (New York: The New Press, 2018).

[273] Richard Wilkinson and Kate E. Pickett, 'Income Inequality and Social Dysfunction', *Annual Review of Sociology*, 35 (2009), 493–511. Kate Pickett and Richard Wilkinson, 'Income Inequality and Health: A Causal Review', *Social Science & Medicine*, 128 (2015), 316–26.

[274] Richard Wilkinson, *The Impact of Inequality: How to Make Sick Societies Healthier* (London: Routledge, 2020).

sterling work across the world.[275] It has been criticized by some for doing so, and particularly by conspiracy theorists. That is an inevitable accompaniment of doing helpful practical work in the real world.

6.6.3 Antidemocratic Activity Powered by Social Media and Misinformation

The accumulation of extreme wealth in conjunction with the availability of social media such as Twitter/X has underlain a major rise in antidemocratic activity around the world. Although it has happened in many countries, I will concentrate on the USA because it is so clearly documented in that case.

Social media and captured narratives. The dynamic in the USA is a combination of, on the one hand, very rich people who wish to manipulate the political system in order to further enrich themselves, no matter the cost to the environment or to others, and on the other, of religious idealogues who regard themselves as superior to the rest of humanity and are determined to control the country no matter what others may want. Being able to control social media channels enables multibillionaires such as Musk to capture narratives and thereby literally buy control of the country. The ideological version is sanctioned by religious belief systems of the MAGA fundamentalist churches, as well as many Catholic churches and the machinations of Opus Dei. The nature of their hoped-for dictatorship is explicitly set out in the *Project 2025 Playbook*.[276] If carried out, it will impact all aspects of society: health, social, economic, political and legal. Neither group believes in democracy: they want to establish a dictatorship where their ideas are imposed on all the others. Peter Thiel, from the first group, is explicit that he does not believe in democracy, and it is implicit in the way Elon Musk behaves. Racism and othering is central to them. Caring extends only to your immediate circle—not to the poor or foreigner. The Good Samaritan doesn't figure in their worldview. One of the paradoxes is the way that poor people can be persuaded to vote for political parties that act against their own interests[277]—a magic worked by propaganda programmes based in misinformation and stirring up hatred of the other.

Social media information wars: systematic misinformation as a key policy. Manipulation of social media by automatic systems—bots—and the formation of echo chambers of like-minded people, leading to extreme brainwashing and ignorance, is a central feature of what has been going on. An in-depth analysis of how misinformation works is given by Roozenbeek and van der Linden in their book *The*

---

[275] See the Wikipedia articles Amazon (company), Alphabet, Inc, Apple Inc., Meta Platforms, Microsoft and Tesla, Inc. See also the Wikipedia article Bill and Melinda Gates Foundation, and e.g. Bill and Melinda Gates Foundation, 'How Do Better Toilets Lead to Better Attendance?' (25 March 2024).

[276] See the Wikipedia article Project 2025, the Project 2025 website, and Project 2025 Playbook.

[277] John Roozenbeek and Sander van der Linden, The Psychology of Misinformation (Cambridge, UK: Cambridge University Press, 2024).

Psychology of Misinformation.[278] They consider how deliberately organized misinformation('disinformation') works, leading to echo chambers, filter bubbles, and conspiracy theories.

Conspiracy theories shaped the 2024 US election: 'As voters headed to the polls this year, they confronted an avalanche of deceptive and false assertions that went well beyond matters of election integrity: Democrats are importing immigrants to vote for them illegally. Haitians are eating pets in Springfield, Ohio. Elites are controlling the path of hurricanes'.[279] Post truth politics was dominant, where the *Oxford English Dictionary* defines post-truth as 'relating to or denoting circumstances in which objective facts are less influential in shaping public opinion than appeals to emotion and personal belief'.

Misinformation by Trump and Musk. Fact-checkers at the *Washington Post* documented 30,573 false or misleading claims by Donald Trump during his first presidential term, an average of 21 per day. Telling lies was central to his 2024 election campaign.[280] Elon Musk used his power as owner of X to post many lies that made a key difference to the election outcome.[281]

In the midst of all the lies and falsehoods, where can one turn to for a reliable version of events? Someone one can solidly rely on is historian Heather Cox Richardson.[282] She has a deep knowledge of US history, and solid values based in a deep belief in democracy. Her take on the firehose of threats and false information by Donald Trump is given in this key post from 6 December 2024, after he unexpectedly threatened to take over Greenland, Canada and the Panama Canal:

> It is starting to seem like the best way to interpret social media posts from President-elect Donald Trump is through the lens of professional wrestling. Never a true athletic competition—although it certainly required athletic training—until the 1980s, professional wrestling depended on 'kayfabe', the shared agreement among audience and actors that they would pretend the carefully constructed script and act were real. . . . in the mid-1990s, wrestlers and promoters began to mix the fake world of wrestling with reality, bringing real-life tensions to the ring in what might or might not have been real. 'Suddenly', Reisman wrote, 'the fun of the match had everything to do with decoding it.' Nothing was off-limits, and the more outrageous the storylines, the better. '[F]ans would give it their full attention because they couldn't always figure out if what they were seeing was real or not.' This 'neokayfabe' 'rests on a slippery, ever-wobbling jumble of truths, half-truths, and outright falsehoods, all delivered with the utmost passion and commitment.' Reisman concluded that producers and consumers of neokayfabe 'tend to lose the ability to distinguish between what's real and what isn't.' In that, they echo the world identified by German-American

---

[278] Roozenbeek and van der Linden, *The Psychology of Misinformation.*

[279] Sarah Ellison, 'Conspiracy Theories Shaped the 2024 Election', *Washington Post* 4 November 2024.

[280] Wikipedia article False or misleading statements by Donald Trump.

[281] Anwar Ashraf, 'Fact Check: Elon Musk Spreads US Election Lies' DW, *In Focus* (2 November 2024). Kanishka Singh and Sheila Dang, 'Musk and X Are Epicenters of US Election Misinformation', Reuters (5 November 2024).

[282] Heather Cox Richardson, *Letters from an American* at Substack.com.

historian and philosopher Hannah Arendt in her 1951 book *The Origins of Totalitarianism.*[283] 'The ideal subject of totalitarian rule is not the convinced Nazi or the convinced Communist,' she wrote, 'but people for whom the distinction between fact and fiction . . . and the distinction between true and false no longer exist.'

The Big Lie. The central lie that has been promoted by Trump and his allies has been to reject the results of the 2020 presidential election: as part of their attempts to overturn this election, Trump and his allies repeatedly falsely claimed there had been massive election fraud and that Trump had won the election.[284] Crucially, this lie, part of a carefully planned strategy, was central to the January 6th insurrection.

The explicit attack on democracy: the January 6th insurrection. The violent attack on the Capitol in Washington, DC, on 6 January 2020,[285] instigated by then-President Donald Trump, was centrally motivated by repeated use of this Big Lie beforehand, using social media such as Twitter.

An extensive careful investigation of the attack was carried out by a January 6th Committee set up by then House Speaker Nancy Pelosi, despite attempts by the Republican Party to prevent this happening.[286] The result is an impressive report that investigated the insurrection in depth, interviewing many witnesses including members of Trump's staff.[287] The introduction to the report starts like this:

> In our work over the last 18 months, the Select Committee has recognized our obligation to do everything we can to ensure this never happens again. At the outset of our investigation, we recognized that tens of millions of Americans had been persuaded by President Trump that the 2020 Presidential election was stolen by overwhelming fraud. We also knew this was flatly false, and that dozens of state and federal judges had addressed and resolved all manner of allegations about the election. Our legal system functioned as it should, but our President would not accept the outcome. What most of the public did not know before our investigation is this: Donald Trump's own campaign officials told him early on that his claims of fraud were false. Donald Trump's senior Justice Department officials—each appointed by Donald Trump himself—investigated the allegations and told him repeatedly that his fraud claims were false. Donald Trump's White House lawyers also told him his fraud claims were false. From the beginning, Donald Trump's fraud allegations were concocted nonsense, designed to prey upon the patriotism of millions of men and women who love our country.
>
> Among the most shameful findings from our hearings was this: President Trump sat in the dining room off the Oval Office watching the violent riot at the Capitol on television. For hours, he would not issue a public statement instructing his supporters to disperse and leave the Capitol, despite urgent pleas from his White House

---

[283] See the Wikipedia article *The Origins of Totalitarianism*.
[284] Ellison, 'Conspiracy Theories Shaped the 2024 Election'.
[285] See the Wikipedia articles January 6 United States Capitol attack and Attempts to overturn the 2020 United States presidential election.
[286] See the Wikipedia article United States House Select Committee on the January 6 Attack.
[287] See the Senate Committee to Investigate the January 6th Attack on the United States Capitol report and the summary, Introductory Material to the Final Report of the Select Committee.

staff and dozens of others to do so. Members of his family, his White House lawyers, virtually all those around him knew that this simple act was critical. For hours, he would not do it. During this time, law enforcement agents were attacked and seriously injured, the Capitol was invaded, the electoral count was halted and the lives of those in the Capitol were put at risk. In addition to being unlawful, as described in this report, this was an utter moral failure—and a clear dereliction of duty. Evidence of this can be seen in the testimony of his White House Counsel and several other White House witnesses. No man who would behave that way at that moment in time can ever serve in any position of authority in our nation again. He is unfit for any office.

In a particularly telling event, Senate Majority Leader Mitch McConnell made this statement in the Senate: 'There is no question that President Trump is practically and morally responsible for provoking the events of that day' (the attack on the Capitol that tried to prevent the popular vote prevailing). He then refused to vote for Trump's impeachment for doing so. Holding on to political power at all costs is demonstrably more important to him than acting in a moral way. Despite its falsity, the Big Lie continues to be propagated today by many in the Republican Party with Donald Trump the President again. Indeed supporting the Big Lie is being taken as a test of loyalty. Falsehood is now central to the nature of the Republican Party and Trump supporters such as Elon Musk, who used his social platform X to propagate the Big Lie.[288]

6.6.4 Capitalism and Its Discontents

Capitalism has been a source of enormous prosperity, and provided the populations of many countries in the world with the many kinds of objects, amenities and services that make their lives safe and comfortable.[289] However, it has also been the cause of many problems, because its goals have often been seen via very narrow financial terms. Central to promoting the idea has been the influence of Nobel Prize-winning economist Milton Friedmann, who famously stated,

> There is one and only one social responsibility of business—to use its resources and engage in activities designed to increase its profits so long as it stays within the rules of the game, which is to say, engages in open and free competition without deception or fraud.[290]

This has been seized upon by businessmen to justify rapacious capitalism, but a fight back is taking place that strives to place communal values back at the centre of the functioning of the capitalist system. An example is commenting that directors should

---

[288] Heather Cox Richardson, *Letters from an American* (6 January 2025) at Substack.com.
[289] See the Wikipedia article Capitalism.
[290] Milton Friedman, 'The Social Responsibility of Business is to Increase Its Profits', *New York Times Magazine*, 32 (13 September 1970).

view themselves not as shareholders' servants, but as trustees for great institutions that serve not only shareholders but other corporate stakeholders, including customers, creditors, employees and the community.[291] A 'people-centred' economics system is possible.[292] I will turn to this after giving four examples of the damage caused by the Friedmann doctrine.

6.6.4.1 Example 1: Gun Manufacturers, the NRA and USA Congressmen

Gun makers in the USA make a huge profit by manufacturing assault rifles designed for the military, but in most states are sold to anyone whatever without any background check.[293] The buyers are often allowed to carry these weapons openly in public places such as stores because of 'open carry' laws. There is no reason that the general public should own weapons such as an AK-47, which can fire 600 rounds per minute.[294] If they really need deadly weapons for self-defence, a non-automatic handgun will more than suffice. The National Rifle Association (NRA) pays large sums of money to Republican Congressmen to ensure that the law stays this way, no matter what the public at large wants.[295] The result is hundreds of gun deaths per year in school classrooms while gun manufacturers and congressmen make their profits.[296] This is the Friedmann doctrine writ large. Whatever the law says, it would be possible for gun manufacturers to impose such checks if they had any concern about the outcome of their sales. They choose not to do so. School children, teachers and parents pay the deadly price.

6.6.4.2 Example 2: Denial of Health Care

Because there is no generally available public health care system in the USA, adequate health care is only available through health insurance schemes. However, in order to maximize profit, despite medical aid contributions being up to date, these schemes often simply don't pay for the needed health care when it is needed—if they pay it at all.[297] They operate under a policy of deny and delay, resulting in debt and often premature death.[298] This predatory practice has been confirmed by two commissions that have looked into it.[299] In December 2024, Brian Thompson, CEO of UnitedHealthcare (UHC), was shot and killed by Luigi Mangione because of such denials of health

---

[291] Lynn Stout, 'The Dumbest Business Idea Ever: The Myth of Maximising Shareholder Value', *The European Financial Review* (April/May 2013).

[292] Milton Friedman and the shareholder value myth, *People Centred Economic Development* (13 May 2023).

[293] See the Wikipedia articles Gun Law in the United States and Gun Laws in the United States by State.

[294] See the Wikipedia article AK-47.

[295] Wikipedia, List of congressional candidates who received campaign money from the National Rifle Association.

[296] Wikipedia: Gun violence in the United States.

[297] "United Healthcare has faced scrutiny, criticism and lawsuits over its denial of medical claims". *The New York Times* 5/12/2024.

[298] Michael Sainato, 'Delays, Denials, Debt and the Growing Privatization of Medicare', *The Guardian* (3 June 2024).

[299] Christ A. Grimm, 'High Rates of Prior Authorization Denials by Some Plans and Limited State Oversight Raise Concerns about Access to Care in Medicaid', Managed Care Office of the Inspector General (July 2023), OEI-09-19-00350. Richard Blumenthal, 'Senate Permanent Subcommittee on Investigations Releases Majority Staff Report Exposing Medicare Advantage Insurers' Refusal of Care for Vulnerable Seniors' (17 October 2024).

claims. Under Thompson's leadership, UHC's profits increased from $12 billion in 2021 to $16 billion in 2023.[300] Profits are being made by choice of predatory policies that violate the agreement made with the clients and are known to cause distress, poverty and deaths.

6.6.4.3 Example 3: Boeing 737 Max

The Boeing Company has an amazing record of building jet passenger aircraft, particularly the Boeing 747 double deck airliner which transformed long haul flights around the world.[301] But it also built a series of very successful short-haul jetliners, particularly the Boeing 737.[302] The problem began when Boeing merged with McDonnell Douglas in 1997 and a clash of corporate cultures ensued: 'Boeing's engineers and McDonnell Douglas's bean-counters went head-to-head and the latter won.' Consequently, in 2001 Boeing's corporate headquarters were moved from Seattle, where the engineers were based and the aircraft were built, to Chicago. Engineers were replaced by accountants at corporate headquarters, and the corporate culture changed from making the best engineering product to maximizing profit. One way this happened was that as later improved versions of the Boeing 737 aircraft were designed, the same designation (737) was used so that the expense of new type rating validation by pilots was avoided—even though the aircraft was in reality evolving into something significantly different.

The 737 MAX first flew in 2016 and was certified in March 2017. However, because of a more powerful engine and its higher mounting, there was a tendency for the aircraft to pitch up in certain flight circumstances. A **Maneuvering Characteristics Augmentation System** (MCAS) that was intended to make the aircraft mimic the flight characteristics of previous Boeing 737 versions was introduced.[303] When a single angle of attack sensor indicated that the angle was too high, MCAS would automatically move the horizontal stabilizer in a nose-down direction and prevent a stall happening. The company claimed that this eliminated the need for pilots to have simulator training on this specific aircraft type. Presumably having only one angle of attack sensor was another move to save costs—despite the dangers this created when the single detector got stuck or iced up. Incredibly, with FAA approval, Boeing removed references to MCAS from both the flight crew operations manual and quick reference handbook.

This is outrageous. **The pilots did not know the system existed and that it could take away control of the aircraft from them**. Consequently they were not trained to handle a system they did not know was there. The result was that 346 people died in two tragic accidents before the aircraft was grounded.[304] In July 2024, Boeing pleaded guilty to criminal charges regarding the fatal accidents. This is a clear example of ethical failure.[305] The accidents and grounding cost Boeing an estimated $20 billion

---

[300] See the Wikipedia article Killing of Brian Thompson
[301] See the Wikipedia article Boeing.
[302] See the Wikipedia articles Boeing 737 and Boeing 737 MAX.
[303] See the Wikipedia articles Maneuvering Characteristics Augmentation System and Boeing 737 MAX groundings.
[304] Wikipedia: Boeing 737 MAX groundings.
[305] Katie Shonk, 'Learning from Ethical Leadership Failures at Boeing', Program on Negotiation, Harvard Law School (17 December 2024).

in fines, compensation and legal fees, with indirect losses of more than $60 billion from 1200 cancelled orders. They are trying to recover by changing back the corporate culture. But the dead remain dead. Their families mourn.

6.6.4.4 Example 4: Environmental Regulations

Erin Brockowich assisted attorney Edward L. Masry in a legal suit against Pacific Gas & Electric Company, alleging contamination of drinking water with cancer-causing hexavalent chromium in the southern California town of Hinkley, resulting in severe illnesses for many.[306] In 1996 the case was settled for US$333 million plus a further $295 million a bit later.

Waters downstream from 60 industries discharging waste from Baton Rouge to the Mississippi River's mouth in New Orleans have had high concentrations of 66 chemicals and toxic metals. Consequently, this area of Louisiana became known as 'Cancer Alley'.[307] Heather Cox Richardson explains:[308]

> Congress set out to safeguard the lives of Americans from toxins released by corporations into the nation's water supply. The Safe Drinking Water Act, the first law designed to create a comprehensive standard for the nation's drinking water, was Congress's answer.
>
> The new law dramatically improved the quality of drinking water in the U.S., making it some of the safest in the world. [. . .] When he signed the Safe Drinking Water Act in 1974, President Ford added simply: 'Nothing is more essential to the life of every single American than clean air, pure food, and safe drinking water.'

Richardson also points out:

> President-elect Donald Trump has vowed to increase production of oil and gas—although it is currently at an all-time high—and such projects are often slowed by environmental regulations. On Tuesday, December 10, he posted on social media, "Any person or company investing ONE BILLION DOLLARS, OR MORE, in the United States of America, will receive fully expedited approvals and permits, including, but in no way limited to, all Environmental approvals. GET READY TO ROCK!!!"[309]

In each of the cases outlines here, capitalist imperatives are knowingly being pursued even to the extent that they lead to deaths. This is going to get worse in the USA, because of the policies of Donald Trump, the Republican Party and his billionaire friends. This puts this kind of capitalism squarely in **value state V3** in the spectrum of values (**Figure 6.11**).

---

[306] See the Wikipedia article *Erin Brockowich*.
[307] See the Wikipedia article Cancer Valley, and Human Rights Watch 'Louisiana's "Cancer Alley": Dire Health Crisis From Government Failure to Rein in Fossil Fuels' (25 January 2024).
[308] See Heather Cox Richardson, *Letters from an American* (15 December 2024).
[309] Richardson, *Letters from an American* (15 December 2024).

Can capitalism be fixed? Is capitalism irremediably flawed, or can it be fixed so? If so how? This is discussed by Paul Collier and Colin Mayer.[310] Each of them believes that capitalism is central to driving a thriving economy, and that it can be reformed so as to work better for the common good. The foundation is this:[311]

> Consider first Friedman's erroneous belief that shareholders 'own' corporations. [. . .] [C]orporations are *legal entities that own themselves,* just as human entities own themselves. What shareholders own are shares, a type of contact between the share-holder and the legal entity that gives shareholders limited legal rights. In this regard, shareholders stand on equal footing with the corporation's bondholders, suppliers, and employees, all of whom also enter contracts with the firm that give them limited legal rights.

Thus shareholders are just one of the interested parties involved in a corporation. Furthermore, the corporation is responsible to behave ethically in relation to the society in which it operates. That is the burden of Colin Mayer's book:[312]

> We have forgotten why we are creating and running businesses. It is not to profit. It is to solve problems. To solve problems that you, I, societies, and the natural world face. And in the process to produce profits. And profits are essential. They are the lifeblood of business [. . .]
>
> What is needed is to have ethical capitalism that primarily aims to produce a public good, The Moral law is that we should only profit from producing solutions to problems of others and not profit from imposing detriments to others. It derives from our purpose, our reason why we exist and are created—to solve each other's problems, to do so profitably and not to profit from producing problems.

This agrees with Mark Carney in his book *Values: An Economist's Guide to Everything That Matters*, where Chapter 13 discusses 'Values-Based Leadership', Chapter 14 'How Purposeful Companies Create Value' and Chapter 15 'Investing for Value(s)'.[313]

But how does that work in practice? It needs a clear choice of positive values, such as a commitment to Meyer's **reformulated Golden Rule**,[314] namely

> Do unto others as they would have done to them.

This has to be incorporated into corporate leadership and culture. Three specific issues then arise.

Firstly, there must be a commitment to limiting environmental damage caused by a firm's operations as much as possible, and to restore damaged environments after operations cease. This requires, as pointed out by Mayer,[315] that any environmental

---

[310] Paul Collier, *The Future of Capitalism: Facing the New Anxieties* (New York: HarperCollins, 2018). Mayer, *Capitalism and Crises.* Mayer, *Prosperity.*
[311] Stout, 'The Dumbest Business Idea Ever', 4.
[312] Mayer, *Prosperity*, 73–4.
[313] Carney, *Values: An Economist's Guide to Everything That Matters.*
[314] Mayer, *Capitalism and Crises.*
[315] Mayer, *Prosperity.*

costs are included in the firm's balance sheets so that the true costs of operation are clear. Suitably structured accounting systems are central to success.

Secondly, ruthless monopolistic practices which lock other firms out from competing need to be curbed. This is both because they have an equal right to compete to provide the relevant services and because it prevents price gouging such as in the case of the provision of insulin in the USA.

Thirdly, there should be an agreement that everyone who either is employed or is wealthy pays a reasonable share of taxes in order to support the common good. Apart from anything else, capitalists benefit from public facilities paid for by taxes: roads, traffic lights, national parks, air traffic controllers, etc.

Ideally these would arise internally in a reformed capitalist system where this is understood to be what capitalism, properly conceived, is about. However, this will not always occur, so they have to be enforced by appropriate laws passed by every government that desires to work for the good of all people in the country. In practice it must be fought for in the face of the political power of the multibillionaires, who can buy senators, presidents and judges. That is where the political, legal and moral struggle arises today.

Positive. Are there examples of firms where positive values of the kind I advocate hold? Indeed there are. Colin Mayer gives as examples Novo Nordisk, manufacturer of insulin; the German bank Handelsbanken; and the Indian company Mahindra with its approach to the used car market.[316]

I add to this list the employee-owned Scott Bader Commonwealth[317]—a chemical company that makes structural adhesives, composites, functional polymers and 3D printing resins. Their website proclaims,

Our Commonwealth is our difference

The Scott Bader difference is underpinned by the fact we are a Commonwealth, which means we are owned by our colleagues, who as Trustees of Scott Bader Commonwealth Ltd, are responsible for the wellbeing of the business.

This rare ownership structure brings great benefits:

- The business is totally independent and cannot be taken over or sold, creating more sustainable business relationships.
- Colleagues have an instrumental role in running the business, a role that is much greater than their everyday tasks and empowers them to learn and grow.
- Our charitable status brings a responsibility to the wider community, to make a positive difference in everything we do.

It has a charitable wing that gives out grants. I very much approve of the idea of commonwealth. It is like the idea of **ubuntu** that I mentioned in relation to the spectrum of values. The relation of the organization to capitalism is laid out in a brief document: *The True Significance of the Scott Bader Commonwealth.*[318]

---

[316] Mayer, *Prosperity*.
[317] *See Quakers in the World, 'Scott Bader Commonwealth'.*
[318] The True Significance of the Scott Bader Commonwealth (4 January 2021).

The existence of the organization is a proof of principle: commercial organizations with such positive values can indeed exist and prosper.

6.6.4.5 Organizations that Are Not Businesses

A key point is that there are many organizations out there that are not businesses: their whole purpose may be to improve the lives of others, or for cultural purposes, sport, religious purposes or building community.[319] Many of the expenditures at level 1 in the QOL feedback system (**Figures 6.4 and 6.5**) are of this nature. Mayer's point about the importance of profit for businesses does not apply to them: they just need sufficient money coming in to keep going. Their motive is not financial; it is service.

Of course some organizations masquerade as something they are not: many 'churches' are actually money-making political organizations. They need to be recognized for what they are, and taxed.

6.6.5 Other Options: Socialism and Marxism

Capitalism by itself is not an adequate foundation for a state or nation that has a caring nature. If one wishes to ensure an adequate quality of life for all citizens, no matter who they are, some aspects of socialism need to be incorporated.[320] This might perhaps be called enlightened capitalism. There are many kinds of common facilities that need to be provided and maintained in any country that aims to look after the welfare of its inhabitants. However, in my view, full-blooded socialism is unlikely to do the job. The core issue is that in the capitalist system, decision-making and agency are assigned to the populace at large, whereas in socialism proper it is assigned to a centralized bureaucracy, with all the problems that then occur. While some bureaucrats are very helpful, there are far too many who are not. This is due to poor training, lack of motivation or engagement in corrupt practices that can so easily occur in this context. If you control public expenditure, you are open to corruption. This has been all too clear in South Africa, where corruption has reached industrial scales in recent decades.

What are the aspects of society where something like socialism is needed? I suggest there are five. First is the issue of health. The state should provide some kind of public health service and prevent price gouging by pharmaceutical manufacturers as in the case of insulin in the USA. The contrast between the National Health Service in the UK and the health situation in the USA could not be more dramatic.[321] The USA is the only developed country in the world without a universal healthcare system. Many people lack adequate health insurance to compensate, and proper medical treatment is not forthcoming when it is needed (see §6.6.4.2). Consequently, medical debt is rife:

---

[319] Kenneth E. Boulding, 'Notes on a Theory of Philanthropy', *Philanthropy and Public Policy*, NBER (1962), 57–72. National Academy of Science, *Launching Lifelong Health by Improving Health Care for Children, Youth, and Families* (2024). Kenneth E. Boulding, 'Toward the Development of a Cultural Economics', *Social Science Quarterly* (1972), 267–84.

[320] See the Wikipedia article Socialism.

[321] See the Wikipedia articles National Health Service and Healthcare in the United States.

it has been found by a 2009 study to be the primary cause of personal bankruptcy in the USA.[322]

Second is the issue of education. Every child in every nation has a right to quality education, as this is the foundation of welfare and prosperity throughout life. Private schools provide this for the wealthy, but every child should have access to adequate public education in a caring state. Particularly important is early childhood education, because this is the foundation of all future education.[323] Crucial are learning to read and learning basic mathematics (particularly compound interest relating to loans). A competitive nation will have adequate secondary and tertiary education available to everyone. This education should promote foundational and innovative thinking, not rote learning.

Third is the issue of public housing. Each person needs some kind of home—a place that provides physical shelter with basic services (water, flush toilets, electricity and now the Internet). If, for some reason, they cannot provide it for themselves, the state needs to intervene to prevent homelessness. Now there is key danger here: promoting a feeling of entitlement. 'We'll just sit back and wait while you provide us with all our necessities.' But it does not have to be like that. Housing policy can be shaped to promote the agency of those involved. There are vast squatter communities living in shacks in most third world countries, particularly in Cape Town. The viable policy to tackle this is to provide the infrastructure needed to build a simple house—a concrete platform, water, flush toilet and an electric connection. The squatters can construct the houses themselves, particularly if provided with training in brick laying, plastering, roofing, safe electrical wiring and so on.[324]

Fourth is the issue of some kind of income in order to survive in the case of those who do not have adequately well-paid employment. This leads to the provision in some countries of a social welfare grant, which provides the support needed for survival. However, it must be acknowledged that this is often problematic, as it can discourage the recipients from looking for work and making a positive contribution to society. Consequently the provision of such a grant is sometimes conditional on proof that the applicant is actually looking for work. In the end this is an issue where social attitudes matter: how local communities perceive such grants and their recipients. It seems to me that where unemployment is rife, for example because coal mines or manufacturing plants have been closed down, some kind of grant should be provided in a caring society. The way it is done needs to be carefully negotiated during the political process that leads to its implementation.

In my country of South Africa, there are a huge number of unemployed people, and in many cases they don't have the education needed to do anything but menial work. In that case, I strongly advocate a **universal bsasic income** (UBI), also known

---

[322] David Himmelstein et al., 'Medical Bankruptcy in the United States, 2007: Results of a National Study', *American Journal of Medicine*, 122(8) (2009), 741–6.
[323] See the Wikipedia article Early childhood education.
[324] David Dewar and George Ellis, *Low Income Housing Policy in South Africa with Particular Reference to the Western Cape* (Urban Problems Research Unit, University of Cape Town, 1979).

as a **basic income grant**.[325] The idea is that every single citizen, no matter how old or young they are, be provided with a small but steady amount of money each month, with no preconditions whatsoever. This can be done, for example, by a microchip in an ID card, which can be used to draw cash or to pay for goods directly in a shop. Because of the small size of the grant it will make no difference to the rich, and can be clawed back from them via their income tax payments. But for the very poor, this can make all the difference in the world. In a family with two infants and two children, there will be available a total of six UBI payments each month, which can save the family from total destitution. It can be used to go to town to look for work, to start a small business such as selling chickens, to buy seeds to plant vegetables or material to sew and so on.

This is likely to be met with howls of indignation from the rich, who do not have the compassion required to endorse such a move. But they too will benefit directly via reduced rates of crime by people who no longer have to steal in order to survive. The idea has been tried in one or two places, and it has been experimentally shown that the funds mainly do not get wasted.[326] Of course it takes a large sum of money to finance it nationwide, but the key point is this: **the money doesn't vanish**. It gets circulated throughout the country thereby creating a multiplier effect. It stimulates small businesses in remote rural areas where every individual in every family in a small village gets such an income. I regard the idea as so important that in **Appendix B,** I reproduce a representation made to the South African Government by the South African Yearly Meeting of the Quakers.

The crucial point is this: this is the only possible action that would completely transform the lives of all the poorest people in the country within a year.

Fifth is public ownership of essential services. Provision of water, electricity, waste removal, sewerage facilities and transport such as buses and trains may be done by either private companies or public organizations such as municipalities. There can be problems with either. The private company will aim to provide a good service so as to attract customers, but may engage in price gouging, and can do so if they have a monopoly. Furthermore it is a sad fact that private water companies have been releasing raw sewage into lakes, rivers, and the sea in the UK.[327] This is completely unacceptable: the ugly face of capitalism is evident as the owners profit on the basis of destroying recreational facilities and even the health of others. This will be less likely to happen if the facility is publicly provided, but in that case the service may be low quality because a publicly owned entity does not have to make a profit, so there is no incentive for the employees to provide a high-quality service. An example of the decay that can result is the state of the Johannesburg public library as I write.[328] Usually

---

[325] Guy Standing, The Case for a Basic Income, The Great Transformation Initiative (2020). Guy Standing, *Basic Income as Common Dividends: Piloting a Transformative Policy* (London: The Progressive Economy Forum, 2019).

[326] Guy Standing, *Basic Income: And How We Can Make It Happen* (London: Penguin UK, 2017). Nyasha Trivayi et al., *Cash transfers: Past, Present and Future Evidence and Lessons Learned from the Transfer Project Innocenti Research Brief* (New York: Unicef, 2021).

[327] BBC, How much sewage is released into the sea and rivers in England? (7 October 2024); Shaoni Bhattacharya, 'The UK's Rivers Are Riddled with Sewage Pollution—New Wetlands Could Help Clean Them Up', BBC (5 July 2024).

[328] Rob Rose, 'Johannesburg Library Symbolises City's Decay', *Financial Times* (4 January 2025).

the decision to privatize is made politically, as when water or transport services are privatized. In that case public pressure and monitoring is essential in order to ensure that quality is maintained. Such a monitoring process can easily be arranged.

6.6.6 Decentralization of Agency: Local Power

Decentralization of agency is the central problem of organizations: how to get cooperative action between independent actors with agency. This is discussed by Stafford Beer in *Brain of the Firm* and *Platform for Change*,[329] and in depth by Jensen and Meckling in their foundational paper.[330] That paper inter alia talks about the key concept of *agency relationships* as follows:

> We define an agency relationship as a contract under which one or more persons (the principal(s)) engage another person (the agent) to perform some service on their behalf which involves delegating some decision making authority to the agent.

They continue:

> If both parties to the relationship are utility maximizers, there is good reason to believe that the agent will not always act in the best interests of the principal. The principal can limit divergences from his interest by establishing appropriate incentives for the agent and by incurring monitoring costs designed to limit the aberrant activities of the agent. In addition in some situations it will pay the agent to expend resources (bonding costs) to guarantee that he will not take certain actions which would harm the principal or to ensure that the principal will be compensated if he does take such actions. However, it is generally impossible for the principal or the agent at zero cost to ensure that the agent will make optimal decisions from the principal's viewpoint. In most agency relationships the principal and the agent will incur positive monitoring and bonding costs (non-pecuniary as well as pecuniary), and in addition there will be some divergence between the agent's decisions and those decisions which would maximize the welfare of the principal.
>
> The problem of inducing an 'agent' to behave as if he were maximizing the 'principal's' welfare is quite general. It exists in all organizations and in all cooperative efforts—at every level of management in firms, in universities, in mutual companies, in cooperatives, in governmental authorities and bureaus, in unions, and in relationships normally classified as agency relationships such as those common in the performing arts and the market for real estate.

But the assumption made here is that all concerned are nothing but utility maximizers. In terms of the spectrum of values (**Figure 6.13**), this amounts to the assumption that all concerned would be located at most at **value state V = +2**. But that need not

---

[329] Beer, *Brain of the Firm*. Beer, *Platform for Change*.
[330] Jensen and Meckling, 'Theory of the Firm'.

be the case. Indeed Colin Mayer's book proposes it is realistic to advocate for **value state V = +3**.[331] In my discussion of **Figure 6.13**, I propose one could even go better, as suggested by Stucke and Ezrachi.[332] That is what transformational leadership is about, as explained by James McGregor Burns.[333] It is related to sense-making in organizations:[334] if the leader can instil a sense of important purpose and sense-making, the members of the organization can be lifted to heights of motivation that transcend the kind of utility maximization discussed in the quote above. In short, not everything has to be motivated by profit, as emphasized by Martha Nussbaum in her important book *Not for Profit: Why Democracy Needs the Humanities*.[335] A specific example of the successful decentralization of agency in a commercial organization is Mayer's discussion of how it is done by delegation in Handelsbanken.[336] This is a positive example and a role model.

I close with two key examples where decentralization is not taking place, with very negative results. The first is the practice of **scripted teaching**.[337] Here some education authorities lay down to the finest detail what will be taught and when and for how long in a primary school classroom, often in relation to the context of phonics lessons. They involve 'highly structured lessons, often with specific time allotments for teaching specific skills, and often word-for-word scripts of what the teacher is to say'.

This is a classic example of refusing to allow either the child or the teacher any autonomy. Both are expected to behave like robots, never deviating from the prescribed script. The project completely fails to relate to what might be going on inside the child's head. The presumption involved is extraordinary—children are little machines whose understanding and capabilities can be predicted to the last detail; we don't have to waste time trying to determine what they actually understand. This soul-destroying methodology allows teachers no agency—so they are resigning in droves. The children are bored out of their minds. If any of them displays some initiative, it will be stopped. This is in strong contrast to the approach to education laid out in **Appendix A**.

The second is the current power of centralized bureaucrats in the UK. They are based in London but have been given the power to determine details of what officials should do in far-off places like Birmingham or Nottingham or Norwich. They don't know important details of what is happening there, as reporting systems can only deal with a finite number of items, and those alone are what are brought to the bureaucrat's attention. Nevertheless they are given the power to determine the details of action and outcomes in far-off places. This is a key reason so many rural people in the UK are disaffected: it results from them being left behind.[338] There is a great need to devolve

---

[331] Mayer, *Capitalism and Crises*.
[332] Stucke and Ezrachi, *Competition Overdose*.
[333] Burns, *Leadership*.
[334] Karl Weick, *Sensemaking in Organisations* (London: Sage Publications, 1995).
[335] Martha Nussbaum, *Not for Profit: Why Democracy Needs the Humanities* (Princeton, NJ: Princeton University Press, 2012).
[336] Mayer, *Capitalism and Crises*.
[337] See the Wikipedia article Scripted teaching.
[338] Paul Collier, *Left Behind: A New Economics for Neglected Places* (London: Public Affairs, 2024).

power and agency to the periphery, as explained by Collier, and in broad agreement with the understandings of Staffor Beer.[339]

6.7 The Great Struggle: Values and Views of Humanity

This chapter has been about the central theme of this book: it is our values that ultimately shape our individual and joint actions, and so are foundational to what kind of society we live in.[340] They do so by shaping our goals in the QOL feedback system (**Figures 6.5 and 6.6**).

The big scene is set by Peter Godfrey Smith in his book *Living on Earth: Life, Consciousness and the Making of the Natural World*, writing about how our consciousness and agency arose and the responsibilities that come with them.[341] Moral issues arise, and an article that deals with this from a similar view to mine is Kenneth Boulding's *Economics as a Moral Science*.[342] He emphasizes how standard economic theory is based on a view that all human beings are purely selfish. But this is not always the case, and one can move economics toward a theory that also takes positive values into account.

Boulding also provides a well-thought out large-scale view of how society works in the paper 'The Relations of Economic, Political and Social Systems' and book *Economics as a Science: Morality, Ecology, Evolution, Sociology, Mathematics*.[343] His integrative view of the relation between economic and organizational contexts is similar to that given in this book, although developed differently. The chapters are 1, 'Economics as a Social Science'; 2, 'Economics as an Ecological Science'; 3, 'Economics as a Behavioural Science'; 4, 'Economics as a Political Science'; 5, 'Economics as a Mathematical Science'; 6, 'Economics as a Moral Science'; and 7, 'Economics and the Future of Man'.

How this all works out is greatly influenced by the leadership in a society. As discussed by James MacGregor Burns, one can have transactional leadership (Part IV of his book) or transformational leadership (Part III).[344]

Transactional leadership follows the classic economic model: leaders relate to followers by exchanging things: jobs for votes, subsidies for campaign contributions. Because it is transactional and in the end based in economics alone, it makes the mistake of equating value with prices. That which is not priced—nature, community, diversity—is not valued. Furthermore if we value the present over the future, then

---

[339] Beer, *Brain of the Firm*. Beer, *Platform for Change*.

[340] Ellis and Noble, 'Economics, Society, and the Pre-eminent Role of Values'.

[341] Peter Godfrey Smith, *Living on Earth: Life, Consciousness and the Making of the Natural World* (Glasgow: William Collins, 2024), reviewed by Alan Love, 'Great Power and Great Responsibility: How Consciousness Changes the World', *Nature*, 636 (2024), 32–34.

[342] Kenneth E. Boulding, 'Economics as a Moral Science', *The American Economic Review*, 59(1) (1969), 1–12.

[343] Kenneth E. Boulding, 'The Relations of Economic, Political and Social Systems', *Social and Economic Studies*, 11 (1962), 351–62. Kenneth E. Boulding, *Economics as a Science: Morality, Ecology, Evolution, Sociology, Mathematics* (London: McGraw Hill, 1970).

[344] Burns, *Leadership*.

we are less likely to make the investments today that will reduce risk tomorrow. This functioning is enabled by markets, which Carney calls 'the most powerful instrument we have ever created'.[345] This is because it is a powerful form of decentralization of agency. He elaborates, 'markets don't have values, people do . . . and we must close the gap between what we value and market prices', and develops this in more depth on in the pages that follow.[346] The ethical value represented is that we can work together to further our personal interests, and should do no more. Burns discusses in his book's Part IV the following aspects of transactional leadership: opinion leadership, group leadership, party leadership, legislative leadership and executive leadership.

Transforming leadership is much deeper. Burns characterizes it as follows:[347]

> The Transforming leader recognizes an existing need or demand of a potential follower, but, beyond that, the transforming leader looks for potential motives in followers, seeks to satisfy higher needs, and engages the full person as a follower . The result of transforming leadership is a relation of mutual stimulation and elevation that converts followers into leaders and may convert leaders into moral agents. . . . I mean the kind of leadership that can produce social change and will satisfy the followers authentic needs . . .
>
> Transformational leadership occurs when leaders and followers raise one another to higher levels of motivation and morality.

Burns discusses in his Part III the following aspects of transforming leadership: 'Intellectual Leadership (Ideas as Moral Power)', 'Reform Leadership', 'Revolutionary Leadership' and 'Heroes and Ideologues'. However, the central chapter from the viewpoint of this book is Chapter 2, 'The Structure of Moral Leadership', talking about the power and sources of values.[348] Examples of such transforming leadership are the lives of Mahatma Ghandi, Martin Luther King Jr., Desmond Tutu and lesser known examples such as Philip Noel-Baker.[349]

The great crisis facing us today has many interlocked dimensions. Peter Willis names it a Holocrisis with many dimensions—but it is all one whole crisis.[350] He claims these many sub-crises are not as discrete or separate as they appear on the surface, but in fact form a living, dynamic whole, whose roots are in the materialistic and individualistic assumptions of modern civilized humanity.

I see the core of this crisis as being the significance of values on social outcomes: the great struggle to keep something like democracy in place, as it is under attack worldwide.

Heather Cox Richardson wrote as follows on 19 December 2024:[351]

---

[345] Carney, *Values: An Economist's Guide to Everything that Matters*, xiv.
[346] Carney, *Values: An Economist's Guide to Everything that Matters*, xv, 4ff.
[347] Burns, *Leadership*, 4–5.
[348] Burns, *Leadership*, 141–256, 29–48.
[349] See the Wikipedia articles Mahatma Gandhi, Martin Luther King Jr., Desmond Tutu and Philip Noel-Baker.
[350] Peter Willis, 'How the Polycrisis Challenges Us to Confront Fear, Failure, and Mortality', *Daily Maverick* (2 December 2024).
[351] Heather Cox Richardson, *Letters from an American* (19 December 2024), on Substack.com.

[In July 1776] members of the Second Continental Congress . . . adopted the Declaration of Independence, explaining to the world that . . . their vision of human government was different from that of Great Britain. In contrast to the tradition of hereditary monarchy under which the American colonies had been organized, the representatives of the united states on the North American continent believed in a government organized according to the principles of natural law.

Such a government rested on the 'self-evident' concept 'that all men are created equal, that they are endowed by their Creator with certain unalienable Rights, that among these are Life, Liberty and the pursuit of Happiness.' Governments were created to protect those rights and, rather than deserving loyalty because of tradition, religion, or heritage, they were legitimate only if those they governed consented to them.

This is what is under attack worldwide. It is nothing other than a battle for control of countries and the world of people with different values expressed in the different ways they see humanity.

There are those who believe all human beings are equal and should be cherished, as in the quote above, and all should be provided equal opportunities. Then there are those who believe human beings are not equal, and of course they (whoever they are) are the superior ones who have the inalienable right to run society, whatever the cost to others and to the environment, and irrespective of what others want.

This is a worldwide movement, but it is particularly clear in the case of newly arrived tech bros funding the Republican Party in the USA such as Elon Musk and Peter Thiel.

A key issue then is how does a genuine democracy work? This is laid out by A. C. Greyling in his book *The Good State: On the Principles of Democracy.*[352] But those are large-scale issues. In most countries, whatever the national situation, one can take local power and make positive things happen. An example is the creation of food gardens in Bellville, Cape Town.[353] There is agency possible at these scales in every country.

Further Reading

Basic aspects of Society

Peter Berger, *Invitation to Sociology: A Humanistic Perspective* (New York: Anchor Books, 1963).

Kenneth Boulding, *Economics as a Science: Morality, Ecology, Evolution, Sociology, Mathematics* (New York: McGraw Hill, 1970).

David Elder Vass, *The Causal Power of Social Structures: Emergence, Structure and Agency* (Cambridge, UK: Cambridge University Press, 2010).

---

[352] A. C. Greyling, *The Good State: On the Principles of Democracy* (New York: Oneworld Publications, 2020).

[353] Abigail Baard, 'Urban Food Garden in Bellville: Transforming Lives through Sustainable Agriculture and Job Creation', *Daily Maverick* (17 December 2024).

Organizational Issues

Herbert A. Simon, *Administrative Behaviour: A Study of Decision-Making in Administrative Organizations* (New York: Simon and Schuster, 1947).
Peter Drucker, *The Practice of management* (London: Routledge, 2012).
Stafford Beer, *Brain of the Firm* (London: Wiley, 1981).

The issue of technology

Vaclav Smil, *How The World Really Works:* A Scientist's Guide to Our Past, Present, and Future (London: Penguin, 2022).
W. Brian Arthur, *The Nature of Technology: What It Is and How It Evolves* (New York: Simon and Schuster, 2009).

The Issues of Values and Leadership

Mark Carney, *Values: An Economists Guide to Everything That Matters* (Glasgow: William Collins, 2021).
James MacGregor Burns, *Leadership* (New York: Harper Perennial Political Classics, 2010).

7

Is There Meaning in the Universe?

The Deep Structure of the Cosmos

Note to the reader: this final chapter is where I express my personal views on the deeper underlying issues regarding the existence and nature of the Universe. Undoubtedly some of what I write will be rejected by some, particularly hardline reductionist physicists and philosophers. So be it. This is how I see it.

Famously, a number of physicists have stated that they can see no meaning in the Universe, a particular case being Nobel Prize winner Steven Weinberg. At the end of his book *The First Three Minutes*,[1] he wrote

The more the universe seems comprehensible, the more it also seems pointless.[2]

This is an extraordinary example of tunnel vision. It results from seeing as meaningful only the stuff that physicists study when doing their professional work. But physicists' personal lives contradict this. Firstly, they are involved with family and friends in meaningful ways. Secondly, their dedicated pursuit of scientific results is obviously seen by them as a very meaningful activity.

In fact the world is teaming with purpose at individual and social levels (see **Figure 7.1**), ranging from personal events—birthday parties, sports events, buying and selling, writing and reading books, devoting one's life to being a doctor or nurse or engineer or scientist or airplane pilot—to the political level such as the Republican–Democrat conflict in the USA.

These are all undeniably meaningful interactions, despite what some science communicators may say.[3] This meaning is, of course, not visible if one only looks at the particle physics or atomic level (**Figure 7.2**), which is Steven Weinberg's main area of concern. Purpose is also not visible on the largest scales either, such as shown in amazing images made by the James Webb Space Telescope (**Figure 7.3**). But purpose is certainly there on the scale of human life (**Figure 7.1**), where it shapes our existence. Function, purpose, meaning and agency occur and underlie daily life as we experience it.

---

[1] Steven Weinberg, *The First Three Minutes* (New York: Basic Books, 1993).
[2] Elaborated by John Horgan in 'Steven Weinberg's Pointless Final Theory', *Cross-Check* (27 January 2025).
[3] Chris Ellis, 'Science Communicators Need to Stop Telling Everybody the Universe is a Meaningless Void', *The Conversation* (28 November 2023).

How We Come to Be. George F. R. Ellis, Oxford University Press. © George Ellis (2026).
DOI: 10.1093/9780198950189.003.0009

Figure 7.1 Trusting and care. An amazing interaction between a young child and Dutch red cross worker Eric Corton in Malawi. Picture taken on behalf of the Dutch radio station 3FM Serious Request. [4]

The issue of **how** purpose emerges is essential in understanding how our mind/brain comes into existence. This happens firstly via evolutionary processes acting over geological timescales, selecting a genome that can lead to existence of a brain capable of meta-reflection. Secondly, via the developmental processes whereby each individual brain comes into being, developing from a fertilized egg. Thirdly, via functional processes leading to specific intentions being stored in neural networks during our ongoing interaction with the society around us. [5] In short, summarizing previous chapters in this book:

> Purpose emerges because evolution has led to our existence and functioning, based in the underlying physics, that ensures we will indeed become beings with agency enabling us to carry out the purposes we choose for ourselves in our lives. [6]

But this only occurs because the Universe has a special nature that allows it to happen. If it had been very different, perhaps with very different laws of physics, or

---

[4] Eric Corton photograph by Ben Houdijk, reproduced with permission. See EricCorton keert terug naar Malawi |3FM Serious Request for a moving video of this visit to Malawi.

[5] Michael Tomasello, *Becoming Human: A Theory of Ontogeny* (Cambridge, MA: Harvard University Press, 2019).

[6] Kevin Mitchell, *Free Agents: How Evolution Gave Us Free Will* (Princeton, NJ: Princeton University Press, 2023).

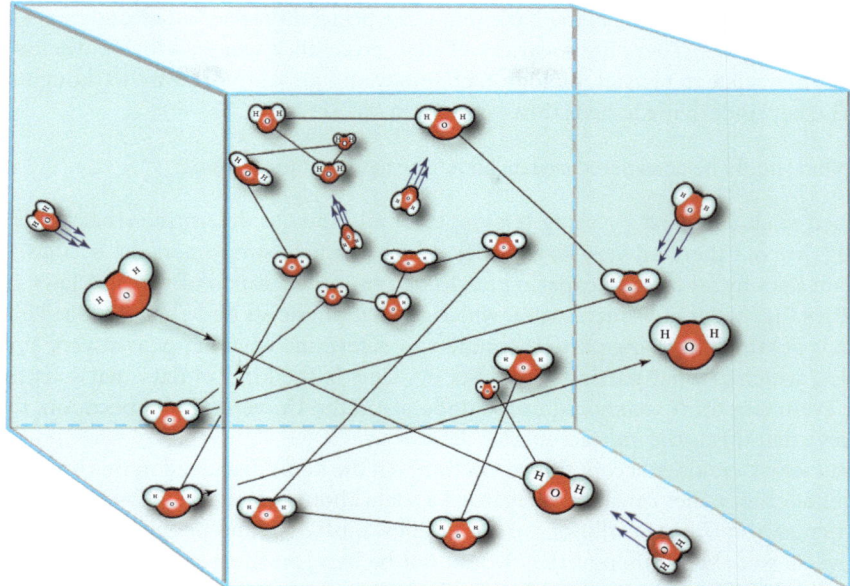

Figure 7.2 Molecular collisions.

Figure by Mauro Carfora.

Figure 7.3 James Webb Space Telescope image. Every dot in the image is a galaxy (the bright lights with crosses are stars).

From NASA.

with a much greater rate of expansion after decoupling, or a background radiation temperature that never dropped below 3000 K, none of this would have happened. Can we explain why this was the case?

At a deeper level, I have emphasized in previous chapters the way that purpose relates to ethics and values. Can any of that discussion of ethics and values relate to the deep structure of the cosmos? And what is that deep structure? I claim that the best way to consider this is to characterize the possibility spaces that underlie what actually happens. There are a variety of such possibility spaces, which I discuss in §7.2. They relate to physics, biology, technology and also, importantly, to thoughts.

The key underlying issue is this:

What should be considered as data about the nature of the cosmos?

We get such data first by using telescopes of all kinds to determine what kinds of things are out there on very large scales, which is what astronomers do. Second, we obtain data by laboratory experiments to determine the kinds of physical laws and entities that occur at human scales, which is what physicists and chemists do. Third, data is obtained by using particle colliders to determine what happens at very small scales, which is what particle physicists do. This is the kind of data that scientists like Weinberg think we should use to understand the Universe. By implication, they believe that this is the *only* acceptable data.

But why should we accept this restriction? All the items discussed in this book that occur at other scales can also be regarded as data about the nature of the Universe. In particular, the Universe allows camels, giraffes, birds, fish and people to exist. If the Universe were very different they would not be here. So their existence is also data about the Universe. More than this, the Universe allows love and hate, courage and fear, faithfulness and treachery, kindness and cruelty to exist. Surely this also tells us something about the Universe? They do indeed occur, and play a key role in what happens at human scales on Earth. That needs explanation that goes beyond a purely materialist view, assuming that their existence is significant—which is my position. In summary,

> Everything that exists in the Universe—electrons, atoms, molecules, teapots, books, digital computers, planets, stars, black holes, gravitational waves, galaxies, love and hate, courage and fear, faithfulness and treachery—gives us data about the Universe. It is its specific nature that allows them all to exist. The nature of what can possibly exist is characterized by abstract possibility spaces, which can be regarded as the deep underlying structure of the cosmos.

That is the topic of this chapter. I discuss, in turn, §7.1, 'Why Are the Laws of Physics What They Are? The Anthropic Issue'; §7.2, 'The Possibility Spaces That Are the Deep Structure of the Cosmos'; §7.3, 'Mathematical Realism and Moral Realism'; §7.4, 'Why Anything Exists with Its Specific Nature: The Four Existential Options'; §7.5, 'The Great Events and Transitions in Individual Life'; and §7.6, 'There is Indeed Meaning in the Universe'.

7.1 Why Are the Laws of Physics What They Are?
The Anthropic Issue

The profound question underlying our existence has two parts:[7] why does anything exist? Why does this specific thing exist? I postpone discussion of why anything

---

[7] Derek Parfit, 'Why Anything? Why This?', *London Review of Books*, 20 (1998), 24–7.

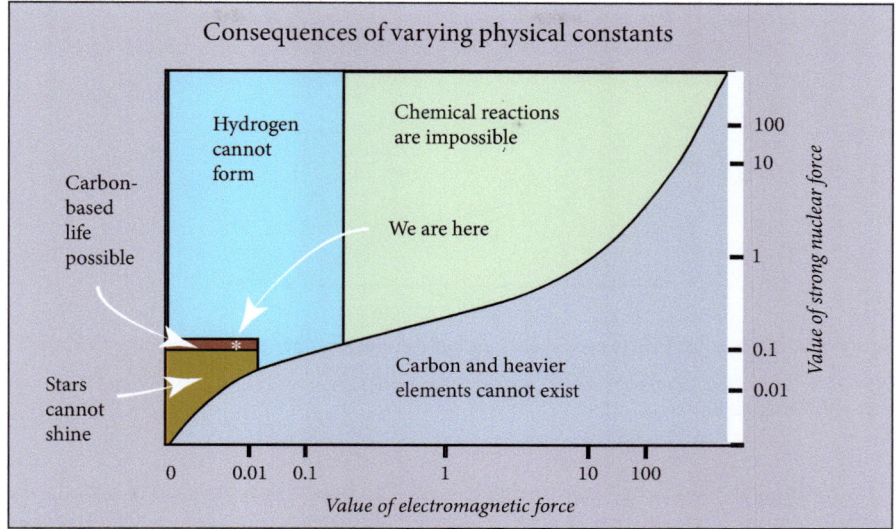

Figure 7.4 Anthropic limits on physical constants (strong nuclear force, electromagnetic force).

Redrawn by Mauro Carfora from J Schombe (University of Oregon), Cosmo Lectures: Anthropic Principle

exists to §7.4, and consider here the fine tuning that allows our existence as human beings (see also §1.2). This was first proposed in a coherent form by Brandon Carter,[8] then commented on by many others,[9] and summarized in depth in a major book by John Barrow and Frank Tipler.[10] Unfortunately this book went over the top and proposed a whole variety of such principles as the weak anthropic principle, strong anthropic principle, participatory anthropic principle and final anthropic principle, which resulted in the book being met with incredulity and ridicule.[11] Nevertheless the anthropic principle is an important idea that has led to interesting scientific investigations. An up-to-date summary is given in a recent volume edited by Sloan et al.[12]

The key issue from the present point of view is that there are two kinds of such fine-tuning issues. The first are to do with physics (**Figure 7.4**), and the second with cosmology (**Figure 7.5**). The nature of the fundamental constants is discussed

[8] Brandon Carter, 'Large Number Coincidences and the Anthropic Principle in Cosmology', in *Confrontation of Cosmological Theories with Observational Data: Proceedings of the Symposium, Krakow, Poland, September 10–12, 1973* (Dordrecht: Reidel, 1974), 291–8.

[9] For example, Bernard Carr and Martin J. Rees, 'The Anthropic Principle and the Structure of the Physical World', *Nature*, 278 (1979), 605–12.

[10] John Barrow and Frank Tipler, *The Anthropic Cosmological Principle* (Oxford: Oxford University Press, 1986).

[11] Martin Gardner, 'WAP, SAP, PAP, and FAP', *New York Review of Books*, 23 (1986), 22–5. John Earman, The SAP Always Rises: A Critical Examination of the Anthropic Principle', *American Philosophical Quarterly*, 24 (1987), 307.

[12] David Sloan, Rafael Batista, Michael Hicks and Roger Davies, *Fine Tuning in the Physical Universe* (Cambridge, UK: Cambridge University Press, 2020).

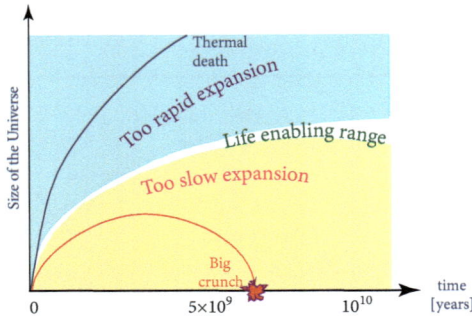

Figure 7.5 Anthropic limits on cosmological dynamics (rate of expansion with time).

Redrawn by Mauro Carfora from Vojtech Ullmann (Ostrava-Poruba): Anthropic principle and the existence of multiple universes.

illuminatingly in a book by Uzan and Leclerq.[13] The bounds on physical constants are very interesting, such as those shown in **Figure 7.4**: the ratio between the electromagnetic force and the strong nuclear force. You can see that there is a variety of different reasons why only a very small range of values of these two constants allows life to exist. Of course if we didn't exist we wouldn't be around to ask this question—so it can be rephrased: why does the Universe allow us to exist, when a random universe with other constants would not?

One possible answer is that we live in a multiverse with many expanding universe patches like that we live in, but with some hypothetical mechanism producing different values of those physics parameters in each one.[14] Then just by luck, things will work out all right in at least one expanding domain in the multiverse (see the discussion in §1.3 and §7.4), and that explains why things work out fine. There are a variety of other such physical relations that are relevant to the existence of life.[15]

Bounds on the cosmos are equally important (**Figure 7.5**). Given the physics that allows life to exist, the cosmos must be of such a nature as to allow galaxies, stars and planets to come into existence. But whether this will happen depends on the rate of cosmological expansion. Too rapid expansion, for example if the cosmological constant is too large, will prevent structure forming.[16] If the rate of expansion is too slow, the Universe will collapse into black holes before stars and planets have had time to form. A further important parameter is the degree of spatial inhomogeneity at the start of structure formation. Again a multiverse with many varying domains with different values of these cosmological parameters is a potential solution to the puzzle.

---

[13] Jean-Philippe Uzan and Bénédicte Leclerq, *The Natural Laws of the Universe: Understanding Fundamental Constants* (Berlin: Springer, 2005).

[14] See the Wikipedia article Multiverse.

[15] Carter, 'Large Number Coincidences and the Anthropic Principle in Cosmology'. Barrow and Tipler, *The Anthropic Cosmological Principle*.

[16] Steven Weinberg, 'Anthropic Bound on the Cosmological Constant', *Physical Review Letters*, 59 (1987), 2607–10.

However, is the multiverse concept a part of standard cosmological theory? I believe the answer is no. The recent book by Nobel Prize winner Jim Peebles mentions the idea briefly early on, and then quietly drops it.[17] The canonical observational cosmologists' view of the Universe is represented by the parameters determined by the Planck satellite, in concordance with all other observations.[18] After acknowledging a few small discrepancies between the various sources of data, the Planck team state, in their definitive summary of our observational knowledge of the Universe,

> [we] conclude that the 6-parameter ΛCDM model provides an astonishingly accurate description of the Universe from times prior to 380 000 years after the Big Bang, defining the last-scattering surface observed via the CMB, to the present day at an age of 13.8 billion years.

There is no mention of any parameter related to an alleged multiverse. There are, however, many constraints placed by these observations on the inflationary era in the early Universe.[19] They are consistent with single-field, slow-roll inflation with minimal kinetic term, such as the Starobinsky model, which indeed results in a multiverse kind of situation occurring.[20] The problem then is how different physics can occur in different local universe patches. There is currently no satisfactory proposal as to how this could happen.

Nevertheless it is clear that, in principle, the existence of a multiverse can indeed provide an answer as to how both the local physical constraints and the cosmological ones might be satisfied, which is the position of my distinguished colleague Martin Rees.[21] I return to just how satisfactory an explanation this is in §7.4.

7.2 The Possibility Spaces That Are the Deep Structure of the Cosmos

The physical Universe is made up of particles and atoms, arranged as molecules, rocks, planets, stars and galaxies, but also as living beings, including you and me. However, underlying what actually occurs are Platonic possibility spaces, which are the deep structure of the cosmos. They govern what is possible, not just in terms of physics and astronomy but also in biology—giraffes and elephants, mosquitos and human beings—and indeed in terms of thoughts.

I consider, in turn, §7.2.1, 'Physical Possibility Spaces'; §7.2.2, 'Biological Possibility Spaces'; §7.2.3, 'Mathematical and Logical Possibility Spaces'; §7.2.4, 'Mental Possibility Spaces'; and §7.2.5, 'The Deep Structure of the Cosmos'.

[17] P. James E. Peebles, *Cosmology's Century: An Insider History of Our Modern Understanding of the Universe* (Princeton, NJ: Princeton University Press, 2020).

[18] N. Aghanim et al., 'Planck 2018 Results—VI. Cosmological Parameters'. *Astronomy and Astrophysics*, 641 (2020), A6.

[19] Jerome Martin, Christophe Ringeval and Vincent Vennin, 'Encyclopædia Inflationaris' (opiparous edition, 2024), *arXiv preprint* arXiv:1303.3787; Jerome Martin, Christophe Ringeval and Vincent Vennin, 'Cosmic Inflation at the Crossroads' (2024), *arXiv preprint arXiv:2404.10647*.

[20] See the Wikipedia article Starobinsky inflation.

[21] Martin Rees, *Just Six Numbers: The Deep Forces That Shape the Universe* (London: Hachette UK, 2008).

7.2.1 Physical Possibility Spaces

The classic example of possibility spaces in physics are phase spaces in classical mechanics.[22] An example is that for a pendulum that can go over the top (**Figure 7.6**). The bob is held at a constant distance from the point of suspension by a rigid rod. The horizontal axis is angular position, which recurs every 2π, so it is actually rolled up like a cylinder. The same physical position is represented multiple times.

The vertical axis represents angular speed. The possible orbits are represented by the lines shown, characterized by their energy E. Because the pendulum is assumed to be frictionless, energy E is conserved on each orbit, and is always equal to the initial value E_0. As shown by the arrows, the lower half of the cylinder represents the pendulum moving to the left (angular position decreasing) and the upper part represents the pendulum moving to the right (angular position increasing).

The phase space shows all possible orbits for such a pendulum, separated by a critical value E_p. If $E_0 < E_p$ then motion takes place on one of the closed ellipses in phase space. This is just the usual way a pendulum swings back and forth between maximum angular values to the left and the right. If $E_0 = 0$, the phase point lies at the centre of the closed ellipses and the pendulum is just hanging down in a stationary state. If $E_0 >$

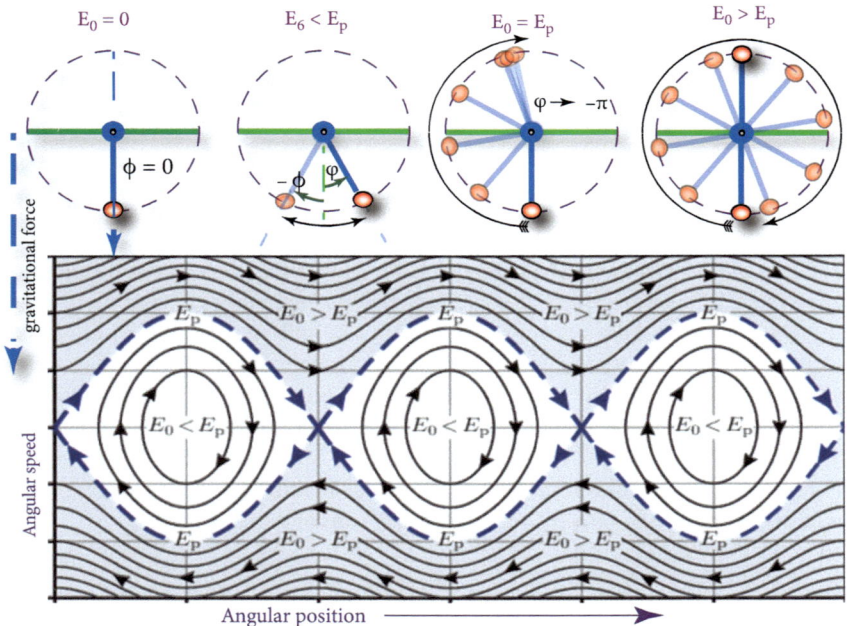

Figure 7.6 Pendulum phase space. E_p is a critical energy; E_0 is the initial energy of the pendulum.

Phase space diagram by Karlheinz Ochs, 'A Comprehensive Analytical Solution of the Nonlinear Pendulum', *European Journal of Physics*, 32 (2011), 479. Top figures by Mauro Carfora.

[22] See the Wikipedia article Phase space.

E_p then the pendulum swings right over the top, either moving to the right (an angle ever increasing but repeated every 2π), as in the top half of the diagram, or to the left (an ever decreasing angle repeated every 2π), as in the bottom half. The interesting situation is the critical case $E_0 = E_p$. There are then two cases. First, it may lie at the critical point where the two orbits intersect, with arrows coming in on one set of orbits and out on the other. This is the critical case of the pendulum being unstably balanced vertically with the bob directly above the point of suspension. It won't stay there: the slightest disturbance will set it off to either the left or the right. This is a hyperbolic unstable point. Finally, it may move on one of the orbits approaching that critical orbit, when in principle it takes an infinite amount of time to approach the vertical state. It would be diverging from that vertical state as represented by one such hyperbolic intersection and then approaching it again as represented by the next such hyperbolic intersection.

I have gone into this case in detail because it is an excellent example of a possibility space in classical mechanics, showing all the possible positions and states of motion of the relevant system. Of course in practice there will be friction, which will modify this phase space. Instead of the ellipses, there will be ingoing spirals ending up at the stable state at the centre with the pendulum hanging vertically down.

A useful discussion of physics possibility spaces in general is given by Bishop and Ellis.[23] They state (§2.5),

> Contexts, determined by stability conditions, large-scale constraints, symmetry breakings and regularities, are related to possibility space. The most basic laws of physics define what is physically possible in the world. Yet, not all of these possibilities are actualizable through the basic laws and particles of elementary particle physics by themselves. It is the specific, concrete contexts that make particular regions of possibility space accessible. Think of contexts as specifying accessibility conditions for particular subspaces of the physical possibility space defined by the most elementary laws.

In the case above, the relevant context is the rigid rod constraining the movement of the bob.

There are also general principles that need to be represented in any possibility space for a physical system. First, energy must be conserved, so perpetual motion is impossible in any system that does work. This is embodied in the modified pendulum phase space that takes friction into account, with the orbits modified to converge to the fixed point $E_0 = 0$. Second, because of special relativity theory, no physical system can exceed the speed of light. The pendulum phase space also needs to be modified to take this into account. Because of the huge value of the speed of light ($c = 3 \times 10^8$ m/sec), this will make no perceptible difference to the phase diagram shown.

---

[23] Robert Bishop and George Ellis, 'Contextual Emergence of Physical Properties', *Foundations of Physics*, 50 (2020), 481–510.

In quantum physics, possibility spaces are Hilbert spaces, which can, for example, represent the possible states of a vibrating string, and are used to represent the idea of a quantum state.[24]

7.2.2 Biological Possibility Spaces

These occur for both microbiological states (molecular biology) and macrobiological states (physiology).

Molecular biology. A wonderful book by Andreas Wagner characterizes a number of Platonic abstract possibility spaces for microbiology.[25] They are for proteins (§4), and for a variety of genotype–phenotype maps: that for metabolism at the biomolecular level (§3), characterized by genotype networks,[26] and for gene expression patterns (§5): a library of regulator circuits and their expression as genotypes.

It is crucial for evolutionary progress that all these possibility spaces are highly degenerate: neutral evolution therefore allows for genotype networks to navigate over very large distances and so to arrive at neighbourhoods where very different phenotypes can be nearby (§6). This solves the key riddle of how evolution has been able to explore so much of these possibility spaces since the origin of life, given the huge dimensions of these spaces. Genotype networks accelerate evolution, as discussed by Wagner.

Possibility spaces for DNA underlie the idea that DNA is the genetic code.[27] All the different possibilities for DNA in the end underlie all the different kinds of plants and animals that exist on Earth. These possibilities correspond to all possible sequences of the four nucleotides A, G, C and T in the DNA coils. DNA profiling locates an individual within this possibility space.[28]

Physiological possibilities. Possibilities, such as how to make a heart work at interlocking scales,[29] are limited by physical laws that underlie what is physiologically possible.[30] The limits on physiological possibilities mean that evolution will, on multiple occasions, find similar solution to problems such as seeing (there are a limited number of possible kinds of eyes) or flying (there are a limited number of possible wings and ways of controlling them). Thus this leads to the occurrence of convergent evolution.[31]

---

[24] See the Wikipedia articles Hilbert space, Mathematical formulation of quantum mechanics and Quantum state.

[25] Andreas Wagner, *Arrival of the Fittest: Solving Evolution's Greatest Puzzle* (London: Penguin, 2014).

[26] João Rodrigues and Andreas Wagner, 'Evolutionary Plasticity and Innovations in Complex Metabolic Reaction Networks', *PLoS Computational Biology*, 5 (2009), e1000613.

[27] Maël Montévil, 'Possibility Spaces and the Notion of Novelty: From Music to Biology', *Synthese*, 196 (2019), 4555–81. See also the Wikipedia article DNA.

[28] See the Wikipedia article DNA profiling.

[29] Denis Noble, 'Modelling the Heart—From Genes to Cells to the Whole Organ', *Science*, 295 (2002), 1678–82.

[30] Steven Vogel, *Cats' Paws and Catapults: Mechanical Worlds of Nature and People* (New York: W. W. Norton, 2000).

[31] See the Wikipedia article Convergent evolution and George McGhee, *Convergent Evolution: Limited Forms Most Beautiful* (Cambridge, MA: MIT Press, 2011).

A key feature of physiological possibilities is scaling laws, for example how energy demand expressed by metabolic rate varies with the mass of an organism.[32] Empirical evidence suggests an allometric law: a power-law relation between them. Various explanations have been given as to why this is the case. There are others, such as the fact that a mouse will not work if you scale it up to be six feet tall. It's legs then won't support it: it won't be able to stand or walk.

The relation between them. Both micro- and macrobiological possibility spaces in the end depend on the underlying physics, but with a strictly biological aspect added in: occurrence of biological function at every emergent physical scale. In brief, physical possibilities underlie biological ones, and the biological possibility spaces characterize key emergent aspects of the way physics underlies life.

7.2.3 Mathematical and Logical Possibility Spaces

Completely different are Platonic spaces associated with mathematics and logic. These are not determined by experiment, as in the physical and biological cases, but by abstract reasoning. Examples are the discovery of Pythagoras' theorem (**Figure 7.7, left**) and the existence and value of the number π (**Figure 7.7, right**), including the fact that it is an irrational and transcendent number. Together with essentially all research mathematicians, I believe in mathematical Platonism. Such abstract relations are truths that hold everywhere in the Universe. They will be known to advanced-enough mathematicians everywhere—on a planet circling Alpha Centauri, or in the Andromeda Galaxy. These relations can be thought of as living in abstract Platonic spaces that we explore with our minds.

There are two key points related to this. The first is that some have queried how, as humans with our limited perceptual capacities, we could possibly have access to the content of such abstract Platonic spaces. A complete answer has been in a remarkable book by philosopher Paul Churchland explaining how the neural network structure of our brain is indeed able to determine the nature of such relationships: *Plato's Camera: How the Physical Brain Captures a Landscape of Abstract Universals*.[33] This book resolves that issue convincingly.

The second is this: some will comment, 'But mathematics is a social construct that changes with time. We did not know π was irrational until Lambert proved it in 1761. Now we do.'[34] But this is confusing mathematics itself (the abstract entity) with the mathematics known to a particular society at a particular place and time (which is indeed a social construct). This is illustrated in **Figure 7.8**. The abstract entity (mathematics itself, on the right) is timeless and true at all places and all times and is not a social construct. Mathematics known to a particular society at a particular time (on the left) is indeed a social construct. It is related to mathematics itself by a projection

---

[32] Geoffrey B. West, James H. Brown and Brian J. Enquist, 'A General Model for the Origin of Allometric Scaling Laws in Biology', *Science*, 276 (1997), 122–6; Fabiano Ribeiro and William Pereira, 'A Gentle Introduction to Scaling Laws in Biological Systems' (2021), *arXiv preprint arXiv:2105.01540*.

[33] Paul M. Churchland, *Plato's Camera: How the Physical Brain Captures a Landscape of Abstract Universals* (Cambridge, MA: MIT Press, 2012).

[34] See the Wikipedia articles Pi and Proof that π is irrational.

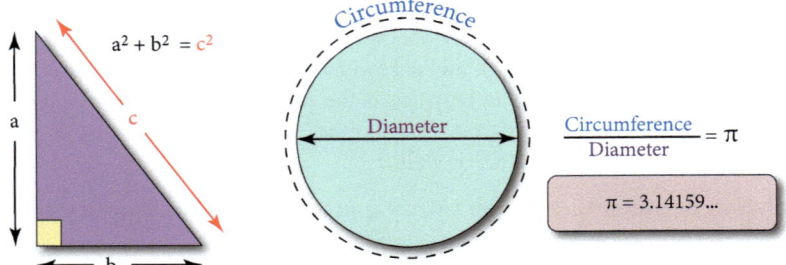

Figure 7.7 Mathematical realism: some examples. Pythagoras' theorem (left); the number π (right), which carries on forever because it is an irrational and transcendent number. The number written down is just an approximation. Notwithstanding its deceivingly simple characterization the number π is extremely important in mathematics. It has a large number of distinct defining properties, making its role surprisingly ubiquitous. These relations will be true everywhere in the Universe, and have been true since time began—or perhaps before.

Figure by Mauro Carfora.

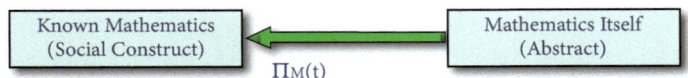

Figure 7.8 Mathematical realism: the basic relationship. The relation between the abstract space of mathematical relations and the shared mental space of mathematics understood by a society.

Figure by Mauro Carfora.

operator $\Pi_M(t)$ that changes with time. Similarly there are abstract spaces for logic, including formal logical systems and Boolean logic.

7.2.4 Mental Possibility Spaces

The deeper underlying possibility space is the space of all possible thoughts. *You can only think a thought if it is possible to think it!* That at first sounds like a tautology, but in fact is a profound statement about the nature of our existence. It is presented in the remarkable short story 'The Library of Babel' by Jorge Luis Borges.[35] This describes a vast library containing all possible 410-page books of a certain format and character set. The Wikipedia article referring to the book states,[36]

> Though the vast majority of the books in this universe are pure gibberish, the laws of probability dictate that the library also must contain, somewhere, every coherent

---

[35] See the Wikipedia article Jorge Luis Borges.
[36] See the Wikipedia article The Library of Babel.

book ever written, or that might ever be written, and every possible permutation or slightly erroneous version of every one of those books. The narrator notes that the library must contain all useful information, including predictions of the future, biographies of any person, and translations of every book in all languages.

Thus it contains all possible thoughts, expressed in some language or other. Underlying this is another fascinating possibility space: that of all possible languages.[37]

A key point is this. The books are all of finite length. However, this contradicts statements by some generative linguists, stating that, because of recursion, infinitely long sentences can occur.[38] They would not be contained in the library. This is a classic example of being led astray by not taking realistic context into account. There are two key points. The first is that a sentence is intended to convey meaning between two people. That means you must remember the beginning of the sentence when you reach the end. But short-term memory has limits: we can only store about seven items in short-term memory.[39] Consequently a sentence of about 490 words is the maximum that can effectively convey meaning.[40] The second is that you'd literally die before you could finish reading an infinitely long sentence, if one were to exist.

Thus I claim that there is a finite possibility space for thoughts that characterizes all the thoughts we can have. It is smaller than the Library of Babel: it does not include all the meaningless stuff that exists there. But, crucially for the argument that follows, it not only includes all possible thoughts about science and engineering, politics and economics, biology and evolution, groceries and cooking, music and art, and so on. It also includes thoughts about love and hate, right and wrong, good and evil.

7.2.5 The Deep Structure of the Cosmos

I claim that these possibility spaces are the deep structure of the cosmos, underlying our physical and mental existence. They have existed unchanging since the beginning of time, or maybe even before. They are the same at all times and all places in the universe. They include possibilities for the physical structures such as stars and galaxies and planets and rocks and crystals, for biological existence such as plants and birds and fish and mammals, for engineering creations such as tunnels and dams and aircraft and digital computers, for political and economic systems, for board games and physical games and for pictures and music.

We choose from these possibilities and explore them. They include possibilities of behaving selfishly or kindly, hatefully or lovingly. Possibility spaces don't determine the outcomes that actually occur; they determine what is possible.

---

[37] Nikhil Mahant, 'Extraterrestrial Tongues', *Aeon* (9 May 2025).
[38] Paul Ziff, 'The Number of English Sentences', *Foundations of Language*, 11 (1974), 519–32.
[39] George Miller, 'The Magical Number Seven, Plus or Minus Two: Some Limits on Our Capacity for Processing Information', *Psychological Review*, 63 (1956), 81.
[40] See George Ellis and William Stoeger, SJ, 'Appendix', in George Ellis and Mark Solms, *Beyond Evolutionary Psychology* (Cambridge, UK: Cambridge University Press, 2018).

Why they exist and have the nature they do is the deep issue underlying the nature of the physical cosmos. This is a metaphysical issue: science is unable to determine the answer. I return to it in §7.4.

7.3 Moral Realism

Crucial is the distinction between moral realism and moral relativism.[41] The issue is this:

> Is it possible to state that the genocide that occurred in the Holocaust was evil as a matter of fact?[42]

That is the position of moral realism. Or is this only a matter of opinion, because different cultures have different ethical values, and none is more valid than any other? Was the Holocaust just one way of behaving that is no more evil or good than any other? This is the position of moral relativism.

It is my view that it is imperative that one is able to say that both the Holocaust and Apartheid were evil as a matter of fact, not just a matter of opinion. To make this clear, one must make the same distinction as I made in **Figure 7.8** between mathematics itself and mathematics as known in any particular society. Here one must make the distinction (**Figure 7.9**) between morality as it is—what is, in fact, good or bad as characterized by moral realism—and ethical understanding of what is good or bad in any particular society at some specific time and place. The latter is indeed a social construct, which will to a greater or lesser degree reflect the nature of true morality—what is in fact good or bad. They are related by a projection $\Pi_E(t)$ that changes with time.

A key point is that ethical or moral values cannot be determined by science, no matter what some reductionist scientists claim.[43] This violates the basic principle 'You can't get an ought from an is', as stated by David Hume.[44] Any time a scientist claims this is possible, whether through social forces, evolution or psychology, they always

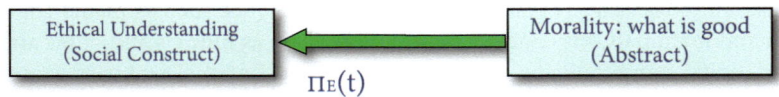

Figure 7.9 **Moral realism: the basic relationship.** The relation between the abstract space of moral realism and the shared mental space of ethics understood by a society at a particular time and place (cf. Figs. 7.7 and 7.8).

Figure by Mauro Carfora.

[41] See the Wikipedia articles Moral realism and Moral relativism.
[42] See the Wikipedia article The Holocaust.
[43] Joel Primack, 'Science Cannot Give Us Values', Princeton valedictory address, *University, a Princeton Quarterly*, 30 (1966), 1.
[44] See the Wikipedia article Is-ought problem.

bring in unjustified assumptions about what is good and what is bad by the back door.[45] There is no purely scientific measure for being good or bad.

This does not mean one cannot attain such a measure; indeed that is precisely what I proposed as regards to the spectrum of values discussed in §6.7. However, it is a moral measure, not a scientific one. Indeed that whole proposal is based in the concept of moral realism.

The **United Nations Universal Declaration on Human Rights** (UDHR) is based in acceptance of moral realism.[46] It discusses its nature as follows:

> Recognition of the inherent dignity and of the equal and inalienable rights of all members of the human family is the foundation of freedom, justice and peace in the world; ... disregard and contempt for human rights have resulted in barbarous acts which have outraged the conscience of mankind. ... the advent of a world in which human beings shall enjoy freedom of speech and belief and freedom from fear and want has been proclaimed as the highest aspiration of the common people; therefore human rights should be protected by the rule of law.

The thirty articles that follow state that all human beings are born free and equal in dignity and right without distinction of any kind, such as race, colour, sex, language, religion, political or other opinion, national or social origin, property, birth or other status and regardless of the political, jurisdictional or international status of the country or territory to which a person belongs.

Those rights included freedom from slavery, torture, degrading punishment, arbitrary arrest, exile and arbitrary interference with privacy, family, home or correspondence, and attacks upon honour and reputation. They include the right to equality before the law and to a fair trial, the right to travel both within a country and outside of it, the right to marry and to establish a family and the right to own property. They include the right to freedom of thought, conscience and religion, freedom of opinion and expression, peaceful assembly, the right to participate in government either directly or through freely chosen representatives and the right of equal access to public service. After all, the UDHR notes, the authority of government rests on the will of the people, expressed in periodic and genuine elections which shall be by universal and equal suffrage.

They include the right to choose how and where to work, the right to equal pay for equal work, the right to unionize and the right to fair pay that ensures an existence worthy of human dignity. They include the right to a standard of living adequate for health and well-being, including food, clothing, housing and medical care and necessary social services, and the right to security in the event of unemployment, sickness, disability, widowhood, old age or other lack of livelihood in circumstances beyond one's control.

They include the right to free education that develops students fully and strengthens respect for human rights and fundamental freedoms. Education shall promote

---

[45] George Ellis, 'Can Science Bridge the Is–Ought Gap? A Response to Michael Shermer', *Theology and Science*, 16 (2018), 1–5. George Ellis, 'A Mathematical Cosmologist Reflects on Deep Ethics: Reflections on Values, Ethics, and Morality', *Theology and Science*, 18 (2020), 175–89.

[46] See the Wikipedia article 'Universal Declaration of Human Rights'.

understanding, tolerance and friendship among all nations, racial or religious groups, and shall further the activities of the United Nations for the maintenance of peace. They include the right to participate in art and science. They include the right to live in the sort of society in which the rights and freedoms outlined in the UDHR can be realized.

This sets out some values as not being acceptable. It is a statement of universal constraints on behaviour which we arrive at by taking moral realism seriously.

7.4 Why Anything Exists with Its Specific Nature: The Four Existential Options

The issue of why anything exists and has the nature it has is a deep philosophical issue: it is discussed in a talk I gave in Beijing many years ago, entitled 'Why Are the Laws of Nature as They Are? What Underlies Their Existence'.[47] As explained at the start of this book, this is a metaphysical issue that cannot be settled on the basis of science per se. It requires philosophical analysis, taking into account all data that might be relevant as regards both physics, chemistry and cosmology and, I submit, the meaningful nature of our social and individual lives, as discussed already.

The four existential options are (a) happenstance, (b) necessity, (c) high probability: the multiverse and (d) underlying purpose.

7.4.1 Happenstance

The first option is that things happened by pure chance (labelled *happenstance* by Paul Davies). This does not mean they were probable: they simply are what happened. There is nothing more to say about it: just accept it.

This is a logically solid proposal. However, it is not popular: we want some kind of understanding or unification of things. This provides neither.

7.4.2 Necessity

The second option is that this is the only way things could be. There was no other option.

Proving this was the physicists' goal when they pursued for decades a Theory of Everything (TOE), by which they mean a physical theory of fundamental particles from which everything else is supposed to follow. But this project has failed. The best candidate for a TOE is String Theory/M theory, which is not only not well formulated, but also has not been shown to lead to the actual physics we know represents what

---

[47] George Ellis, 'Why Are the Laws of Nature as They Are? What Underlies Their Existence?', in Donald G. York, Owen Gingerich and Shuang-Nan Zhang, eds, *The Astronomy Revolution: 400 Years of Exploring the Cosmos* (Boca Raton, FL: CRC Press, 2011), 387.

happens at the particle level—the Standard Model of particle physics (SMPP). Instead there are something like 10^{500} string vacua, each representing different physics than the SMPP. There is a vast number of other possibilities than what we experience, from a physics viewpoint. Necessity fails.[48]

7.4.3 High Probability: The Multiverse

There is a multiverse: millions of expanding universe bubbles something like the expanding domain we can see, each with different physics and expansion histories. If there are enough of them, it becomes highly probable that conditions will be favourable for life like ours to exist in at least one of them.[49]

It potentially can solve the issue of why there is a universe domain that is biofriendly and will provide a home for us (see the discussion in §7.1).

But the key point is this. It does not solve the anthropic issue: rather just displaces it one level up. Why does your multiverse have such a nature as to allow any universes regardless whether they are biofriendly or not? If Starobinsky-type inflation occurs, as seems to be preferred by the data, and leads to a suitable multiverse, then why did that Starobinsky type of inflation occur? The problem recurs at the multiverse level. No ultimate explanation has been given.

7.4.4 Underlying Purpose

The final option is that there may in some sense be intention or purpose underlying the existence and specific nature of the Universe. It was meant to be that way. That is the reason there is purpose in the Universe. It reflects the underlying nature of things at a fundamental level.

As remarked already, physics cannot explain what happened before the beginning, because there was no space time then, no physics then and indeed no 'then' then! One needs to reach for an explanation beyond physics, which inter alia explains why physics exists and allows our existence. More than that, it should explain the existence and nature of the possibility spaces just discussed, and the existence of purpose in the Universe.

Personally I find this the most convincing kind of explanation. Note that one is going beyond explaining things by means of Darwinian evolution, which certainly took place. The issue is why is any Darwinian evolution at all possible? That need not have been the case. Why is agency possible? We know it exists (just look at your computer or iPhone). Note that this does NOT propose some kind of deity intervening in the world, overwriting the laws of physics. It asks, why are there any laws of physics at all? And of such a nature as to allow our existence, including enabling evolution and agency to occur? And why do all those possibilities exist, as outlined earlier? If

---

[48] See Peter Woit, *Not even wrong: The failure of string theory and the search for unity in physical law.* Basic Books (AZ), 2006.

[49] Bernard Carr, ed., *Universe or Multiverse?* (Cambridge, UK: Cambridge University Press, 2007).

you don't want to describe them in terms of possibility spaces, that's fine; the question remains: why do those possibilities exist?

Which of these four ultimate options holds? It's not possible to prove any one of them is true; but I take all the evidence of existence of meaning in the Universe to support the last one. It is the deep reason all that meaning exists. That evidence is the topic of the final two sections of this chapter.

7.5 The Great Events and Transitions in Individual Life

In the end, what matters to us is our individual lives and all the struggles, joys and sadness that we experience. I initially wrote a great deal about this, but have condensed it now into just a few topics that make the point. I consider here, §7.5.1, 'Coming into Being: Our Existence as Individuals'; §7.5.2, 'Development, Learning and the Phases of Life'; §7.5.3, 'The Many Varieties of Good Life'; and, §7.5.4, 'The Inevitable Ending: Time as the Ultimate Resource'.

7.5.1 Coming into Being: Our Existence as Individuals

The most profound aspect of all of the human condition is that we each exist as individuals.

> We did not exist at one time, but then we come into being and are functioning individuals at a later time. This is the single most important fact about our existence. It is truly extraordinary.

The sexual reproduction leading to our existence is an immensely complex aspect of the human condition, involving mental, emotional, physiological and social aspects in a contested political and legal context.

You are born either male or female, and that shapes everything that follows in your life. It determines key features of your physical development, strongly influences mental development, certainly plays a major role in life experiences and the chances of being sexually harassed or abused and of course fundamentally determines your relation to procreation. It is a key shaper of your experience. Yes, surgical or chemical intervention can change some physical aspects of your body, but it can't change your fundamental relation to the process of procreation: your ability to be a mother or father.

And what is important is this: although we understand a great deal about the physiological and developmental processes that result in our coming into being, we simply don't understand the core:

> We have no idea how we come to experience consciousness, the passage of time and qualia (sensations such as seeing red, feeling pain, tasting wine).

Yes, we understand a great deal about the neural correlates of consciousness, in particular through brain imaging studies, but not the phenomenon itself. This is the most profound limit of science at the present time. We understand how action potentials spike chains propagate down axons in the brain, resonant circuits occur at macroscopic levels, neuromodulators such as dopamine alter processes in the brain related to emotions and gene regulatory networks underlie brain plasticity that underlie memory. These tell us necessary conditions for consciousness and qualia to occur, but not how they occur.

But we do know that they are possible, because they do, in fact, occur. Their occurrence is characterized by the most profound possibility space of all: **the possibility of existence of qualia, consciousness and experiencing the passage of time.** This is written into the deep structure of the cosmos. It goes much further than the issues discussed in §7.1.

7.5.2 Development, Learning and the Phases of Life

We all go through all the stages of life:[50] infancy, being a toddler, a child, teenager and puberty, maturity, perhaps becoming a parent, eventually reaching old age and finally death (unless death for some reason occurs before old age, so one's life is cut short).

As we progress through these stages of life we develop abilities and knowledge and use them to greater or lesser effect. Crucially, in accord with the main theme of this book, we develop a set of values that shape what we do. Hopefully we develop an ability to undertake meta-reflection about our actions and their effects, and the nature of the life we are leading.[51]

Life-long learning: We have to learn everything we do—to eat, to control excretion, to stand, to walk, to talk, to read, to understand mathematics, to drive a car. One needs to get a school certificate and preferably a university degree in order to have a chance of a well-off life. But far more is going on than formal education: parents are all the time teaching children what is what, and how to behave (**Figure 7.10**). Not just talking, walking and so on, but how to negotiate social life, cross the street safely, be careful of people who might harm you. Above all a caring parent sets boundaries for their children so that they are well socialized and respectful of others—socialization is key part of this process.

7.5.3 The Many Varieties of Good Life

We are taught many subjects at school and university, but not how to lead a good or worthwhile life. We are also not taught how to interact with others in a positive

---

[50] See the Wikipedia article Development of the human body.
[51] Nancey Murphy and Warren Brown, *Did My Neurons Make Me Do It?* (Oxford: Oxford University Press, 2007).

Figure 7.10   The context of informal teaching. The mother is teaching the child what is safe and what is not safe, how to approach others respectfully and so on all the time in an intense ongoing interaction.

Designed by Freepik, http://www.freepik.com

problem-solving way. I will just mention one process that aims to achieve this, namely *Theory U* developed by Otto Scharmer (**Figure 7.11**).[52]

This involves suspending judgement and seeing with open eyes, sensing what is there and letting go of one's own wishes and desires in order to create a positive outcome, as in the case of kenosis discussed in relation to the spectrum of values. This letting go opens up the way to getting to a common agreement on the way forward, and then letting come, enacting, and embodying what is learnt and agreed.

Such a process of letting go is also essential in learning, because in order to learn you have to give up, for the moment, strong adherence to what you know so far in order to be open to a greater synthesis. Furthermore, such letting go is central to the creation of a community, where individual members give up their own wishes to some

---

[52] See the Wikipedia article Theory U.

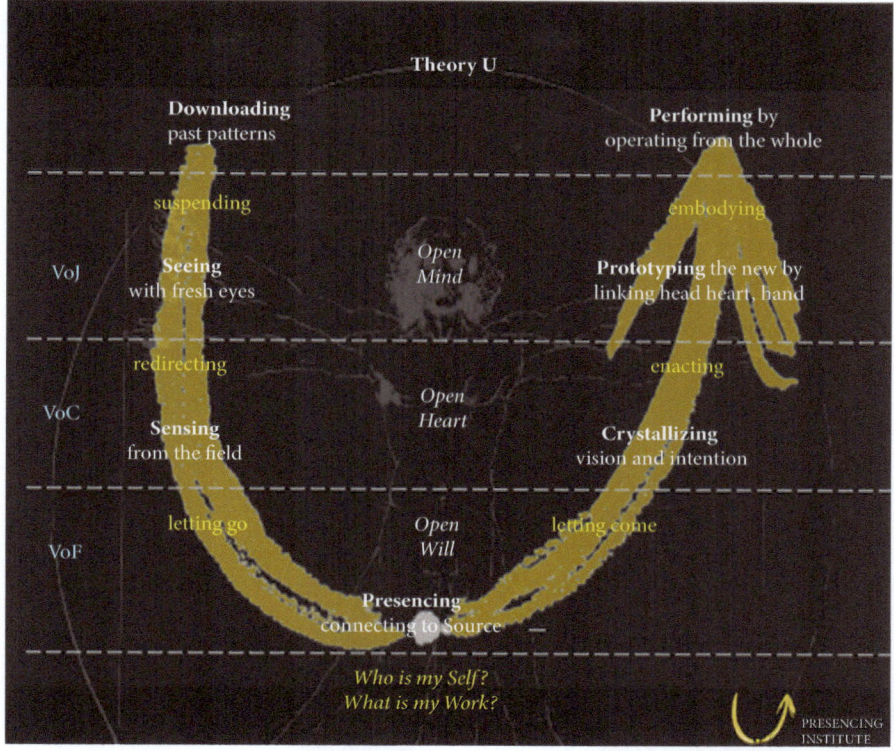

Figure 7.11 The elements of Theory U: learning from the future as it emerges.

Source: O. Scharmer, Presencing Institute; https://commons.wikimedia.org/wiki/File:Theory_U.png

degree in order to create the greater good of a valued community. And of course it is central to the relation of a mother and child, where the mother must eventually be willing to let go of control of the child as she/he develops their own agency and maturity.

Livelihood. The necessity of an adequate livelihood is set out in the UDHR discussed briefly in §7.3. Satisfaction or dissatisfaction with one's life path and career is an important issue in each individual's life. It has two aspects: is the career successful in its own terms—achieving relevant outcomes in terms of its aims, and being paid for doing so? And is it recognized as being successful?

I will make just two comments here. The first is that the values underlying one's work are important, for example in terms of running a business. I discussed this in some depth in Chapter 6, referring in particular to the writing of Colin Mayer.[53] The second is that in academic terms, I regard it as very important to take a holistic approach to whatever issue one is trying to understand, rather than a simplistic reductionist approach, as so often happens. Worse, academic specialization is

---

[53] Colin Mayer, *Capitalism and Crises: How to Fix Them* (Oxford: Oxford University Press, 2024).

quite often accompanied by the arrogance of an insider as against outsiders.[54] This happens particularly in relation to the fundamental sciences, specifically particle physics, as against those working in more applied sciences. An attitude or respect and open-mindedness is called for, and opens the way to greater understanding.

Sport. Both competitive and noncompetitive sports are a key feature of many millions of people's lives, as either a participant or a spectator. In competitive sports such as football or rugby, a major part of what makes it enjoyable for spectators is being in a crowd that are communally taking part in the event and experiencing its emotional ups and downs.

In my own case, although I have taken part in competitive rowing, being part of a Lady Margaret Boat Club team in Cambridge that won an oar in the Bumps, I prefer noncompetitive sport.

Firstly, I have greatly enjoyed flying both small aircraft and gliders. I love the combination of fine motor control, on the one hand, and mental understanding involved in flying, on the other: taking into account the three-dimensional situation of the plane, the forces acting on it and the effect of using the controls to determine its motion. In particular it is very satisfying to make a precise approach and smooth landing at a prespecified spot on a runway. In the case of gliding, additionally one needs to understand the air currents—thermals, ridge lift and standing waves—that provide lift and enable long-distance flights. My longest was 210 km.

Secondly, since I was twelve, I have enjoyed walking and climbing in the beautiful mountains round Cape Town, and continue to do so to this day. I enjoy the physical exercise. Not only is this needed to keep physically healthy, but I believe that it's good for the mind—to relax in the outdoors before getting back to academic work. I enjoy the companionship of my fellow walkers on my weekly hikes. And I love the fynbos vegetation we have in Cape Town and the wonderful views we get from the various peaks we climb **(Figure 7.12)**.[55] These walks—maybe a bit of rock scrambling—are an important part of each week.

Finally, I enjoy doing Pilates exercises once or twice a week to strengthen my physical inner core.[56] What makes this practice so helpful is the precise muscle control involved. It has two key features: careful control of breathing, and moving only one limb at a time while keeping all the others motionless.

Creative endeavours of all kinds. Many forms of art are a creative activity giving satisfaction to those doing them. In performing arts, the art is in the process. This includes theatre, dance, ballet, music of many kinds and gymnastics. In the case of visual arts, the art is in the product: painting, drawing, sculpture, pottery, sometimes photography, films and architecture such as King's College Chapel and the Sagrada Familia cathedral in Barcelona. It includes writing literature and poetry. Such endeavours are a form of self-realization, which is essentially the same as self-actualization

---

[54] Philip H. Abelson, 'Bigotry in Science', *Science*, 144 (1964), 371.
[55] See the Wikipedia article Fynbos.
[56] See the Wikipedia article Pilates.

Figure 7.12 Mountain views. Left, view from the ridge above Silvermine Dam: Elsies' Peak and Simonstown Mountain. Right, view from Kommetje Peak of Kommetje Beach, Noordhoek Beach on the right, Chapman's Peak, Noordhoek Peak behind on the right and the Sentinel on the left.

Photos: George Ellis.

in Maslow's hierarchy of needs.[57] Various reports confirm that engaging in artistic endeavours can be beneficial to health.[58]

The core of great art. Wikipedia characterizes art as 'a diverse range of human activity, and resulting product, that involves creative or imaginative talent expressive of technical proficiency, beauty, emotional power, or conceptual ideas'.[59] A key point is that great art arises by a form of letting go by the artist. She/he starts off with an initial conception of what the outcome should be, and fashions it in that way to start. But as the work progresses, it starts to have a form and character of its own that in effect starts a dialogue with the artist, and should be respected: 'This is the way I should be developed, not the way you first thought!' The grain of the wood, the swirls and veins of a particular coloured marble will offer natural ways of development; a poem written so far forms a context where new lines naturally develop; characters in a novel each develop a particular characteristic nature of their own that needs to be respected by the writer, once that character has taken shape.[60] Great art arises in this way, as emphasized by Harold Pinter in his Nobel Prize lecture:[61]

It's a strange moment, the moment of creating characters who up to that moment have had no existence. What follows is fitful, uncertain, even hallucinatory, although sometimes it can be an unstoppable avalanche. The author's position is an odd one. In a sense he is not welcomed by the characters. The characters resist him, they are not easy to live with, they are impossible to define. You certainly can't dictate to them. To a certain extent you play a never-ending game with them, cat and mouse, blind man's bluff, hide and seek. But finally you find that you have people of flesh and blood

[57] Abraham H. Maslow, 'A Theory of Human Motivation', *Psychological Review*, 50 (1943), 370–96.

[58] Daisy Fancourt and Saoirse Finn, 'What is the Evidence on the Role of the Arts in Improving Health and Well-being?', Health Evidence Network Synthesis Report 67, WHO Regional Office for Europe (2019).

[59] See the Wikipedia article Art.

[60] See Philip Pullman, 'The Writing of Stories', *Daemon Voices: Essays on Storytelling* (Oxford: David Fickling Books, 2017), 23–44.

[61] Harold Pinter, 'Art, Truth and Politics' (Nobel Lecture, 7 December 2005).

Figure 7.13 Pets and humans: an intense relationship.

Photo: George Ellis.

on your hands, people with will and an individual sensibility of their own, made out of component parts you are unable to change, manipulate or distort.

Domestic art. Artistic creativity does not only occur in 'artistic' contexts. It can also occur in the context of the home: cooking ('culinary art'), sewing, crocheting, patchwork, making clothes, interior design and so on.

Hobbies. Hobbies of many kinds have developed and occupy many in their spare time: stamp collecting, embroidery, knitting, painting, woodwork, photography, gardening, making model trains and model train layouts, making and flying model aircraft and so on.[62] I have done many of them—but not knitting and embroidery!

Music. A particularly important aspect of many people's lives is music of all kinds.[63] It has a deep resonance to our emotional nature and with rhythms we respond to. But there is also the element in many cases of a large crowd enjoying the rhythm, skill, beauty and emotion of a great performance.[64] It has always been one of the regrets of my life that I have not learnt to play any musical instrument well, although there was a time many years ago that I had a drum set and learnt to perform it reasonably.

Animal ownership and well-being. Relationships with animals is often important: the joy arising from owning pets, dogs and cats in particular, and interacting with them (**Figure 7.13**). Our empathetic bond with dogs and cats arises because we share

---

[62] See the Wikipedia article Hobby.
[63] See the Wikipedia articles Music and Rhythm.
[64] For one list, see 50 Greatest Musicians of All Time: https://www.imdb.com/list/ls000957075/.

with them the same primary emotional systems, discussed in Chapter 5. Interacting with them, looking after them and receiving affection in return leads to well-being and is central to many peoples' lives.[65] Much distress is often caused by a pet's death, as I have seen in a number of cases.

In the end, friendship with other people—a caring and loving community—is the key resource above all others that makes life worth living.

The best way I can express this is by reproducing a speech that was made by Pulitzer Prize-winning author Anna Quindlen at a graduation ceremony of Villanova University on 23 June 2000 when she was awarded an honorary doctorate:[66]

I'm a novelist. My work is human nature. Real life is all I know. Don't ever confuse the two, your life and your work. You will walk out of here this afternoon with only one thing that no one else has. There will be hundreds of people out there with your same degree: there will be thousands of people doing what you want to do for a living. But you will be the only person alive who has sole custody of your life. Your particular life. Your entire life. Not just your life at a desk or your life on a bus or in a car or at the computer. Not just the life of your mind, but the life of your heart. Not just your bank accounts but also your soul.

People don't talk about the soul very much anymore. It's so much easier to write a resume than to craft a spirit. But a resume is cold comfort on a winter's night, or when you're sad, or broke, or lonely, or when you've received your test results and they're not so good.

Here is my resume: I am a good mother to three children. I have tried never to let my work stand in the way of being a good parent. I no longer consider myself the centre of the universe. I show up. I listen. I try to laugh. I am a good friend to my husband. I have tried to make marriage vows mean what they say. I am a good friend to my friends and them to me. Without them, there would be nothing to say to you today, because I would be a cardboard cut-out. But I call them on the phone and I meet them for lunch. I would be rotten, at best mediocre, at my job if those other things were not true.

You cannot be really first rate at your work if your work is all you are. So here's what I wanted to tell you today: Get a life. A real life, not a manic pursuit of the next promotion, the bigger pay cheque, the larger house. Do you think you'd care so very much about those things if you blew an aneurysm one afternoon or found a lump in your breast?

Get a life in which you notice the smell of salt water pushing itself on a breeze at the seaside, a life in which you stop and watch how a red-tailed hawk circles over the water, or the way a baby scowls with concentration when he tries to pick up a sweet with his thumb and first finger.

Get a life in which you are not alone. Find people you love, and who love you. And remember that love is not leisure, it is work. Pick up the phone. Send an email. Write

---

[65] Katherine J. Bao and George Schreer, 'Pets and Happiness: Examining the Association between Pet Ownership and Wellbeing', *Anthrozoö*, 29 (2016), 283–96.
[66] See Anna Quindlen's Commencement Address at Villanova (23 June 2000).

a letter. Get a life in which you are generous. And realize that life is the best thing ever, and that you have no business taking it for granted. Care so deeply about its goodness that you want to spread it around. Take money you would have spent on beer and give it to charity. Work in a soup kitchen. Be a big brother or sister. All of you want to do well. But if you do not do good too, then doing well will never be enough.

It is so easy to waste our lives, our days, our hours, and our minutes. It is so easy to take for granted the colour of our kids' eyes, the way the melody in a symphony rises and falls and disappears and rises again. It is so easy to exist instead of to live.

I learned to live many years ago. I learned to love the journey, not the destination. I learned that it is not a dress rehearsal, and that today is the only guarantee you get. I learned to look at all the good in the world and try to give some of it back because I believed in it, completely and utterly. And I tried to do that, in part, by telling others what I had learned. By telling them this: Consider the lilies of the field. Look at the fuzz on a baby's ear. Read in the back yard with the sun on your face.

Learn to be happy. And think of life as a terminal illness, because if you do, you will live it with joy and passion as it ought to be lived.'

7.5.4 The Inevitable Ending: Time as the Ultimate Resource

We have a limited time available to do what we wish to do because of the inevitability of death, as the counterpoint to birth. Our available time is precious: it is the ultimate resource we have. As we grow older, health issues and illness intrude more into our daily lives. Particularly distressing are cancer, with its devastating effects, and mental disorders—dementia such as Alzheimer's disease in the elderly. But despite all the progress of medical science, our lives will end in death.

Death of a loved one is a devastating experience, leading to great grief due to the loss, as chronicled by C. S. Lewis.[67] No, it's not going to be possible for us to avoid death, as some billionaires hope. The processes of aging are irreversible.[68]

Will we be reincarnated into some other form of existence, as many religions proclaim? From a scientific viewpoint, life is made possible by the functioning of our physiological systems, and when they cease to function, the brain stops to function and results in the end of consciousness. Near death experiences suggest to some that consciousness continues in some form.[69] This is not the scientific consensus, which believes that consciousness is an emergent process resulting from functioning of the brain as a whole, and will simply cease when the brain ceases to function for any reason.

[67] C. S. Lewis, *A Grief Observed*, Readers' Edition (London: Faber & Faber, 2014).
[68] See the Wikipedia article Aging.
[69] See the Wikipedia article Near death experience.

7.6   There is Indeed Meaning in the Universe

To reinforce the point that there is indeed meaning in the Universe, I close this book with some examples out of many I could have chosen. They are, §7.6.1, 'The General Principle: *Man's Search for Meaning* by Viktor Frankl'; §7.6.2, 'A Lived Experience: Aquilino Gonell's Statement about the January 6th Insurrection at the US Capitol'; §7.6.3, 'A Powerful Novel: *The Prince of Tides* by Pat Conroy'; §7.6.4, A Poem: 'In the Meantime' by Tom Hirons; and §7.6.5, 'A Short Quote from Thich Nhat Hanh, *The Miracle of Mindfulness*.

7.6.1   The General Principle: *Man's Search for Meaning by* Viktor Frankl

This is a profound book that explores the depths of human suffering, resilience and the quest for purpose. Viktor Frankl, a neurologist, psychiatrist and Holocaust survivor, relates his experiences in Nazi concentration camps and emphasizes the power of choice.[70] There is a space between stimulus and response, and there lies our power to choose our response in which lies our growth and our freedom. Even in the direst circumstances, we can choose our attitude toward our suffering. He argues that the primary drive in human existence is not pleasure or power but the pursuit of meaning. Our main motivation for living is our ability to find meaning in life in all circumstances, even the most miserable ones. He suggests that suffering is not meaningless. We cannot avoid suffering, but we can choose how to cope with it, find meaning in it and move forward with renewed purpose.

He states that love is key to finding meaning in life. Thinking of his wife and their love during his time in the concentration camps gave him the strength to endure the suffering. Love transcends the physical presence of the beloved and can provide profound motivation to live. We are fundamentally free to take responsibility for our lives. This focus on responsibility to something or someone other than oneself is a key pathway to finding meaning. Frankl emphasizes that one of the core aspects of human existence is the ability to change. No matter our circumstances, we have the freedom to change ourselves and our attitudes toward our situations. This is crucial for mental health and finding meaning.

Meaning can often be found in potentiality toward something—in being drawn to goals and futures we wish to bring into reality. This forward-looking perspective helps individuals navigate through their suffering by focusing on what they can still achieve or become. Frankl's experiences in the concentration camps illustrate that even under the most inhumane conditions, individuals can retain their sense of humanity and make choices that reflect their innermost values and beings. We might not control our circumstances, but we can always control our attitude toward those circumstances.

---

[70] Viktor E. Frankl, *Man's Search for Meaning* (Boston: Beacon Press, 1959).

7.6.2 A Lived Experience: Aquilino Gonell's Statement about the January 6th Insurrection at the US Capitol

Gonell is an immigrant and former sergeant in the United States Capitol Police, who are based in Washington, DC, and suffered greatly during the insurrection on 6 January 2021. Below is an extract from his recently published op-ed piece in *The New York Times*.[71]

> For my efforts doing my duty as a Capitol Police sergeant, I was beaten and struck by raging rioters all over my body with multiple weapons until I was covered in my own blood. My hand, foot and shoulder were wounded. I thought I was going to die and never make it home to see my wife and young son. Over the last four years, it's been devastating to me to hear Donald Trump repeat his promise to pardon insurrectionists on the first day he's back in office. 'It will be my great honor to pardon the peaceful protesters, or as I often call them, the hostages,' he said in a speech last year. But all of us who were there and anyone who watched on TV know that those who stormed the Capitol were not peaceful protesters. Pardoning them would be an outrageous mistake, one that could mean about 800 convicted criminals will be back on the street.
>
> It could also put me in danger, as I've continued to testify in court and I've given victim statements in cases against dozens of the rioters who assaulted me and my fellow officers.
>
> I was one of the fortunate ones that day; nine people wound up dead as a result of the rampage. Two protesters had fatal medical episodes, one rioter overdosed during the uproar and another was fatally shot by a policeman while forcing her way into the House Chamber. One of my colleagues, 42-year-old Officer Brian Sicknick, suffered two strokes after the trauma of fighting off multiple protesters who sprayed him with a chemical irritant. He didn't survive. Four D.C. policemen harmed in the riots later died by suicide.
>
> I required multiple surgeries, years of rehab and treatment for recurrences of the post-traumatic stress disorder I was diagnosed with in the Army. I was vilified and called 'a traitor', as Mr. Trump and some of his fellow Republicans called the riot a 'day of love' and a 'peaceful protest' by 'warriors', 'patriots', 'political prisoners' and 'mistreated hostages'.
>
> Although I left the Capitol Police force, I remain haunted by that day. Now Mr. Trump's promised actions could erase the justice we've risked everything for. Although I don't blame all Trump supporters—some of my own relatives support him—I do detest what MAGA extremism did to me and my team on Jan. 6. I resent the ongoing whitewashing of the barbarity and the collective amnesia of right-wing politicians who aren't willing to hold Mr. Trump accountable. I can't bear to hear Republicans describe themselves as the 'law and order' party.
>
> Mr. Trump is returning to the presidency at 78, while I had to leave the career I'd worked for my whole life at 42 as a result of injuries suffered while doing my job. I sometimes wonder why I risked my life to defend our elected officials from a mob

---

[71] *The New York Times* Opinion Section (5 January 2025): For Many of us January 6 Never Ended.

inspired by Mr. Trump, only to see him return to power stronger than ever. It's hard to witness a rich white man get rewarded for treachery while I'm punished for fulfilling my duty. Maybe that's why so many people don't do the right thing—because it's hard and it hurts.

At least I get to hear my son call me his hero, as we remember the people who put everything on the line to protect our democracy and continue to tell the truth about Jan. 6.

7.6.3 A Powerful Novel: *The Prince of Tides* by Pat Conroy

I have been told that the greatest novel ever written is *War and Peace* by Leo Tolstoy.[72] So I bought it and read quite a bit of it, and found it disappointing. I get the major message of the book that there is a great deal of randomness in times of war; a lot of what happens is simply meaningless. But I found the characters thin and unsympathetic: why were they worth caring about? It simply did not move me.

By contrast, the book *The Prince of Tides* by Pat Conroy is an extraordinary piece of writing. The descriptions of both South Carolina and New York are lyrical. But the most important thing is the depth he brings to describing relationships, feelings and their development over time as terrible events occur but eventually love arises out of abusive relationships. Conroy has a wonderful capacity for describing all of this. I keep reading it time and again to savour the writing and the depth of understanding of relationships and emotions.

7.6.4 A Poem: 'In the Meantime' by Tom Hirons

My wife Carole Bloch proposed this poem for inclusion:

IN THE MEANTIME
Tom Hirons

Meanwhile, flowers still bloom.
The moon rises, and the sun.
Babies smile, and somewhere,
Against all the odds,
Two people are falling in love.

Strangers share cigarettes and jokes.
Light plays on the surface of water.
Grace occurs on unlikely streets
And we hold each other fast
Against entropy, the fires and the flood.

---

[72] See the Wikipedia article War and Peace.

> Life leans towards living
> And, while death claims all things at the end,
> There were such precious times between,
> In which everything was radiant
> And we loved, again, this world.[73]

7.6.5 A Short Quote from Thich Nhat Hanh, *The Miracle of Mindfulness*.

This is an old favourite of mine: how to realize the amazing nature of the world around us.[74]

> I like to walk alone on country paths, rice plants and wild grasses on both sides, putting each foot down on the earth in mindfulness, knowing that I walk on the wondrous earth. In such moments, existence is a miraculous and mysterious reality. People usually consider walking on water or in thin air a miracle. But I think the real miracle is not to walk either on water or in thin air, but to walk on earth. Every day we are engaged in a miracle which we don't even recognize: a blue sky, white clouds, green leaves, the black, curious eyes of a child—our own two eyes. All is a miracle.

The unique integrity of each person. Each person is different from each other. Each has their own struggles and successes and failures, friendships and losses and their personal integrity and existence as people living and learning and facing life and ultimately death.

The existence of each of us is based in the combination of all the features and inter-actions discussed in the chapters of this book. This immensely complexly interacting whole and its history is truly amazing.

It is how we all came to be.

Further Reading

The anthropic principle

David Sloan, Rafael Batista, Michael Hicks and Roger Davies, *Fine Tuning in the Physical Universe* (Cambridge, UK: Cambridge University Press, 2020).

Possibility spaces: biology

Andreas Wagner, *Arrival of the Fittest: Solving Evolution's Greatest Puzzle* (London: Penguin, 2014).

---

[73] Tom Hirons, 'In the Meantime', in *At the Orphan's Door* (Devon: Feral Angels Press, 2023).
[74] Thich Nhat Hanh, *The Miracle of Mindfulness: An Introduction to the Practice of Meditation* (London: Penguin Random House, 2015).

Comprehending abstract possibility spaces

Paul M. Churchland, *Plato's Camera: How the Physical Brain Captures a Landscape of Abstract Universals* (Cambridge, MA: MIT Press, 2012).

Mental possibility spaces

Jorge Luis Borges, The Library of Babel, trans. Andrew Hurley (Boston: Godine, 2000).

The four existential options

George Ellis, 'Why Are the Laws of Nature as They Are? What Underlies Their Existence?', in Donald G. York, Owen Gingerich and Shuang-Nan Zhang, eds, *The Astronomy Revolution: 400 Years of Exploring the Cosmos* (Boca Raton, FL: CRC Press, 2011), 387.

Is there meaning in the Universe?

Kit Wilson, 'Are We really living in a Materialist Age? Let's just say that I am skeptical', The Hedgehog Review (Spring issue 2025: After Neoliberalism).

Andrew Briggs, Hans Halvorson and Andrew M. Steane, *It Keeps Me Seeking: The Invitation from Science, Philosophy, and Religion* (New York: Oxford University Press, 2018).

Orders of Magnitude

An order of magnitude is a rough estimate of how large things are. It is indicated by powers of 10.

Thus 10^x represents x powers of 10.

$10^0 = 1 =$ one

$10^1 = 10 =$ ten

$10^2 = 100 =$ a hundred

$10^3 = 1,000 =$ a thousand

$10^6 = 1,000,000 =$ a million

$10^9 = 1,000,000,000 =$ a billion

$10^{11} = 100,000,000,000 =$ a hundred billion stars in a galaxy such as Andromeda

$10^{13} = 10,000,000,000,000 =$ ten-thousand billion cells in a human being

$10^{14} = 100,000,000,000,000$ atoms per cell

$10^{23} = 100,000,000,000,000,000,000,000$ stars in the visible Universe

$10^{27} = 1,000,000,000,000,000,000,000,000,000$ atoms per person

$10^{80} = 100,000,000,000,000,000,000,000,000,000,000,000,000,000,000,000,000,000,000,000,000,$ 000,000,000,000,000,000,000 baryons (protons and neutrons) in the visible Universe

Glossary

abiogenesis: processes by which life started.

abstract causation: causal effects due to abstract entities such as the rules of chess and algorithms.

action potential: electrical impulse travelling down axons, conveying information from one neuron to another.

agency: the ability to move as you wish, or to act to change the environment in a desired way.

algorithms: detailed specification of all the steps in a computer program enabling it to solve a specific task.

anthropic principle: the Universe is fine-tuned so as to allow life to exist.

artificial intelligence (AI): the imitation of intelligence by artificial neural networks realized in digital computers.

artificial neural network (ANN): representation of a neural network by running a program on a digital computer.

astronomical unit (AU): the distance between the Earth and Sun.

axons: long thin extension of a neuron conveying information to other neurons via action potentials.

baryon: positively charged protons and uncharged neutrons, out of which ordinary matter is made.

Basic Income Grant (BIG): see universal basic income (UBI).

Big Bang: the era in the expanding early Universe when it was very hot and processes such as cosmic nucleosynthesis took place.

bosons: integer spin particles obeying Bose–Einstein statistics, for example photons.

causal closure: the existence of a sufficient set of conditions whereby outcomes are determined. This only happens when one takes into account all upward and downward causation that occurs. In the real world, interactions at the particle physics level are not, by themselves, causally closed because of these upward and downward interactions.

causality: effects by which some initial set of physical or abstract conditions reliably determine later such states. The outcomes may be unique, or may only be determined statistically. An example is the established causal relation whereby smoking causes lung cancer.

cell: the basic unit of life. All living entities are made of cells, which carry out all the basic functions of life.

cell nucleus: the central part of a cell, inter alia containing genes (DNA).

central nervous system (CNS): the set of interacting neurons that receive and analyse information and enable thought and action.

cold dark matter (CDM): the major part of matter in the Universe, which is quite unlike ordinary matter (baryons) as it does not interact with baryons or with light.

contextual wavefunction collapse (CWC): a process whereby a quantum state becomes a classical state in a way that is dependent on the local context.

cosmic background radiation (CBR): blackbody radiation pervading all intergalactic space left over from the hot early stages of the Universe when it was emitted at 4000 K. It cools down as the Universe expands, and is observable today when its temperature is 3 K = −270°C. Thus it nowadays has a microwave wavelength, and is known as cosmic microwave background radiation (CMB).

cosmic inflation: extremely rapid exponential expansion of the very early Universe when its volume increased by a factor of at least 10^{78} in an extremely short time.

cybernetics: feedback control processes occurring in engineering.

deoxyribonucleic acid (DNA): the double helix genetic material that contains genes coding for proteins. Different DNA will result in different developmental outcomes via developmental processes.

developmental processes: the processes whereby a fertilized egg becomes a functioning organism. It occurs by gene regulatory networks (GRNs) reading specific DNA coding regions in a contextual way.

dopamine: neuromodulator that generates feelings of pleasure.

downward causation: processes whereby higher levels in a hierarchical structure affect what goes on at lower levels. This takes place via time-dependent constraints, and by altering lower-level elements in a contextual way.

electromagnetic spectrum: the full range of electromagnetic radiation, from radio through microwave to infrared, visible light, ultraviolet, X-rays and gamma rays.

electromagnetism: interactions between electricity and magnetism that underlie the existence of electromagnetic radiation and thus radio, TV, cell phones, radar and so on, as well as the functioning of electric generators, electrical motors and underlying chemical bonding.

emergence: the way complex structures arise out of simpler parts, with completely new properties arising at emergent levels. The nonlinearity involved is characterized by the phrase 'the whole is greater than the sum of the parts'.

enzymes: proteins that greatly speed up chemical reactions, without themselves being used up.

epigenetics: processes that control the reading of genes via gene regulatory networks (GRNs).

equivalence class: the class of all sets of entities that obey precisely the same abstract relationships.

evolution (biology): the processes of natural selection over long timescales that result in specific forms of DNA coming into existence, because their developmental outcomes enhance relative survival rates of specific organisms and species.

evolutionary–developmental biology (EVO–DEVO): the interaction effects between evolution and developmental processes. Neither by itself can result in the evolutionary outcomes we see.

fermions: spin-½ particles obeying Fermi–Dirac statistics, for example an electron.

free will: the ability to act to attain a chosen goal, within the limits set by physical constraints— and the ability to choose those goals.

general relativity: Einstein's theory of how gravity is based on spacetime properties, whereby matter determines aspects of spacetime curvature, and spacetime curvature determines how matter moves.

gene regulatory network (GRN): interacting network of RNA and proteins that determine what genes (segments of DNA) will be read at a specific time and place in an organism.

global climate change: changes to global climate patterns due to the enhanced greenhouse effect caused by emission of greenhouse gases, primarily carbon dioxide, resulting in an increase in average global surface temperature. Effects are increased droughts, fires, floods and tornadoes.

global pandemics: infectious diseases that spread across multiple countries, sometimes causing millions of deaths, including Spanish Flue, HIV/AIDS and COVID-19.

global positioning system (GPS): satellite-based system transmitting radio signals, enabling determination of position with an accuracy of meters. This accuracy depends on the satellites having very accurate clocks, and corrections to account for general relativity effects.

Grand Unified Theory (GUT): a Grand Unified Theory of fundamental physics, but not including gravity.

gravity: force of attraction between any two masses, proportional to the masses and inversely proportional to the distance between them. The force is always attractive because negative gravitational masses do not occur.

greenhouse effect: clouds reflecting radiation back toward the Earth, so that the Earth is hotter than it would otherwise have been, just as in the case of a greenhouse. The greenhouse effect at the right level is essential to our existence (otherwise the Earth would be too cold) but emission of greenhouse gases is now causing global climate change.

homeostasis: physiological feedback processes that keep variables such as temperature, heart rate, concentration of sodium, potassium and calcium ions, and blood sugar levels with safe bounds.

immune system: physiological system that protects an organism from diseases by combatting pathogens such as viruses and bacteria. The innate immune system protects against threats previously encountered in our evolutionary history. The adaptive immune system protects against threats not previously encountered.

kenosis: an attitude of being prepared to sacrifice on behalf of another.

km: kilometre = 1000 metres

ΛCDM: theory of matter in the Universe involving both cold dark matter and a cosmological constant Λ.

last scattering surface (LSS) of radiation in the early Universe: the time when radiation changes from being tightly coupled to matter to freely flowing.

light year: the distance light travels in one year. As light moves at the speed 300,000,000 meters per second = 186,000 miles per second, this is about 9,460,000,000,000 km (about 6 trillion miles).

macromolecules: very large molecules important in biology such as proteins and nucleic acids (RNA and DNA). They are composed of smaller building blocks: nucleotides in the case of RNA and DNA, and amino acids in the case of proteins.

metabolic network: complete set of metabolic and physical processes forming a network that determines the biochemical properties of a cell.

metabolism: set of reactions essential to life by converting energy in food to usable energy, converting food to usable molecular building blocks and eliminating waste.

metal–oxide–semiconductor field-effect transistor (MOSFET): a particular kind of transistor.

moral realism: view that there are moral statements that represent objective features of existence with universal validity.

moral relativism: view that all moral statements are social constructs with no universal validity.

multiple realizability: the way that higher-level emergent features in a hierarchy can be realized in multiple way at lower levels. This applies to both physical and abstract hierarchies,

multiverse: the proposal that there are many different expanding universe domains out there beyond where we can see them, with a variety of different physical and cosmological natures.

network: set of elements (nodes) joined by bidirectional or directed links.

neural network: connected set of neurons that form a network.

neuron: cell in the nervous system that conveys electrical signals (spike chains) to other neurons via axons and synapses. It is composed of dendrites, a cell nucleus, and axons.

organism: any living entity that functions as an integral whole, undergoing reproduction, growth and metabolism.

periodic table of the elements: arrangement of chemical elements into rows and columns, with each column representing elements with a broadly similar chemical nature.

physiology: physical structure of an organism carrying out all the different functions necessary for biological functioning. This structure is hierarchically organized with different functions being carried out at each emergent level.

placebos: medicines with no active ingredient that nevertheless are effective in improving medical outcomes.

possibility spaces: abstract spaces characterizing possibilities in physical, biological and abstract contexts.

predictive processing: brain processes involving continually updating a mental model of the environment, comparing this model with incoming data, and predicting likely outcomes on this basis.

proteins: macromolecules made of a string of amino acids that carry out key functions at a cellular level. Inter alia they act as catalysts for metabolic reactions (enzymes), provide structural elements (actin), act as molecular machines (kinesin and dynein) and respond to light (chlorophyll and rhodopsin).

quality-of-life (QOL) welfare feedback system: view of society as a multilevel feedback control system shaped to enhance the welfare of a particular coherent group in society.

redshift: if emitted by a source moving away from us, light looks redder than if emitted by a source at a constant distance from us.

reductionism: explaining higher-level emergent properties in terms of lower-level interactions, and specifically in terms of interactions between particles. While reductionism is a core aspect of explanation in physics, biology and neuroscience, it is controversial to what degree it can fully explain the nature of all emergent properties in all these cases.

ribonucleic acid (RNA): very versatile macromolecule made of nucleotides that play key roles in the transcription of DNA to proteins (messenger RNA, mRNA; transfer RNA, tRNA; and ribosomal RNA, mRNA), catalysing reactions, and taking a part in cell signalling.

second law of thermodynamics: the statement that the entropy of an isolated system cannot decrease with time, implying heat cannot flow from a colder to a hotter system.

spacetime: Combination of one time dimension with three spatial dimensions to form a four-dimensional spacetime. This underlies both special relativity and general relativity.

special relativity: theory of physics in flat spacetime (i.e. no gravity) characterizing motion of massive objects as limited by the speed of light c and relating mass m and energy E by the formula $E = mc^2$.

spectrum of values: a characterization of the nature of values of an individual or organization as ranging from the extremely cruel and self-centred to the extremely generous and self-sacrificial.

Standard Model of particle physics: theory describing three of the fundamental forces (electromagnetic, strong, weak) and associated particles.

strong force: fundamental interaction combining quarks to form protons and neutrons, and combining protons and neutrons to form atomic nuclei.

synapses: structure allowing an electrical or chemical signal to pass from one neuron to another, thus forming the key structure enabling creation of neural networks out of neurons.

technology: application of knowledge to achieve practical goals in a reliable way via control of physical forces and materials, or by use of information, software and algorithms.

theory of everything (TOE): hypothetical unified theory of all fundamental physics, including gravity. We do not have such a well-established theory at present.

Theory U: change management method developed by Otto Scharmer.

tragedy of the commons: the abuse of common resources when many people use it in such a way that it gets degraded and becomes valueless.

transactional leadership: a leadership style that relies on penalties and rewards to achieve short-term goals. It prioritizes individual interests and extrinsic motivation as the means to obtain a desired outcome.

transformational leadership: a leadership style where a leader's behaviour inspires their followers to work towards common goals because they are seen as valuable. This intrinsic motivation can achieve remarkable results.

universal basic income (UBI): unconditional income grant to everyone in a society.

uroborus: an ancient symbol depicting a serpent eating its own tail.

weak force: a weak interaction between subatomic particles that is responsible for radioactive decay.

zygote: a fertilized cell that will grow into a plant or organism.

Neuroscience and Literacy

An Integrative View

This Appendix reproduces an article I wrote in 2021 with my wife Carole Bloch, an expert in early literacy. In contrast to the globally prevalent reductionist view, it presents an integrative view on how young children become literate.

We think that it is crucial to take the integrative neuroscience view seriously because it offers a broad, inclusive understanding of what is going on. This inclusive view includes, but does not prioritize, the elements that reductionist neuroscientific thinking proposes are the critical ones for beginner readers. It is our view that without an integrative understanding, millions of children in diverse settings around the world will continue to be failed by systems that constrain rather than enable opportunities for them to learn to read for meaning. Three inter-related issues underlie what we present in this article.

Firstly, the reductionist view presents a whole text as if it is nothing more than the sum of its parts. It assumes that a learner has to build up the whole by putting the parts together first, from smallest to big. It claims that comprehension then follows. But this is not true. The whole is much more than the sum of its parts, inter alia because meaningful texts are read in a contextual way, where context often determines not only meaning but how words are pronounced. Learners need to be encouraged to seek meaning in texts from the start. We explain in detail how that happens.

Secondly, a crucial conceptual issue are the purposes people have for reading and writing. These include the contextual sociocultural practices related to written language which children learn through involvement and exploration. Using written language is about making and conveying meaning. It enables you to understand what others think and feel, and enables you to convey what you think and feel to others. This is a form of communication between the reader and writer, beautifully stated by Carl Sagan in a passage quoted in §5.3.5 of this book. These purposes for reading and writing ought to frame and drive the hard work every child does when becoming literate, because it is this which motivates children to want to learn. The reductionist view minimizes the significance of engaging with purposeful, authentic texts in the beginning stages of teaching reading.

Thirdly, the affective dimension is crucial. Because motivation is key to learning anything whatsoever, this includes learning how to read and write. As explained in §5.4 on affective systems, the search for meaning (the SEEKING system) is a key driver of motivation. If the initial stages of being taught seem meaningless because the focus is only or mainly on phonics exercises, many children cannot see the point of what is required of them. They then find it hard, if not impossible, to bring meaning to the task.

The article is reproduced with the permission of the Publisher and of the Royal Society of South Africa.

Neuroscience and Literacy

An Integrative View

George Ellis & Carole Bloch

1. Introduction: the Context for this Paper

The well-known global debates and divisions in relation to the way children become literate have increasingly been influenced by neuroscience evidence on reading (Seidenberg et al., 2020). Research undertaken by academics and researchers in the field, concentrated in the powerful centres of the Global North (especially the USA and the UK), have far reaching impact on those making policy and developing national documents (sometimes legally framed) at government level in relation to literacy teaching in diverse settings (Hoffmann, 2012). This in turn influences pedagogy, curriculum approaches and teaching methods. The longstanding proposed 'science of reading' (Seidenberg et al., 2020; Shanahan, 2020) has increasingly been used by some involved in these debates to claim that neuroscience studies support the *Simple View of Reading* (SVR), a model of reading which views the reading process in unidirectional linear terms. Literacy teaching debates and practices are at present dominated globally by approaches arising from the SVR (see Castles et al., 2018; Clark, 2020 and a special issue of the *Reading Research Quarterly*: Goodwin and Jiménez, 2020). Because of the claimed supporting neuro scientific evidence, there is now a widely perceived gravitas and authority adhering to this particular model, which insists on building skills as a prior step to comprehension. It is used to claim that the 'reading wars' which pitted phonics against whole language should now be over (Castles et al., 2018). Many literacy specialists and teachers find an either-or position misleading and unhelpful; some prefer a 'balanced approach' as a middle ground to ensure children get 'the best of both worlds' in teaching programmes (Willson and Falcon, 2018). In any event there is no consensus over common narrow interpretations of the SVR; several recent papers focus on the complex and multifaceted interplay between decoding and listening comprehension (Bua Lit Collective, 2018; Cervetti et al., 2020; Compton-Lilly et al., 2020) or even propose a *Complete View of Reading* (CVRi) (Francis et al., 2018) or similar (Snow, 2018).

Still, aspects of reductive neuroscience are used to justify the kind of singular teaching focus on skills, which is currently commonly viewed as necessary for all children as they begin their formal schooling. This is understood to be the case irrespective of vastly diverse socioeconomic and cultural contexts and individual experiences, with immense significance for the serious global literacy teaching challenges. South Africa is a case in point. The longstanding systemic problems with teaching early literacy effectively in this particular context (Taylor, 1989; Bloch, 1999, 2000; Alexander and Bloch, 2010; NEEDU, 2013), characterized by historical language inequities and extremes of economic and social inequality, illustrates how urgent this matter is. Several government-led initiatives have been attempted to improve matters since apartheid ended (e.g. DoE, 2008; DoE, 2011; Van der Berg et al., 2016), but children have continued to perform badly on all assessments; and particularly on PIRLS, which is focused on comprehension. The 2016 results claimed that 78% of South African children could not read for meaning by the end of grade 4 in African languages or English (Howie et al., 2017).

This intensified efforts and discussions to both understand why this situation exists and to provide viable solutions to get children 'reading for meaning by age 10'[1] (Reeves, 2017; Hickman, 2018; Bua-lit Language and Literacy Collective, 2018; Fleisch and Dixon, 2019). But the view of reading which dominates research, policy and curriculum-related interventions reflects the Simple View of Reading (Spaull et al., 2020) and early literacy teaching in classrooms is still characterized by the foregrounding of decontextualized exercises.

Recent initiatives include an acknowledgement that learning to read and write should begin in languages children understand; this has led to attention being given to establishing reading benchmarks in African languages (Jukes et al., 2020; Spaull et al., 2020); also increased government-level action on the recognition that the years before formal school are crucial ones for laying firm learning foundations (DoE, 2015; Harrison, 2020). This is in response to the broad global consensus that during their preschool years, children learn best in informal and meaning- and play-based ways, including their oral language foundations and first steps to literacy. Yet there is an increasing push down pressure to consider earlier skills teaching from formal schooling, compromising time for play in the preschool years (Campbell, 2020). And, while the curriculum for the first 4 years of primary schooling also orients toward meaningful teaching and learning (DoE, 2011), a conceptual schism exists between curriculum statements and their teaching implications, on the one hand, and on the other, popular understandings and practices regarding what to prioritize for initial literacy teaching in school. Grade R, which is simultaneously the last preschool year and/or the first primary school year, is caught at the centre of this conceptual and practical conundrum.

In the interests of working toward equity and justice for all children, our concern in this paper is to problematize the validity of the reductive neuroscience view of how reading works. We do this by offering alternative evidence from the growing body of integrative neuroscience which perceives reading as a nonlinear holistic process strongly influenced by affect and involving a foundational search for meaning through prediction, developing through tentative exploration as skills are built. In particular, we dispute the widely claimed view that oral language is natural but written language is not (Shaywitz, 2003: 49–50; Wolf, 2018). Rather we argue that both are cultural inventions that originated at different times through similar evolutionary processes in the long distant past, in order to meet social needs (Harari, 2011).

A note on terminology: in the educational body of literature on teaching literacy, the term 'reading' has been used far more than 'writing'. This reflects how these aspects of literacy tend to be viewed and taught separately. More recently, the term 'literacy' is being used as a conscious umbrella term to bring more integrative sociocultural understandings to bear (Frankel et al., 2016). In this paper, when referring to published work we tend to use the term 'reading' as the authors often do, while in our own writing we use 'writing and reading' and 'literacy' synonymously, unless we are specifically referring to one of them.

1.1 Contrasting Early Literacy Perspectives

The different views about literacy and how it is learnt arose from historical disagreements about the nature of knowledge and how language learning happens (Altwerger et al., 2007: 4). Two contrasting pedagogical perspectives coexist, and have been argued about for hundreds of years (Huey, 1908; Chall, 1967; Pearson, 2004; Kim, 2008; Castles et al., 2018; Miller, 2020). The

[1] These terms became widely used in South Africa since the President of South Africa called for all children to 'read for meaning' by aged 10 in his 2019 State of the Nation Address, after being alerted to the severity of the challenges following these results.

central issue has come to be how and when comprehension comes about. These two perspectives can be related to two models of literacy (Street, 1984). One, which Street (2006) calls the 'autonomous model', views literacy as constituting separate sets of skills, to be taught independent of context. The other views literacy as being based in the social practices of communities. In this 'ideological model', there are different forms and uses for literacy in different sociocultural and linguistic settings, and these form the basis for teaching. Understandings about young children's literacy learning and teaching can be viewed as falling under one or other of these umbrellas.

Early literacy teaching approaches which correspond to Street's broad autonomous model of literacy are underpinned by skills-based[2] models (summarized in Figure A.1). Now supported by reductive neuroscience studies based in a linear view of cognition and action, the automatic decoding of phonics skills (recognizing and sounding out letter-sound relationships) followed by fluency are widely seen as essential prior steps to comprehension, because of how the brain is believed to function. This applies to each language a child is being taught to read. The meaning-based[3] model (summarized in Figure A.2), which corresponds to Street's

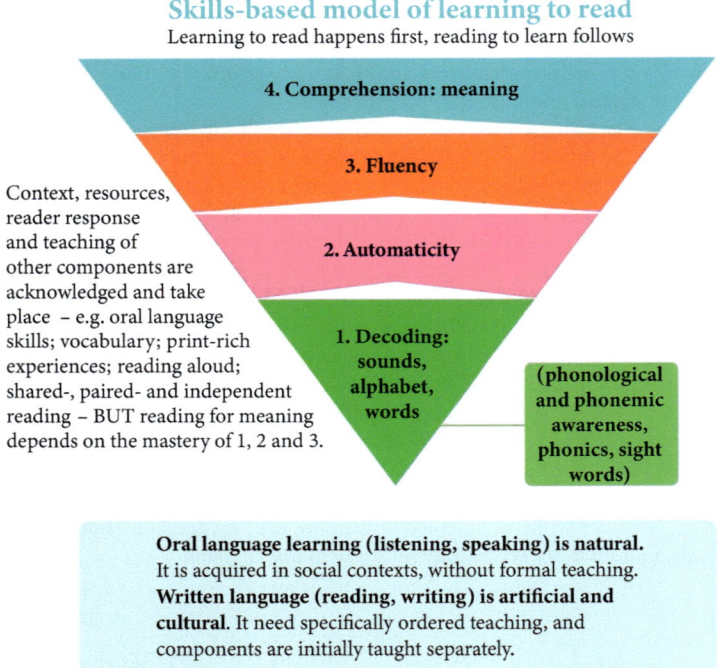

Figure A.1 **Skills-based model of learning to read.**

Source: Carole Bloch.

---

[2] We use the terms *skills based* to refer to 'part to whole' views of early literacy teaching which, may differ in detail, but all start from the 'bottom up' with phonics and other technical skills, also referred to as 'phonics based', or structured literacy.

[3] We use the term *meaning based* to refer to views of early literacy teaching which may differ in detail, but prioritize context, sociocultural practices and meaning making. They are sometimes called 'top down', emergent literacy, whole language or 'social practices'.

ideological model, underpins approaches which see initial and ongoing meaning construction taking place, with alphabetic knowledge and phonics skills being taught in the context of authentic literacy-related experiences. This implies using relevant languages[4] and a focus on

Meaning–based model of learning to write+read
Learning how to write+read happens while print is used for
personally meaningful reasons

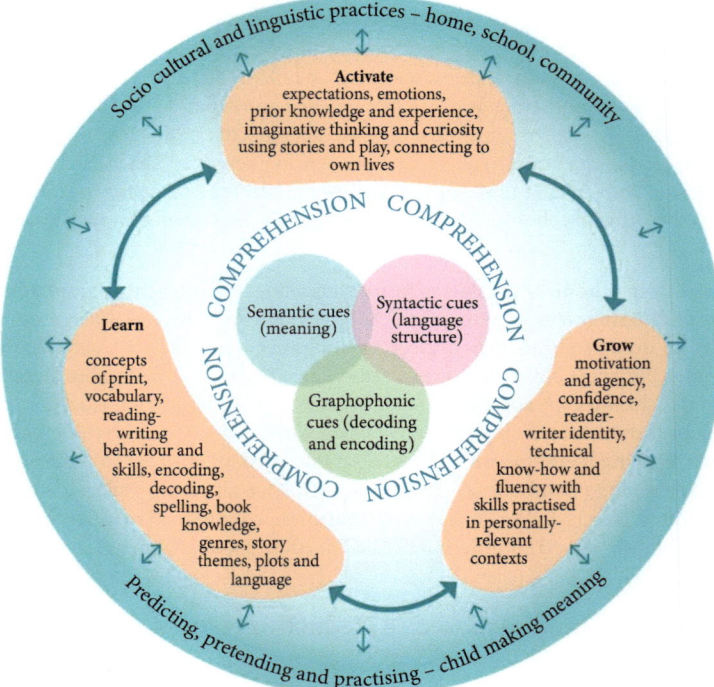

Oral and written language are **learned in social contexts with both informal and formal teaching.** Although oral language is much older, they both evolved as symbolic systems to communicate meaning.

Children apply their knowledge of oral and written language to communicate purposefully from the start. They learn to write+read from knowledgable and interactive role models who teach skills as these are needed to carry out desirable print related activities.

Figure A.2 Meaning-based model of learning to jointly read and write.

Source: Carole Bloch.

[4] Although we do not deal directly with multilingualism, multiliteracies and learning in this paper, we flag this as involving significant pedagogical issues which are impacted on directly by the views of neuroscience which underpin programmes for language and literacy teaching for all children.

motivation, personal agency and meaning, with predictive understandings connecting content to children's sociocultural understandings and practices.

It is important to note that in a meaning-based model[5] there is no 'profound mistake' being made through omitting phonics, as alleged by Seidenberg et al. (2020), because the view is of a complex meaning-based process with graphophonic cues working in concert with semantic and syntactic cues, as indicated in Figure A.2. Phonics is taught as it arises in the texts being read and also as it is required to write.

Informal learning before school. A significant body of international interdisciplinary early literacy research evidence has been conducted into the years before formal schooling.[6] Those with meaning-based perspectives tend to conceive of learning related to written language as forming part of the informally structured foundations of learning processes, consistent with early childhood wisdoms and traditions which value holistic learning and play (Bruce, 2015). The view is that these foundations ought to be deepened and expanded as school begins (Bloch, 2018; Bua lit, 2018). By contrast, from a skills-based perspectives, such learning is usually conceptualized as preparatory: pre-reading and pre-writing activities, the basic building blocks which are needed to be taught to young children so that they are ready for the formal teaching of reading and writing in school. The strong implication is that proper literacy learning begins here.

Formal learning in school. At the start of formal schooling, many teaching programmes follow the narrow skills-based interpretation of the SVR (Gough and Tunmer, 1986; Compton-Lilly et al., 2020). In doing so, they may neglect to emphasize and enable crucial meaning-based elements and experiences children require in the vital early stages of becoming literate, thereby restricting opportunities for appropriate quality learning. We argue that from the early years onward, major features of how the predictive brain works, which are currently ignored in the early literacy teaching literature, need to be taken into account in order to ensure appropriate conditions of learning (Cambourne, 1995, 2000, 2020) for all young children whenever they encounter written language. We do not agree that these conditions are different for children from low SES, poorly served communities (Abadzi, 2006, 2008).

1.2 Understanding Based in Reductive Neuroscience Views

Many studies present the brain regions involved in oral language (Friederici, 2017) and written language (Shaywitz, 2003; Dehaene, 2010; Wandell et al., 2012; Kearns et al., 2019). Some psychologically or cognitive-based texts give a brief presentation of the neuroscience, e.g. Deacon (1998), Wolf (2008), Schnelle (2010), Seidenberg (2017) and Hruby and Goswami (2019). Others model cognitive processes without linking to neuroscience proper, e.g. Tomasello (2003) and Willing-ham (2017). Many link language to evolution, e.g. Tomasello (2000), Donald (2001) and Greenspan and Shanker (2004). We note that much of the literature which focuses specifically on neuroscience and reading has grown out of studies related to dyslexia, e.g. Shaywitz (2003) and Wolf (2008). An important question is thus to what extent studies of dyslexia throw light on normal reading processes?[7] These studies have by definition a deficit

---

[5] We avoid the term 'whole language' as this is a loaded term, with various interpretations and misinterpretations of its meaning.

[6] To date, most of this early literacy research has been done in high socioeconomic status (SES) countries of the Global North. Despite significant recent scientific attention on the importance of the 'first 1000 days', the early years of childhood are still poorly provided for, and are very low in actual status and value in terms of support for quality care and educational provision, except for the children of the elite. Slowly it is being instituted in the Global South; research attention follows in its trail.

[7] We use the term 'normal' here to include diverse SES, cultural and linguistic practices and contexts.

view of the reading process built in, which means their recommendations will necessarily be affected by that view.[8]

In many of these writings, the link to neuroscience is limited to diagrams of active domains and pathways in the brain when phonemes, words or non-words are read, or more accurately, decoded, particularly referring to the Visual Word Form Area (VWFA).[9] These are supported by functional neuroimaging studies. While this gives useful information about neural pathways associated with reading, one should be aware that they are rarely accurate representations of the full functional brain networks operating when a person reads or attempts to read meaningful texts, and hence they only give a very partial picture of what goes on in the brain when such purposeful reading takes place.[10] Furthermore many assume that perception operates in a linear manner from sensory data input to analysis of that data in the cortex, resulting in the SVR. For this reason, we term this reductionist neuroscience.

To focus the discussion, we refer mainly to four bodies of work which encapsulate the reductionist neuroscience view: Shaywitz (2003), because her neuroscience research into dyslexic children's brains informs 'normal' reading. as do her views of natural and unnatural language; Dehaene (2010), as it is in many ways the groundwork on reading and the brain that many others refer back to; Abadzi (2006, 2008, 2017), who has been immensely influential for development aid literacy programmes via her work at the World Bank and the Global Partnership for Education in the Global South;[11] and Castles et al. (2018), as this paper summarizes the SVR and is an up-to-date review of the 'reading wars' between the two positions on literacy and how it should be taught. Castles et al. (2018) summarize the SVR (Gough and Tunmer, 1986) thus:

> The *Simple View of Reading* posits that reading comprehension R is the product of two sets of skills, 'decoding' D and 'linguistic comprehension' C:

$$R = D \times C. \tag{1}$$

> The logical case for the Simple View is clear and compelling: Decoding and linguistic comprehension are both necessary, and neither is sufficient alone. A child who can decode print but cannot comprehend is not reading; likewise, regardless of the level of linguistic comprehension, reading cannot happen without decoding. . . . Early in development, reading comprehension is highly constrained by limitations in decoding. As children get older, the correlation between linguistic and reading comprehension strengthens, reflecting the fact that once a level of decoding mastery is achieved, reading comprehension is constrained by how well an individual understands spoken language.

---

[8] It is pertinent to consider how this might affect both the confidence of young beginning readers who live in poorly served communities and expectations of them as readers. Many are taught by teachers who have been trained to perceive them as already lacking in school readiness skills; a deficit model of reading is added to this.

[9] The role of the VWFA as unique to reading has been called into question recently, inter alia by Vogel et al. (2012, 2014), Moore et al. (2014), Martin et al. (2019) and Vidal et al. (2021). The VWFA is not present on functional MRI scans before learning to read, but appears and enlarges as reading skill is gained. Also noted is that from the outset the VWFA is strongly connected to the dorsal temporoparietal areas which are activated during speaking and listening, and are present as early as 2 months of age. The fMRI scans show that, after the initial visual reception, reading results in nearly instantaneous and simultaneous involvement of widespread ventral, dorsal and frontal areas involved in the sound, shape and meaning of words in skilled readers. We thank Roland Eastman for these comments.

[10] A study which does this is Fedorenko et al. (2016), showing how different brain areas are indeed involved.

[11] Klaas and Trudell (2011), Piper et al. (2016), and including South Africa, see for example Spaull amd Pretorius (2016: 9).

They then use this view that comprehension is initially constrained by limitations in decoding to motivate the imperative of a skills-based model (although they do also emphasize the importance of broader reading experiences). But firstly, decoding as such is only necessary for the dorsal pathway, one of the two neural reading pathways that they describe (see Section 5.3 below): it does not explicitly occur in the ventral pathway because graphophonic as well as other structural language features are always sampled to the extent that they are needed to predict the meaning of the text. Miscue analysis (Flurkey et al., 2008) demonstrates that successful reading involves preserving meaning of the text, though not necessarily word accuracy. So letter-by-letter decoding is not, in fact, necessary in order to read: words can be grasped as a whole in a gestalt way (Section 2.1). Secondly, comprehension early in development is more likely to be constrained when reading is taught with the strong primary emphasis on decoding skills they recommend. This restricts the child's attempts to understand the text directly by drawing on other clues, because of the way teaching focuses the child's attention toward accuracy and fast decoding. For example Spear-Swerling (2019) argues against encouraging students to attend to multiple-cueing systems[12] when reading, which is what a mature reader will do.

Dehaene makes this explicit when he says

> The child's brain, at this stage, is attempting to match the general shape of the words directly onto meaning, without paying attention to individual letters and their pronunciation—a sham form of reading. (Dehaene, 2010: 200)

His is acknowledging that the use of the ventral pathway is possible and indeed young children can do so, but his reductionist perspective leads him to recommend preventing this from happening. He defines reading inadequately. He wants the parts to work rather than the integral process, and characterizes as 'sham' reading that which is both the intention of proficient readers and a profound reading path for young learners. He shuns precisely what children need to do to avoid a possible memory overload, which is a reason given for the need to concentrate children's attention on developing swift and automatic decoding (see Section 5.4). Dehaene also dissuades teachers from encouraging children from making attempts at conventional reading:

> Children need to understand that only the analysis of letters one by one will allow them to discover a word's identity. (Dehaene, 2010: 229)

This contradicts the predictive understanding of perception we highlight below, see for example Friston et al. (2017a), and ignores the other cueing systems which proficient readers use and which should be encouraged and supported in learners. The serious problem is that Dehaene's authoritative advice to educators (Dehaene, 2010: 230), where everything is planned to the last grapheme, is a recipe for rigidity that makes no allowance for prior knowledge and development and social and cultural experiences, or the role of motivation and the drive toward understanding. He makes statements against including illustrations in books (Dehaene, 2010: 229) or posters on the wall. This bleak view of early literacy teaching completely ignores the powerful symbolic life and imagination of young children, their impressive linguistic and intellectual capabilities and the affective dimension of the mind that we emphasize below.

Abadzi (2017: 8) offers a view of comprehension when she claims that, in contrast to its usual prominent position in high SES educational contexts, 'comprehension' need not be the aim of the learning process for poor children (Abadzi, 2017: 8):

> Should instruction focus on reading comprehension early on? Middle-class children often process quickly and have rich vocabulary; so, in high-income countries, literal comprehension

---

[12] We refer to multiple-cueing systems below in Section 5.2.

may be too simplistic. Instead, 'comprehension' is often used to signal inferences or predictions. These require more knowledge than offered in a text. Poorer students have more limited vocabulary and expression, and they may lack the academic language to deal with classroom conversations.

This view suggests reading need not imply comprehension, and that is precisely the problem that can occur when the focus is on teaching skills out of context. Does she believe that it's 'natural' for middle-class children to develop rich vocabularies? This highly problematic position begs the question of how to address deep inequalities in transformative ways to promotes intellectual and affective justice and equity. She continues:

> To teach the poor efficiently, we must make learning easiest on their brains. The research suggests that, when time is scarce, reading components could be taught sequentially. The sequence could roughly follow that of the reading stimuli as they go through the brain. Teachers must focus instruction and practice on the early visual processes and speed those up in order to facilitate complex cognition. Middle-class reading instruction, such as the simultaneous teaching of the 'five pillars',[13] may slow down and complicate the acquisition of this quintessentially visual skill. The answer to the twenty-first-century reading crisis may lie in second-century practices, such as decoding, that apparently most human brains could perform. (Abadzi, 2017: 11)

Visual processes are not the bottleneck, because of the predictive nature of vision which involves the corticothalamic feedback circuits. Much information needed for interpretation is already present before the signal arrives. The extreme position taken by Abadzi implies that the brains of poor (African language speaking?) children are different from those of more affluent ones and are unable to deal with complexity. Apart from being insulting and patronizing, it misleads teachers and learners down imaginative and intellectual cul-de-sacs. In Abadzi (2008) she makes a major issue out of the mind's short-term timeframe and the need to read fast. But there is no need to read fast; the need is to read and to learn to read with comprehension (Dowd and Bartlett, 2019). The problem arises if one insists that **reading starts with tracking and interpreting individual letters in a morass of print** (Abadzi, 2008) and then teach in such a way as to enforce this as the priority. She later admits that **fluency is achieved when an instant word recognition pathway is activated** (this is the ventral pathway). She claims this happens after **much practice in pairing consistently sounds with groups of letters.**

However, teaching approaches which emphasize the authentic language of storytelling expose children to precisely the rich vocabulary and expression which Abadzi claims they may not have. In settings which have been dominated by colonial and post-colonial education systems, reinstating story as a legitimate educational form is worthy in itself, as well as providing an obvious segue to written language; motivation comes as adults and children connect with personal and cultural histories, at the same time as they create some of the texts to read.

Shaywitz emphasizes the view that oral language is natural, and written language is unnatural, for example

> Spoken language is instinctive, built into our genes and hardwired into our brains. Learning to read demands that we take advantage of what nature has provided: a biological model for language. (Shaywitz and Shaywitz, 2004)

She, and many others, have used this as one of the powerful motivators for skills-based reading models (e.g. Wolf, 2018; Spaull and Pretorius, 2019: 5). We strongly critique this understanding in Section 4 below.

---

[13] These five pillars were identified by the National Reading Panel (2000) as phonemic awareness, phonics, reading fluency, vocabulary and reading comprehension.

A: The linear context-free model of reading

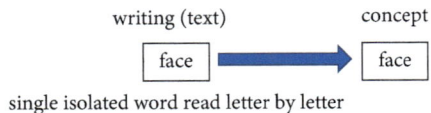

single isolated word read letter by letter

B: The nonlinear contextual model of reading

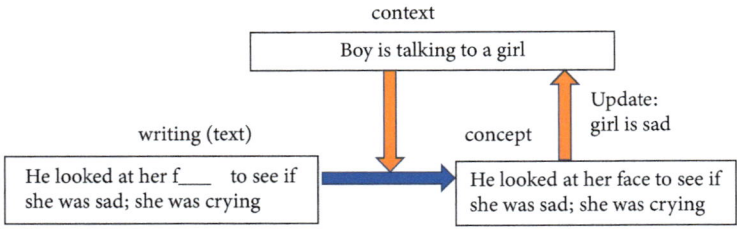

Missing word implied by context; context updated in response to answer

Figure A.3 The linear reading process envisaged by the Simple View of Reading (A), and how contextual information enables word identification in a contextual way (B). When the context is updated in response to the information gained, one has a model of the predictive processing understanding of the reading process. The closed loop makes it nonlinear. The letter ambiguity Castles et al. (2018) discuss ('fact' or 'face'?) can be resolved in this way without reading every letter. The outcome is an interactive model of reading (Rumelhart, 1977) in agreement with Seidenberg et al. (2020).

Source: George Ellis.

1.3 Understanding Based in Integrative Neuroscience Views

Miłkowski et al. (2018) claim that cognitive neuroscience has undergone a silent revolution based in the integration of wide perspectives with the rest of the cognitive neurosciences. These substantial change in neuroscience perspectives on brain function develop from earlier views on how perception works, for example Gombrich (1961), Gregory (1978) and Purves (2010), leading to the hierarchical predictive processing view of action and perception espoused by Friston (2003, 2010, 2012), Clark (2013, 2016), Hohwy (2013), Seth (2013), Fabry (2017) and many others, giving a more integrative view of brain function. In discussing this integrative neuroscience and its relevance for literacy learning and teaching, we point out the importance of five major features of how the brain works:

- First, perception is an active, contextually based predictive process, based in detection of errors in hierarchical predictions of sensory data and action outcomes. Reading and writing are particular cases of this process. Not all text need be read; words can be filled in due to context (Figure A.3).
- Second, emotions play a key role in underlying cognitive functioning. Innate affective systems underlie and shape all brain functioning, including communicating by speech and writing.

- Third, there is not the fundamental difference between listening/speaking and reading/writing that is often alleged on the basis of evolutionary arguments. They are both social and cultural practices learnt through social processes.
- Fourth, brain function is not fundamentally based in a rule-based way of responding to data. It is a neural network of huge dimensions, whose natural mode of operation is statistical pattern recognition and prediction, based in nonlocal storage of data. It is a Bayesian machine.
- Fifth, like listening, reading is a nonlinear contextually shaped psychosocial process of conveying meaning in a specific context, shaped by current knowledge. One of the two neural routes to reading does not involve explicit decoding processes, and can be activated from the earliest years.

This predictive nature of perception is enabled by corticothalamic circuitry (Alitto and Usrey, 2003) allowing the downward passing of predictions from the cortex to the thalamus, as depicted in Figure A.4. We contend that a twenty-first century perspective on literacy must include this evidence about not just unidirectional but bidirectional neural messages passing in hierarchical systems. This processing is affected by affective (emotional) messages passed

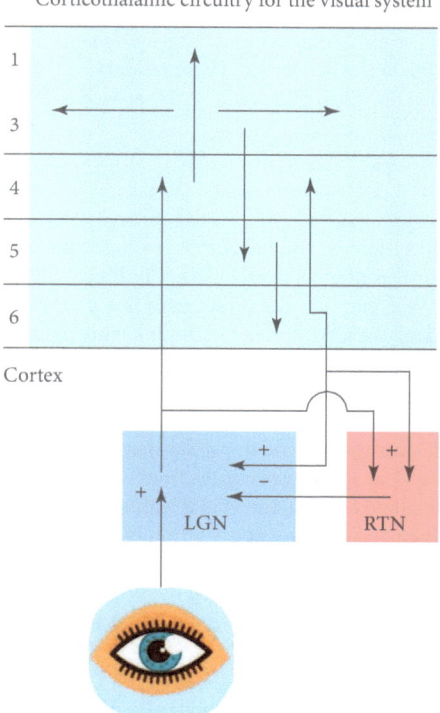

Figure A.4 Corticothalamic circuitry for the visual system. Information flows from the eyes via the optic tract to the lateral geniculate nucleus (LGN) in the thalamus and then via excitatory projections to level L4 in the visual cortex and on to levels L3–L1. Predictive information flows down from L3 to L5 and L6. Neurons in L6 send excitatory feedback to the thalamus and the reticular nucleus (RTN). The feedback axons terminate on relay neurons in thalamic relay nuclei, as do inhibitory projections from the RTN.

Adapted from Alitto and Usrey (2003).

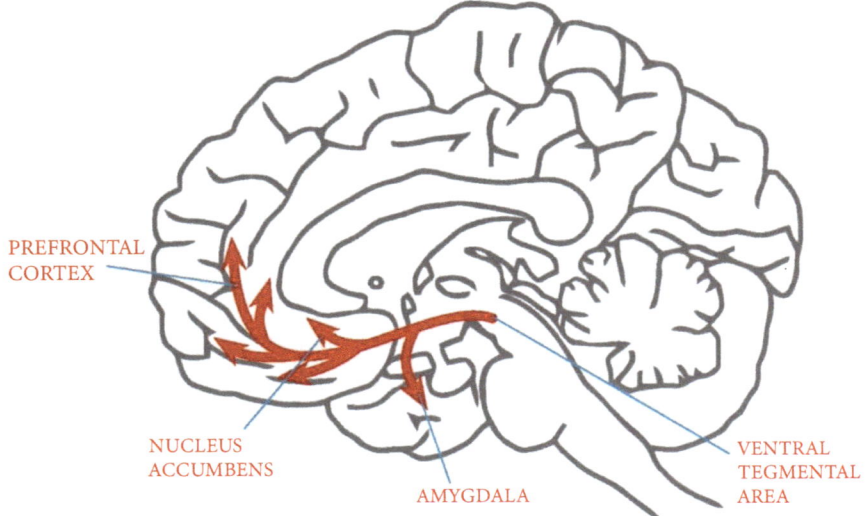

Figure A.5 The SEEKING system is one of the ascending systems that project diffusely to the cortex from nuclei in the excitatory systems, conveying neuromodulators such as dopamine and epinephrine to the neocortex. These reticular activating systems underlie Gerald Edelman's neural Darwinism (Edelman, 1987) as well as Panksepp's primary affective systems (Panksepp, 1998).

Source: Mark Solms.

diffusely from the limbic system to the neocortex via ascending systems (Figure A.5). Crucially, there are two neural routes to reading—the 'indirect' (dorsal) and 'direct' (ventral) pathways (Figure A.6). We argue that the direct path is a powerful biologically natural way by which beginning readers can learn to read without having to explicitly decode. The resulting view of the reading process corresponds with the meaning-based views proposed inter alia by Goodman (1967, 1982), Strauss et al. (2009), Bever (2009, 2013, 2017) and Goodman et al. (2016).

In what follows, we discuss each of the five major features in detail. Section 2 looks at the brain and how perception is an active process, Section 3 at how cognitive function is crucially shaped by affect (emotions) and Section 4 at what is 'natural': what are the innate brain systems? Section 5 looks at how the natural mode of operation of the brain is statistical pattern recognition and prediction, and Section 6 considers the similar neural and psychological processes involved in meaning making and communicating by listening/speaking and reading/writing. Section 7 comments briefly on possible educational implications.

2. Perception is An Active, Contextually Based Predictive Process

The brain works in a complex, nonlinear way. The neocortex is a predictive organ (Hawkins, 2005; Kveraga et al., 2007) based in connectionist principles (Section 5). It is the seat of perception and pattern recognition, learning based in neural plasticity and sensation/action based in prediction and choice (Purves et al., 2008; Gray, 2011). Downward causation takes place in a variety of ways: in relation to perception, attention and motor control

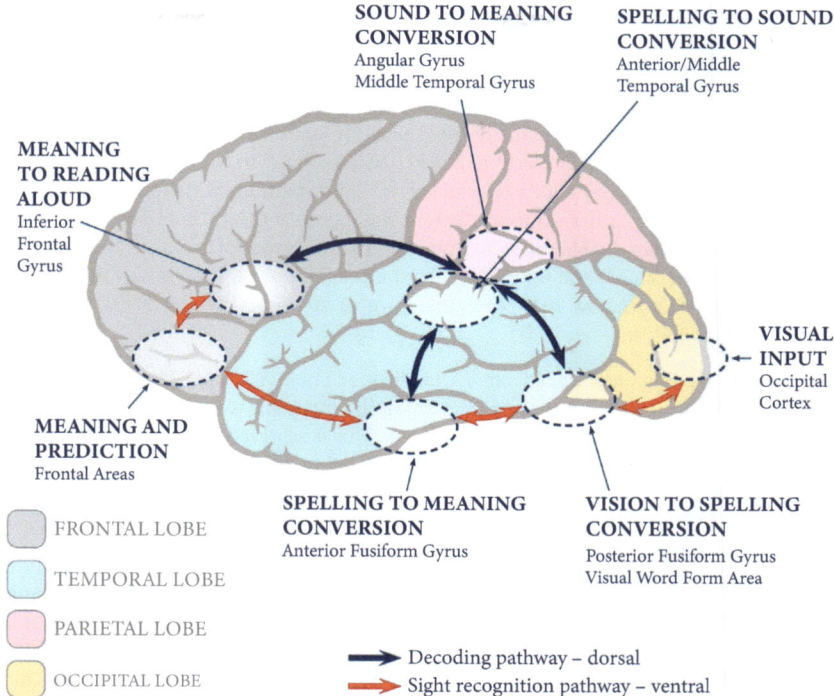

SOUND TO MEANING CONVERSION
Angular Gyrus
Middle Temporal Gyrus

SPELLING TO SOUND CONVERSION
Anterior/Middle
Temporal Gyrus

MEANING TO READING ALOUD
Inferior
Frontal
Gyrus

VISUAL INPUT
Occipital
Cortex

MEANING AND PREDICTION
Frontal Areas

SPELLING TO MEANING CONVERSION
Anterior Fusiform Gyrus

VISION TO SPELLING CONVERSION
Posterior Fusiform Gyrus
Visual Word Form Area

FRONTAL LOBE

TEMPORAL LOBE

PARIETAL LOBE

OCCIPITAL LOBE

Decoding pathway – dorsal
Sight recognition pathway – ventral

Figure A.6 Brain pathways associated with reading. A dorsal pathway underpins phonologically mediated reading, and a ventral pathway underpins direct access to meaning from print. Many further cortical areas will be involved when meaningful reading occurs, for example reading stories with an emotional impact, and the brain engages with that meaning in its social context. Standard neuroimaging studies do not emphasize these further areas because they do not deal with the reading of meaningful texts. For analogous diagrams in the case of oral language, see Friederici (2017: 107, 109, 124, 128, 135).

Adapted from Taylor et al. (2013), Rastle et al. (2001) and Kearns et al. (2019), under the expert guidance of Professor Roland Eastman (former head of the Neurology Department, University of Cape Town).

(Ellis, 2016, 2018). Reading and listening are forms of perception; speaking and writing are forms of action modulated by perception.

2.1 How Perception Works: Hierarchical Predictive Processing

The key point we wish to raise here is the predictive way all sensory systems work as discussed by Gregory (1978) and many others. The brain understands the world in a holistic way on the basis of the clues offered to it (Purves, 2010; Kandel, 2016). It has to do this in order to solve Helmholz's inverse problem, namely we are not provided by our senses with enough data to uniquely determine what the situation 'out there' is. We have to do the best we can with what sensory data is available, even though some needed data is missing.

Consequently vision is an active process (Findlay and Gilchrist, 2003).

Like vision (Frith, 2007; Purves, 2010), reading involves prediction in the light of previous experience and the confirmation or adjustments of such predictions in the light of new information (Smith, 2012). This is nothing other than the process of hierarchical predictive processing[14] (Friston, 2003, 2010; Clark, 2013; Hohwy, 2013; Seth, 2013, 2014), which underlies how reading text with meaning actually takes place (Goodman, 1967; Flurkey et al., 2008; Strauss et al., 2009; Smith, 2012; Goodman et al., 2016). This is indicated by eye-tracking and miscue studies[15] as well as our ability to read scrambled or partially constituted pieces of text.

All perception works in the same contextual way because they are all based in the same cognitive mechanism, applied in different domains. They all proceed by, in advance, predicting what ought to be perceived, and then adjusting the predictions on the basis of incoming data (Bever and Poeppel, 2010; Yon, 2019) so as to minimize surprisal (Friston, 2010). This is stated by Clark (2013) as follows:

> Brains, it has recently been argued, are essentially prediction machines. They are bundles of cells that support perception and action by constantly attempting to match incoming sensory inputs with top-down expectations or predictions. This is achieved using a hierarchical generative model that aims to minimize prediction error within a bidirectional cascade of cortical processing. Such accounts offer a unifying model of perception and action, illuminate the functional role of attention, and may neatly capture the special contribution of cortical processing to adaptive success.

This is a hierarchical process in that it involves multiple levels (Ding et al., 2016) and multiple timescales (Keitel et al., 2018). It is facilitated firstly by the downward passing of information in the cortex (Bar et al., 2006), and secondly by feedback loops of thalamocortical circuitry (Alitto and Usrey, 2003; Kveraga et al., 2007; Briggs and Usrey, 2008) shown in Figure A.4. There is no direct link from either the visual system or the auditory system to the relevant parts of the cortex; in both cases incoming information is first sent to the thalamus for processing. Here cortical predictions generate a difference signal relative to incoming data from the optic nerve, which is then fed back to the cortex as a measure of surprisal (Parr et al., 2018) which is used to update predictions in the cortex.

This is a nonlinear signal processing operation. New data comes in (as it is constantly doing), and you update your current hypothesis on the basis of this incoming data through Bayes' rule, a mathematical relation which your mind automatically implements (Clark, 2013). This happens subconsciously in such a way that these predictions actively and efficiently facilitate the interpretation of incoming sensory information and directly influence conscious experience (Panichello et al., 2013). This updating implements a causal loop that makes the process nonlinear (Figure A.3).

We often fill in what we think is right (based on previous experience) even if it's not what is actually there. A good nontechnical presentation is Yon (2019). Through these processes, vision works in a gestalt or holistic way[16] (Kandel, 2012; Kandel, 2016), whereby one rapidly sees the whole. As explained by Orbán et al. (2008), humans extract chunks from complex visual patterns by generating accurate yet economical representations and not by encoding the

---

[14] Much of the literature refers to 'predictive coding'. However, we do not limit ourselves to schemes designed to predict continuous variables, like the acoustic properties of a voice. Instead, we mean all forms of predictive processing, including those that deal in categorical variables like phonemes, words and sentences, so will refer in the following to 'predictive processing'. We thank Thomas Parr for this comment.

[15] Miscues are 'window on the reading process' (Goodman and Burke, 1973). They uncover both the lower- and higher-level processes readers undertake as they read (decoding phonological and graphic information, as well as predicting. sampling, confirming and correcting).

[16] For up-to-date views on gestalt psychology in perception, see Wagemans et al. (2012), Wagemans et al. (2012a) and Isaac and Ward (2019).

full correlational structure of the input. Thus our brains do not have to notice the parts first in order to construct the whole; rather the whole is perceived first and the parts are usually perceived later. We have all experienced how in a new environment, we tend to notice the big picture first: we see the general outline of things before we start taking account of the details. So babies and young children who are still learning what things are also do this. Babies consciously recognizing their mother for the first time take in and respond to the whole face and the eyes; they get to know the other parts gradually later. A toddler first sees a dog or cat in its entirety; they don't have to first identify and learn the parts of the animal before they can assemble them into a whole. A plastic doll with movable parts is not perceived as a doll only once the child has learned it is made up of arms, legs, a head with many strands of hair and a torso. A real or toy car is understood first as a whole, not by building it up from wheel to windscreen wiper; indeed in general one does not know (or need to know) what all the parts are.

A physical action aspect. In many cases, this process involves physical action: the nature of the world is tested by acting on it and seeing if the outcomes are as predicted (Friston et al., 2017). There is a cycle:

$$\text{predict} \rightarrow \text{perceive} \rightarrow \text{act} \rightarrow \text{predict(repeat)}, \tag{2}$$

where the boundary between the brain and the world can be characterized as a Markov blanket (Friston, 2003, 2010). Parr et al. (2019) state that the variational perspective of cognition formalizes the notion of perception as hypothesis testing, and treats actions as experiments that are designed partly to gather evidence for or against alternative hypotheses. Thus expectations come from experience (Yon, 2019). In fact

> Brains construct hypotheses and test them by acting and sensing... Brains sample information, hold it briefly, construct meaning, and then discard the information. (Freeman, 2004)

Social context. All of this takes place in social contexts, and constitutes sociocultural and linguistic practices involving social engagement with role modelled behaviour (Longres, 1990), leading to a social Bayesian brain (Otten et al., 2017). Intentions and meanings of others drive the understanding of implied features and linkages of a text (Donald, 2001; Frith, 2007; Friston and Frith, 2015).

2.2 Listening and Reading are Forms of Perception

Speech can be regarded as a form of perception; it is a predictive correction process based on prior knowledge (Sohoglu et al., 2012). Written language is perceived in this way too. So if a young child has a word pointed out to her and is told that this says 'cat' or 'giraffe' or 'Granny', she will perceive the entire word, just as she perceives an entire toy doll, car or train. Gestalt imagery is a critical factor in language comprehension (Bell, 1991). This can initially happen before a child understands the alphabetic principle. It depends on experience and context, and is why very young children are sometimes able to read brand names such as McDonalds, Coca Cola or KFC—they are seeking the meaning of the writing they encounter in its context, and are reading the sign as a whole (Harste et al., 1984; Bua Lit, 2018).

For competent readers, reading is fundamentally a contextual, holistic process. Sense making of words and sentences occurs: they are generically understood through contextual dependence on meaning, rather than by stringing together the component parts to reach a cumulative point of comprehension. So contextual word recognition occurs. In English and in other languages, it is common for words to have meanings and pronunciations that are contextually dependent such as 'wound' and 'wind': she wound the clock, his wound hurt; wind the clock, the wind is blowing hard, he planned to wind up his opponent. Reading always involves filling in implied contextual information on the basis of prior experiences and cultural expectations. This happens both at a local level (Who is 'she'?, 'What hurt him?', 'Why did the clock stop?' and so on) and at a more global level (Does mention of an owl imply bad luck or wisdom? What does the phrase 'The Holocaust' mean?, etc.). This is a key part of understanding when reading (Donald, 2001; Box 1 in Castles et al., 2018); see Figure A.3. The brain subconsciously corrects errors and fills in missing words through the predictive processing process. This is the major reason that proofreading a text you have written is so difficult: you literally don't see what is there; you see what ought to be there because that is what your brain expects to see.

Decoding words 'accurately' with phonics rules (Shaywitz, 2003) has extremely limited application in languages with opaque orthographies like English: 'It is tough having a thought that sounds off colour'; there is often a silent 'e' as in 'eye', 'bye', 'were', 'queue', 'quite' and so on (Strauss, 2004; Strauss and Altwerger, 2007). Decoding in languages with transparent orthographies is potentially easier to do from a memory perspective, their spelling being more regular and predictable (Goswami, 2008). This does not, however, detract at all from the predictive nature of the reading process. Indeed, as Seidenberg states, the research shows that **there is no free orthographic lunch and that . . . there is little evidence that precocious knowledge of spelling–sound correspondences confers a comprehension advantage or that the irregularities in written English present an especial burden** (Seidenberg, 2013).

As mentioned above, in many cases this process involves action. Talking and listening are conjoint processes learnt together by an infant as the sounds he hears and makes move from immature babbles to conventional speech (unless deaf, where a range of other cues lead to signing). Alongside this, writing and reading what is written are conjoint processes of active perception which allows movement from immature attempts to ever better approximations to mature reading and writing (Bissex, 1980; Ferreiro and Teberosky, 1982; Bloch, 1997). Learning to read and comprehend can happen without learning to write; it is, however, not possible to learn to write without reading. When learning to read and write in ways based in integrated understandings which centre on purposeful uses of print, attention is on meaning as texts are written and read in concert. These processes reinforce and support each other symbiotically.

2.3 Critiquing the Neuroscientific Basis for the Current Reading Orthodoxy

The view outlined in Section 2.2 differs from current reading orthodoxy, influenced by the neuroscience work done by Shaywitz (2003) studying the brains of children with problems learning to read. This view is strongly represented by the writings of Helen Abadzi. She states,

> To read and make sense of a text, our brains must first link together lines perceived by our eye receptors. The visual areas of the brain register these individual features, and, with practice, they combine them into the letter shapes used in various cultures. (Abadzi, 2017: 4)

But she then states as regards mathematics,

> we group and automatize Arabic numerals. Thus, we see the number 2 365 678 not as a mere sequence of numbers but as chunks in a group that gives a sense of magnitude. Similarly we assemble letters and numbers into complex mathematical equations . . . And how does meaning arise from these grouped shapes? The brain interprets them according to needs in the environment. (Abadzi, 2017: 4)

This is correct. She does not, however, draw the corollary that the same thing happens in reading text. In general, as in the case of mathematics, the cortex chunks the text and interprets it on the basis of environmental context, seeing whole words and phrases rather than strings of letters.

Abadzi makes the following statement: 'The neuronal pathways originate from the visual cortex and move forward, linking sounds and subsequently linguistic processes' (Abadzi, 2017: 5). This is contradicted by the studies we have mentioned above of how sensory processes work. Contrary to her view, prediction and filling in takes place both between cortical layers (Bar et al., 2006; Rauss and Pourtois, 2013) and via thalamocortical pathways (Alitto and Usrey, 2003; see Figure A.4) whereby downward feedback signals affect what one sees and hears. They are omitted from Abadzi's Figure 10.1 (Abadzi, 2017).

The process is not a one-way process from sensory organs to the cortical layers, and it is not a one-way process from incoming sensory data to output. That is a basic misrepresentation of how the brain actually works. Curiously she states in the next paragraph: 'The evidence points to a hierarchical, cascaded, interactive model of word recognition, in which top-down feed-back consolidates fast feed-forward influences via recurrent processing loops' (Abadzi, 2017: 5). Indeed so. This is what underlies the real reading process. This correct statement contradicts her previous one.

She then goes on to say,

Thus, reading involves closely timed sequences, where performance at each stage must be optimized to give reliable and timely input to the next. The meaning-related areas are at the end of this path. It is necessary to lift the print off the page before interpreting a text. (Abadzi, 2017: 5)

In reality (Bever, 2009; Bever and Poeppel, 2010; Bever, 2017), we predict what will be there as we read the words on the page in any detail—in essence interpretation precedes lifting the details of print off the page. This is confirmed by detailed EEG studies (Monsalve et al., 2014).

Abadzi states later, 'Instead, "comprehension" is often used to signal inferences or predictions. These require more knowledge than offered in a text' (Abadzi, 2017: 8). Precisely so. That is why reading is a contextual process of interpretation, extending to a psycholinguistic guessing game in the case of complex texts (Goodman, 1967; Bever, 2009).

2.4 Reading and Predictive Correction: Jumbled Words

A famous illustration of this predictive property is on our ability to read jumbled words (Rayner et al., 2006; Seidenberg, 2017 :85–99): 'yu cn raed this evn thogh wdrs wonrg and messd up.' This is the subject of an informative comment by Matt Davis[17] and the thesis work by Rawlinson (1976). It is significant because it gets to the heart of the predictive reading process. It is summarized by Rawlinson[18] as follows:

My conclusions, and these are open to question of course, were that: Letter features are processed through a route of letter classification/identification. Middle letter identification proceeds largely independently of position. Higher level units seem to be significant only for the beginnings and endings of words. Information from the middle letters may operate via a sampling/probability system (rather than absolute accuracy). That is, you can have sufficient letters, even though in the wrong position, for the brain to 'recognise' the word. My end model was of a multiple access system 'allowing some direct use of features without precise letter identification, use of word length information, and some structuring of phonemic or syllabic units, as well as incorporating a sampling recognition system using letters or their attributes

---

[17] See https://www.mrc-cbu.cam.ac.uk/personal/matt.davis/Cmabrigde/
[18] See https://www.mrc-cbu.cam.ac.uk/personal/matt.davis/Cmabrigde/rawlinson.html

directly.' I suggest the experiments 'demonstrate the considerable flexibility of the reading process'. Stimulus sampling theories seem to apply more than simple phonetic theories of word recognition. As regards learning to read, 'when the child is beginning to learn to read (s)he already has a highly refined set of skills not only for dealing with the known world but also for selecting and using information from the unknown world'. 'Word recognition skills develop which are not only not taught but which develop despite sometimes fairly specific teaching in alternative skills'.

This key evidence strongly supports the predictive processing understanding of reading.

2.5 The Centrality of Social Context

This predictive process is always shaped by social context (Donald, 2001; Frith, 2007). Friston and Frith (2015) explain that in the case of speaking and listening, communication is centred on inference about the behaviour of others:

> We are trying to infer how our sensations are caused by others, while they are trying to infer our behaviour . . . . This produces a reciprocal exchange of sensory signals that, formally, induces a generalised synchrony between internal (neuronal) brain states generating predictions in both agents.

This is what many call 'mindreading' (Donald, 2001: 59–62; Frith, 2007: 16; Heyes and Frith, 2014). Fabry (2017) gives an account of prediction error minimization that is fully consistent with approaches to cognition that emphasize the embodied and interactive properties of cognitive processes. Constant et al. (2019) give the predictive processing view of cognition extending beyond skulls. In short, the brain is a social Bayesian brain (Otten et al., 2017): social knowledge can shape visual perception. Literacy essentially involves the same issues, and is therefore a social practice (Street, 1984; Barton et al., 2000).

2.6 The Nature of Language Processing across Modes

Farmer et al. (2013) summarize how the predictive processing view extends to language processing across modes. It applies equally to spoken, written and sign language, the latter being an important form of language where no phonemes occur. There is no divergence as to how these various language modes are handled by the brain. Berent (2020) summarizes as follows:

> Linguistic principles themselves transfer across modalities. An early exposure to sign language helps because some of its rules are relevant to the later acquisition of English. Language is neither speech nor sign, but an abstract algebraic system that can emerge in either system.

The same applies to spoken and written language. They are all realizations of the same abstract relations (Huybregts et al., 2016). Similarly, significant aspects of learning to read and write are transferred to learning new languages (Bialystok et al., 2005). A key point is made by Seidenberg et al. (2020):

> Reading depends on speech. Students do not relearn language when they learn to read; they learn to relate the printed code to existing knowledge of spoken language. Writing systems are codes for representing spoken language. The structure of spoken words in English—the fact that they consist of sequences of phonemes, syllables, and morphemes that are associated with meaning—is reflected in their alphabetic representations. Learning about the written code is easier for students who know more about characteristics of spoken words that it represents.

Individual differences in knowledge of such properties of spoken language at the start of formal instruction have an enormous impact on students' progress.

Significant implications of this are both the value of enriched language input and the value of ensuring comprehensible input (Krashen, 2017; Krashen and Mason, 2020) for all children, with particular attention to children learning in difficult conditions and learning multilingually.

2.7 An Integrative Predictive Processing View of Reading

Strauss et al. (2009) summarize the predictive processing view of reading as follows:

> Whereas the classical neuroanatomic view is most consistent with a bottom-up, information processing model, the emerging view supports an interactive, constructivist model. The cortex either promotes or inhibits the very input being transmitted to it from the eyes, ears, and other sensory receptors. The psychological interpretation of this neuroanatomic arrangement is that the cortex selects evidence to confirm or disconfirm its predictions. It anticipates what will be seen and heard using knowledge stored in memory. Both this new neuroanatomical view and its psychological reflection are consistent with a transactional socio-psycholinguistic model of reading. Drawing on extensive comparisons of expected and observed responses from oral reading miscue studies, this model of reading emphasizes the fundamental importance of effective and efficient prediction and confirmation in the construction of meaning.

This holistic, meaning-construction view of reading and writing is confirmed by eye movement analysis, miscue studies and the ability to read partly hidden or garbled text.

3. Emotions Play a Key Role in Underlying Cognitive Functioning

Emotions play a key role in underlying normal cognitive functioning from birth onwards. Innate affective systems underlie and shape all brain functioning, including communicating in speech and writing. Genetically determined inbuilt emotional systems functioning via reticular activating systems (Figure A.5) stimulate and guide all cognition and learning from birth (Panksepp, 1998; Panksepp and Biven, 2012; Ellis and Solms, 2017), and so play a key role in particular in oral and written language learning.

3.1 The Key Role Played by Emotions in Normal Cognitive Functioning

A key factor in all brain function is the emotional systems that underlie motivation in life in general (Panksepp, 1998; Damasio, 1999, 2000; Panksepp and Biven, 2012; Ellis and Solms, 2017), and in particular for children in the classroom (Willis, 2006). They are also key in language development (Greenspan and Shanker, 2004: 210). Railton (2017) states

> Recent decades have witnessed a sea change in thinking about emotion, which has gone from being seen as a disruptive force in human thought and action to being seen as an important source of situation- and goal-relevant information and evaluation, continuous with perception and cognition. . . . The affect and reward system—affective system, for short—is the central locus of the learning processes, evaluative representations, and spatial mapping and simulation essential for the reasons-sensitive action guidance.

An important feature is that all memories have an emotional tag, either positive or negative.

Because of their great significance for learning, we will discuss the affective systems in more detail in the next section. A crucial distinction exists between the primary (genetically determined) affective systems and associated emotions and the secondary (socially determined) emotions.

3.2 The Primary Emotional Systems

Innate affective systems (Panksepp, 1998; Davis and Montag, 2019) are 'hardwired emotional systems' that all babies are born with. They underlie and shape all brain functioning, and result in felt emotions. These are our evolutionary inheritance, genetically determined to be what they are because they were essential for our survival in the distant past (Panksepp and Biven, 2012; Ellis and Solms, 2017). They are also the initial and ongoing propensities which all babies and young children bring to any learning.

These primary emotional systems function via the ascending reticular activating system: diffuse projections to the neocortex from nuclei in the arousal system (roughly: the limbic system) that spread neuromodulators such as dopamine and serotonin to the cortex. A particular example (the SEEKING system) is shown in Figure A.5. These primary affective systems both affect immediate behaviour and underlie brain plasticity by shaping neural connections because they form the 'value system' for Gerald Edelman's neural Darwinism (Edelman, 1987; Ellis and Toronchuk, 2005) whereby neural network weights are affected by experience. Panksepp (1998) lists seven such primary emotional systems; Ellis and Toronchuk (2013) suggest a further two, agreeing with claims by Stevens and Price (2015). We will now briefly review those that are most important for early learning.

(A) **The search for meaning:** A core feature of psychology is the search for meaning (Frankl, 1985). This drive is associated with the SEEKING system (Panksepp, 1998), which is the primary hardwired emotional system all babies are born with. It is a prime motivator for all they do: exploring the world around and trying to understand it so that it becomes predictable (and this is what the predictive processing model is about).

In particular they want to understand the meaning of what their primary caregiver does (Greenspan and Shanker, 2004). The SEEKING system and the search for meaning play a key role in all cognitive learning, and in particular learning to speak, because our brains are wired to search for meaning and intention (Frith, 2007), and it is language that enables the joint construction of meaning (Evans, 2015). This leads us to question what happens to young children's impetus to learn when, on entering formal education, they are expected to set aside their expectation (active since birth in informal settings) that seeking and making meaning drives learning, and replace this with working out how to give the teacher what she asks for, irrespective of the sense it makes to them.[19]

(B) **The need for community and belonging:** The second core primordial emotional need is that of belonging to a community (Stevens and Price, 2015), because we have a social brain (Dunbar, 1998). In the case of babies and young children, Panksepp labels this the PANIC/DISTRESS system, which has to do with the strong need to be in the secure presence of the primary caregiver, and the panic and distress experienced when this support is removed (Panksepp, 1998). In the broad context of society, it should more properly be labelled the BELONGING/AFFILIATION system, which includes both mother/child bonding and the deep need to belong to social groups (Ellis and Toronchuk, 2013; Stevens and Price, 2015).

---

[19] This is not to imply that informal learning does not involve working out what the mother or other wants, but that informal learning has a strong self-motivated voluntary aspect.

This interaction between mother and child involving the development of relationship facilitates the emergence of spoken language in the child. The importance for language development of the emotional need to interact intensely with the primary caregiver is explained clearly in the First Idea (Greenspan and Shanker, 2004). Such intense interaction provides rich stimulus for language use in purposeful contexts, contrary to Chomsky's claims of lack of sufficient stimulus to enable language learning. The key contextual feature in early childhood, shaping this all, is the relationship with the caregiver. Tomasello states:

> The glue that holds this all of these factors together is always the child's attempts to understand the communicative intentions of other persons as she interacts with them socially and linguistically... children learn words most readily in situations in which it is easiest to read the adult's communicative intentions... usage based linguistics holds that the essence of language is its symbolic dimension, that is, the ways in which human beings use conventional linguistic symbols for purposes of interpersonal communication. (Tomasello, 2003: 44, 49, 283)

The kind of informal learning which is stimulated through this need for community and belonging is determined and shaped by situated cultural practices used and valued in particular environments (Rogoff et al., 2016). This suggests the strong case for encouraging and enabling learning written language in similar ways, as discussed in depth in Ferreiro and Teberosky (1982), and demonstrated in Bissex's *Gnys at Wrk: A Child Learns to Write and Read* (1980) and *Chloe's Story* (Bloch, 1997). We return to this in Section 6.

(C) **The role of play:** Play is one of the primary emotional systems (Panksepp, 1998; Ellis and Toronchuk, 2013; Ellis and Solms, 2017), leading to rough and tumble play in all mammals, and to various forms of play, including imaginative/symbolic play, in humans. Play, which evolved tens of millions of years before language, has great significance for learning (Gray, 2017: 120–122). Gray states that the varieties of play match the requirements of human existence.[20] As stated by Boyd (2018: 19),

> [It] offers a way of learning species-typical skills by detaching them from serious mode, testing them in safe circumstances in exuberant fashion so that trial and error can refine them at low risk. Play has been so beneficial in the young of so many species that it has evolved to become self-motivating, irresistible—sheer fun.

It involves children in symbolic thinking, exploring and discovering alternative options and their outcomes, and hence leads to creative thinking and understanding (Bruce, 1991).

Behaving symbolically (Deacon, 1998) as children do in pretend/imaginative play underpins literacy learning, a second-order symbolic system (Vygotsky, 1978; Stone and Burriss, 2016). This imaginative/symbolic play arising from the PLAY system has fundamental and ongoing relevance from babyhood onwards: early word play and action play using songs, rhyme and alliteration (Bryant et al., 1990) are all practiced voluntarily by toddlers and young children as they develop a feel for the repetitions and rhythms of their languages. Such behaviour contributes to learning to read, especially when bridging connections can be made from oral to written forms, for instance with rhymes. Children come to sense the 'tune on the page' (Meek, 1988) by encountering these oral wordplays in print, with illustrations to provide initial clues to meaning. Moreover, intrinsically motivated, self-directed child exploration and discovery of written language through play (Bruce, 2015) and story (Gussin Paley, 1990) connected to children's current concerns and interests leads to deep engagement (Roskos et al., 2003; Cooper, 2009; Roskos and Christie, 2011).

---

[20] https://www.psychologytoday.com/za/blog/freedom-learn/200810/the-varieties-play-match-requirements-human-existence

3.3 Emotions, Play and Stories

Language processing involves salience and attention in accord with the predictive processing paradigm (Zarcone et al., 2016). Reading and writing in authentic contexts involves conveying and negotiating meaning, facts, stories and emotions between authors and readers (Meek, 1988).

Play is described as story in action by Gussin Paley (1990). This reveals the significance of stories for early learning in both spoken and written language (Nicolopoulou et al., 2015). As storytelling animals we make sense of our lives through stories (Gottschall, 2012): it is a powerful form of meaning making and social sharing (Redhead and Dunbar, 2013; Wissman, 2019), and we feel compelled to share our stories, factual and fictitious, with one another. Children's attention, imaginations and thinking are activated when immersed in formal or informal contexts in stories (Stanley, 2012)—life stories, history of families and communities or imaginative stories. Authentic language use and learning through stories (Egan, 1989; Sugiyama, 2017) offers adults and young children power and voice. Encouraging children to tell and compose their own stories, and valuing these, makes important connections to children's home funds of knowledge and identity (Moll et al., 1992; Esteban-Guitart and Moll, 2013). Prediction, emotion and the embodied mind are fruitfully entangled together in these contexts (Miller and Clark, 2018).

3.4 The Secondary Emotions: Extrinsic Motivation

Secondary (social) emotions such as pride and shame are also important in mental life. They are not genetically determined due to evolutionary processes, as the primary emotions are. This is because unlike the primary emotions, there are no associated ascending systems in the brain. They are socially determined as a result of social processes and play an important role in shaping sociocultural interactions. They piggyback off the BELONGING/AFFILIATION system which underlies socialization.

Extrinsic rewards tend to be used very early in school through marks, stars, competitions, prizes and so on. The emotional outcome can be both positive (praise, high marks) and negative (tests failed, low marks). Affirmation is indeed a strong motivator that leads to positive behavioural outcomes, but overly competitive or punitive aspects can have either positive or negative behavioural outcomes: they may result in greater effort, but they may also result in humiliation, anger, despair and demotivation.

3.5 Emotion, Reading and Literacy Learning

We do not necessarily consciously acknowledge this, but negative emotional tags are one of the most serious stumbling blocks to learning. This is a well-established fact in the case of mathematics education (Carey et al., 2017). In the case of reading assessments, the Early Grade Reading Assessment (EGRA) has been developed for wide use in the Global South. It is based on the Diagnostic Interpretation of Basic Early Literacy Skills (DIBELS) in the USA, which has been criticized for the emotional upset it causes some young children (Goodman, 2006). Once demotivated, it is very difficult for children to succeed.

We need to pay much more attention to this as it is a potentially critical factor in literacy learning problems in classroom contexts (Meyer and Turner, 2006; Immordino-Yang et al., 2019). However, emotional or affective systems are not mentioned by Shaywitz (2003), Dehaene

(2010), Abadzi (2017) or Castles et al. (2018), although the latter mentions the closely associated features of boredom (14) and motivation (26). Emotion is, however, mentioned by Hruby and Goswami (2019). We regard the emotive aspect of learning to read and write as critical to creating proficient readers, starting at the earliest ages via the caregiver/infant interaction (Greenspan and Shanker, 2004), and illustrated when young children learn to write and read together as in a six-year biliteracy project (Bloch, 2002; Bloch and Alexander, 2003: 104–114) and a small home-based early literacy project (Alexander and Bloch, 2010: 204–210) carried out by one of us with colleagues; this was as much the case with isiXhosa as it is with English only (Bloch, 1997).

The great importance of motivation for learning to read and write (Wigfield et al., 2016) can be understood as arising from and developing these primary and secondary emotional systems.

4. There is not the Fundamental Difference between Oral and Written Language That is Often Alleged

The claim is made frequently by many involved in literacy education that it is an evolutionary fact that oral language, i.e. listening and speaking, represents the only 'natural language', acquired in social contexts without teaching (Shaywitz, 2003). Written language, i.e. writing and reading, is understood to be a cultural and artificial invention needing specifically structured teaching, with components initially simplified and taught separately (Wolf, 2008; van Rooy and Pretorius, 2013; Spaull and Pretorius, 2019: 5).

This claim comes in many forms. Gough and Hillinger (1980) describe reading as an **unnatural act**. It is captured in the following statement by Willis:

> Reading is not a natural part of human development. Unlike spoken language reading does not follow from observation and imitation of other people. (Willis, 2008: 2)

This remark illustrates the contradictory nub at the heart of this debate: the author holds this foundational view, which leads her to believe in the necessity for young children to 'crack the code' in decontextualized ways. Like some others who hold authority as neuroscience experts (e.g. Wolf, 2008), once this is achieved, she reverts to a meaning-based understanding and approach.

Wolf (2018) states new neural circuitry was necessary for reading because reading is neither natural nor innate; rather, it is an unnatural cultural invention that has been scarcely 6000 years in existence. By contrast, we view both oral and written language as equally 'natural' (Goodman and Goodman, 2013). This is because both are social constructs, developed in evolutionary terms as successive modes of symbolic communication, the latter piggybacking on the former, when the need arose as part of human development. They can both be learnt by essentially the same social processes (Section 6.2), with the symbolism of thought realized in different ways (oral and written), and the same is true for sign language and Braille.

4.1 Naturalness of Oral and Written Language: An Innate Language System?

Shaywitz (2003: 45, 49–50) states,

> Reading is more difficult than speaking. . . . Spoken language is innate. It is instinctive. Language does not have to be taught. All that is necessary is for humans to be exposed to their mother tongue. Although both speaking and reading rely on the same particle, the phoneme,

there is a fundamental difference: speaking is natural and reading is not. Herein lies the difficulty. Reading is an acquired act, an invention of man that must be learned at a conscious level. And it is the very naturalness of speaking that makes reading so hard.

She justifies her views of the difference between reading and writing as follows (Shaywitz, 2003: 50):

Profound differences distinguish reading from speaking . . . Reading is not built into our brains. There is no reading module wired into the human brain. In order for children to read, man has to take advantage of what nature has provided: a biological module for language.

That is, she is claiming the key to the difference between oral and written language is innate properties and how they underlie brain development.

Shaywitz is relying on Chomsky's idea (Chomsky, 1965, 1975) of an innate language module in the brain: a **Language Acquisition Device** (LAD). But there are, in fact, no innate cortical modules in the brain representing evolutionary-based hard-wired knowledge of any kind; this is not possible for evolutionary, developmental, information theoretic and physiological reasons (Ellis and Solms, 2017). Rather we are provided with brains that are highly plastic and able to adaptively learn through ongoing experience with the physical, ecological and social environment. We have learning-ready brains.

What is preset is the primary emotional systems that guide action (Section 3.2). But above all, Shaywitz fails to recognize that the process of learning to listen and speak is just as much a learning process as is learning to read and write. The implication of this view is that such 'natural', oral language is acquired effortlessly. We question and contest this. Babies cannot talk when they are born. They learn through a complex, extensive and persistent process involving social interactions, during the first few years of life, as stated by Kuhl:

The learning processes that infants employ when learning from exposure to language are complex and multi-modal, but also child's play in that it grows out of infants' heightened attention to items and events in the natural world: the faces, actions, and voices of other people. (Kuhl, 2010: 716)

And the same is true for written language. Both have to be learned at a conscious level and both involve teaching. This is discussed in Section 6.

4.2 Language Readiness versus a Language Acquisition Device

There is no LAD as envisaged by Chomsky on behavioural grounds. As Evans (2020) states,

Everyone agrees that our species exhibits a clear biological preparedness for language . . . What is in dispute is the claim that knowledge of language itself—the language software—is something that each human child is born with . . . a 'language organ' . . . containing a blueprint for all the possible sets of grammar rules in all the world's languages.

Pinker (2003) called this a 'language instinct'. The problem is that Chomsky proposed his LAD without taking into account the biological processes whereby the brain comes into being. If you bring biological reality into the picture by considering this, such a LAD cannot exist for developmental, genetic and evolutionary reasons, as explained in depth in Ellis and Solms (2017). We summarize the main reasons thus:

First, there is no way that the precise details of the billions of neural connections in the neocortex can be guided by developmental processes: the refined detailed nature of the connections make that impossible. Rather the detailed synaptic connections are initially made randomly, and then refined on the basis of experience (Wolpert et al., 2002). They are not directly genetically determined.

Second, there is not a fraction of the genetic information available in the human genome needed to shape such detailed neuronal connections. It contains about 30 000 genes, which are needed to code for the entire body: heart, lungs, liver, digestive system, skeleton, skin, etc., and in particular to set up the large-scale brain structure. There simply are not enough genes to determine the detailed cortical structure with billions of connections. In any case only a fraction of those genes are specifically human genes that can conceivably be associated with grammar.

Third, setting aside these two critical issues, it is not remotely plausible that the kind of detailed grammatical structures investigated by Chomsky would have been of such a vital importance that they would have resulted in evolutionary selection because they affect survival probabilities so crucially. Selection for an overall language capacity, yes that is critical: but for this kind of detailed grammatical structuring, no way. Because of the predictive processing nature of language perception (Section 2), minor grammatical errors do not harm understanding of the message being conveyed and are not needed for survival. As discussed above, the brain automatically makes the needed corrections.

These considerations are decisive (Ellis and Solms, 2017): there is no genetically determined LAD. The real situation is that we possess a language-ready brain with a generic symbolic capacity (Deacon, 2003) which in suitable social contexts learns to understand both spoken and written language, or sign language in the case of deaf people. Evans (2014, 2020) develops this all in a clear way, emphasizing how, as more data has been collected, the claims of grammatical universals have weakened over time.

There are, however, two further arguments to consider: Chomsky's poverty of stimulus argument, and the issue of where language universals come from.

4.3 The Poverty of Stimulus Argument

There are three counters to this claim made by Chomsky that there is not sufficient evidence provided to children for them to be able to learn the grammatical rules of their home language as a social process.

First, as pointed out by Lewis and Elman (2001), Chomsky's poverty of stimulus argument (1975) fails to hold once stochastic information is admitted. The properties of language in question is shown by them to be learnable with a statistical learning algorithm. They show that simple recurrent networks are able to provide the correct generalizations from the statistical structure of the data. Pullum and Scholz (2002) detail how the linguistic nativist position noted above is not supported by the data. Amodei et al. (2016) show how statistical learning can be done in practice via an end-to-end deep learning approach. This is in line with the predictive processing view. Friston et al. (2020) propose that the neuronal correlates of language processing and functional brain architectures should emerge naturally, given the right kind of generative model. The basic issue is that language processing is not, in fact, rule based; it is based in statistical correlations (see Section 5).

Second, is there, in fact, a poverty of stimulus? We claim there is not in normal situations, where massive stimulus is provided by the main caregivers, as emphasized by Greenspan and Shanker (2004). Rogoff (2003: 69) describes human beings as 'biologically cultural' and states,

> Whether or not they regard themselves as explicitly teaching young children, caregivers routinely model mature performance during joint endeavours, adjust their interaction and structure children's environments and activities in ways that support local forms of learning.

The stimulus which occurs for language learning crucially involves the strong emotional link discussed in Section 3, as well as continuous demonstrations of (culturally) conventional or mature speech in action, to which children gradually adjust their immature speech attempts. These are the basis for statistical learning processes.

Third, this ability to learn either spoken or written language through such interactions is significantly strengthened when these interactions are laden with positive affect, as discussed in Section 3. This enhances the motivation to transact with and understand the message being conveyed, and hence also to grasp the grammatical patterns by which it is conveyed.

4.4 Language Universals

Where then do language universals come from? A plausible view is that they are due to essential syntactic limitations that must necessarily apply to any language whatever due to the requirement that it be an adequate symbolic system for representing the world around. They arise due to fundamental semiotic constraints on any symbolic representation of our experiences and environment, as explained in detail by Terrence Deacon:

> Many of these core language universals reflect semiotic constraints, inherent in the requirements for producing symbolic reference itself ... these constraints shape the self-organisation and evolution of communication in a social context . . . combinations of words inherit constraints from the lower order mediating relationships that give words their freedom of mapping. These classes of constraints limit the classes of referentially consistent higher order symbol constructions. (Deacon, 2003: 112, 118)

That is, they arise because language must provide a meaningful representation of the world around us in order to be useful. Tomasello reinforces this view (Tomasello, 2003:18).

4.5 Naturalness and Evolution of Reading and Writing

We have asserted that there is not the fundamental difference between listening/speaking and reading/writing that is often claimed on the basis of evolutionary arguments and the alleged existence of a LAD in the brain. Oral and written language are both social practices driven by the communication imperative of the social brain (Dunbar, 1998), learnt through sociocultural processes. They both evolved in similar ways through the social processes of cultural evolution.

An examination of the historical record will show that oral language first evolved as a crucial cultural invention between 70 000 and 30 000 years ago (Harari, 2011: 23–28), and writing evolved as a second cultural invention piggybacking on the first between 3500 and 3000 BC (Harari, 2011: 137–148). Neither is hardwired in the brain, as discussed above; both are socially transmitted down the generations. Spoken language evolved to enable efficient human bonding (Dunbar, 1993), in particular enabling communication among kin (Fitch, 2005); Tomasello (2000) and Donald (2001) give broadly consistent viewpoints. Writing later evolved to solve the problem of cooperation in large groups by transcending the severe limitations of our evolved psychology through the elaboration of four cooperative tools—(1) reciprocal behaviours, (2) reputation formation and maintenance, (3) social norms and norm enforcement and (4) group identity and empathy (Mullins et al., 2013). As a major extension of oral language, writing evolved to allow communication over space and time and record keeping over time in unparalleled ways.

While there can be contestation about the details, the fundamental issue is clear: both oral and written language evolved in broadly similar ways to enable the cultural evolution of human cooperation (Fitch, 2005). Thus we suggest that the statement 'reading is unnatural' could usefully be replaced by

Reading and writing are both cultural practices, and culture is natural.

This proposal is strengthened if one looks at the case of sign language (Trettenbrein et al., 2021). This is obviously also a cultural invention, and does not involve phonemes as it is a communication means for deaf people. By looking at the brain areas involved in sign language, the authors show that the human brain evolved a lateralized language network with a supramodal hub in Broca's area which computes linguistic information independent of speech. It can be realized in sound, writing or sign.

5. Natural Brain Operations: Statistical Correlations and Predictions, not Logical Rules

The foundational issue is what is the natural mode of cortical function. This underlies a question: what really is the nature of linguistics? In an important paper, Seidenberg et al. (2020) raise this after considering the problems underlying a rule-based view of language.

5.1 Linguistics: Rules and Exceptions

Seidenberg et al. (2020) summarize the dual-route theory of reading as follows: it consists of

- Rules to produce patterns such as **save-pave-gave**, which are used in sounding out unfamiliar words (or, in research studies, pseudowords such as mave),
- A list of 'exception' or 'sight' words whose pronunciations violate the rules (e.g. have, said, bear) and must be memorized.

They state, 'The instructional implications of the theory are straightforward: teach children the rules (or enough to allow them to "break the code"), and help them memorize the exceptions.' But they then ask, 'What are the rules for pronouncing written English?', and conclude 'No one knows.' The key problem is that the dual-route model does not provide a meta-rule for determining when the standard patterns apply, and without that, you cannot reliably apply the rules. Examples in English are the well-known problems in pronouncing 'ou' (wound, sound, cough, tough, ought) and the problem of silent 'e' (Strauss, 2004), which already afflicts the standard patterns cited by Seidenberg et al. They then ask the key question:

> What if it is difficult to state the rules and how they are learned and decide on the sight words because the system **isn't rule-governed**? What if 200 years of phonics instruction has been based on a false dichotomy?

That is exactly the right question to ask.

5.2 Connectionist Models: The Functioning of Neural Networks and Language Learning

Seidenberg et al. then propose using connectionist models of the brain as providing the basis of speech and reading. Such models (Buckner and Garson, 2019) are not based in following logical rules but in learning and generalizing the statistics of presented text, which trains the weights of the neural network.

This is set out in Seidenberg and McClelland (1989), Seidenberg (2005), Plaut (2005) and Bybee and McClelland (2005). This has to be correct, because the brain is, in fact, a vastly complex neural network (Nicholls et al., 2001) with memory enabled by neural plasticity allowing statistical pattern recognition and active prediction (Carpenter and Grossberg, 1991; Bishop, 1995; Churchland and Sejnowski, 1999; Rolls, 2016), resulting in the brain in effect employing Bayes'rule at a psychological level (Hohwy, 2013). The brain does not in neural terms implement a strict set of logical rules such as occur in computer programs as envisaged by Turing and von Neumann. Rather it is a Bayesian brain (Friston, 2012; Seth, 2014; Otten et al., 2017) that learns statistical associations such as collocations and colligations (Hoey, 2005) underlying active perception. They are developed from embodied experience as ways of conveniently describing those experiences symbolically (Feldman, 2008), often in effect using metaphor as mental models (Lakoff and Johnson, 1980), later generalized to abstract thought and logic.

Note that learning these statistical patterns is not the same as a learning a set of rigorous logical rules such as grammatical rules as envisaged by Chomsky (1965, 1975), which in the end are the source of the alleged problem. Rather, statistical dependencies are learnt by experience—repeated presentation of many examples—and these then form the foundation of prediction of what is to be expected, which are then used in the predictive processing way discussed in Section 2. In particular, the processing mechanisms involved in the visual recognition of novel words occur through the visual system capturing statistical regularities in the visual environment (Vidal et al., 2021). Their relation to the Parallel Distributed Processing (PDP) of reading is examined in Laszlo and Plaut (2012).

Thus, while we can indeed think in a logical rule-based way (how this can occur on the basis of neural networks is discussed by Marcus (2019)), this is not the brain's natural way of functioning. Our brain is a connectionist Bayesian brain whose natural mode of operation is statistical pattern learning and prediction.

As a result, the pattern-matching way of reading presented by Seidenberg et al. (2020), summarized in their Figure A.1, is exactly right. Children pick up the structure of grammar by statistical learning. They conclude:

> Readers do not pronounce words by explicitly applying rules; doing so would be a conscious, slow effortful process (the opposite of 'fluent'). Teaching phonics by teaching rules and memorizing exceptions leaves out the statistical patterns that permeate the system and drive the fast, implicit learning process.

This results in the 'rules' being applied when they are valid, but avoids the problem of trying to determine when they apply and when they do not.

What if the language is an agglutinating language such as isiXhosa, if it does indeed have a highly regular structure? Our comment is that in this case (unlike English) it may indeed be possible to describe the language adequately via a rather strict set of rule. That will not, however, change the natural way the brain operates, as just outlined. It will make it possible to efficiently learn that language in a rule-based way, because the brain's statistical predictions will be well-correlated with the outcome of that set of rules, but this will not imply that that is the best way to do so. Furthermore it should be noted that it is a matter of fact that English is the dominant world language in terms of commerce and science, and hence access to the modern economy is greatly increased by being fluent in English—where phonics 'rules' are highly fallible (Strauss, 2004). The problems pointed out by Seidenberg et al. will arise when children who operate multilingually try to learn English.

5.3 The Development of Rule-Based Understandings in Individual Lives

An interesting issue that arises from this discussion is that, given that rule-based logic is not the natural mode of operation of the brain, how does it arise in developmental terms? A plausible answer is that it arises through taking part in human cultural activities of singing and games.

Music has hidden rules, embodied in the structure of rhythm: this leads to an expectation of what will come next (Huron, 2008), which is in essence a rule played out over time. Children make up verse that involves rhythm (Chukovsky, 1968: 61, 87). All play and games involve rules and an expectation they will be obeyed, conveyed in the statement 'I'm not going to play with him: he cheats' (Elkind, 2007: 119). Vygotsky (1978: 94) confirms this by saying, 'There is no such thing as play without rules.' This applies equally across cultural communities and to all kinds of games.

So our hypothesis is that the connectionist brain learns the basis of rule-driven thought through partaking in songs, rhyme, poetry and games of many kinds. Once that understanding has taken root, it can be developed in terms of logic and then mathematics and science.

5.4 Formal Linguistic Theories Embodying this Viewpoint

The statistical associations of language occur as collocations and colligations which allow lexical priming (Biber et al., 1998; Hoey, 2005) and so underpin predictive understanding of text. How this functionality arises through embodied experience is detailed by Feldman though his neural theory of language (Feldman, 2008). The outcome can be formalized in terms of **Systemic Functional Linguistics** (Halliday, 1977, 1993, 2003). These alternative views of the nature of linguistics are summarized by Peter Fries in Ellis and Solms (2017: 125–133), based on the work of Feldman, Halliday, Hoey and others.

The key outcome for this paper is that the rule-based view of linguistics espoused by Chomsky is not the only game in town. The other approaches briefly mentioned here are far closer to what is validated by biological reality, and have been formalized in alternative views of the nature of linguistics.

6. The Neural Route to Reading with No Explicit Decoding Can be Activated from the Earliest Years

Language includes listening, speaking, signing, reading and writing. Because oral and written language both evolved, it is not a coincidence that there are important similarities in the way each of them function to make and convey meaning. Both receptive aspects of language (listening and reading) and productive ones (speaking and writing) are nonlinear, neurolinguistic-psychosocial processes of understanding, shaped by current knowledge and context. The previous sections related to how the corticothalamic circuits help underlie the way the brain predictively searches for meaning, and the innate emotional systems that power that search. This section looks at aspects of how the cortex enables the link between writing and meaning.

6.1 Oral Language: Meaning Making in Context

The first and foremost point about oral language is

LAN(o): through speech, patterned sounds convey information, meaning, and emotion.

This enables complex communication in sociocultural contexts, where listening and speaking is a joint socially based interaction involving shared attention, prediction and modelling other people's minds (see for example Frith, 2007; Heyes and Frith, 2014).

The basic problem is how we understand a linear stream of symbols representing a hierarchical structure. We have to flatten the hierarchical structure into a linear structure.

Thus, '[s]entences are externally serial (i.e., "horizontal"): derivations are internally hierarchical, (i.e., "vertical"). That is, the computational domain of a derivation can embrace entire clauses and sentences, while the immediate processing appears to be one word after another' (Bever, 2013). We have to learn how to handle this for both oral and written language, where the issue is the same. In the case of oral language, Bever (2017) states it thus:

> A sentence in everyday use combines a stream of sound, with rhythm and pitch variations, with memorized units of meaning, an organizing structure that recombines those meaning units into a transcendental unified meaning that includes informational representations, general connotations, and specific pragmatic implications unique to the conversational context. In other words, each sentence is a miniature opera of nature.

Ding et al. (2016) explain that in speech, hierarchical linguistic structures do not have boundaries that are clearly defined by acoustic cues and must therefore be internally and incrementally constructed during comprehension. This is the predictive processing process that underlies listening to speech.

Cortical activity at different timescales concurrently tracks the time course of abstract linguistic structures at different hierarchical levels, such as words, phrases and sentences. This is how the brain handles the problem flagged by Castles et al. (2018):

> The segmentation of an acoustic signal does not correspond in any straightforward way with segmentation at the phoneme level: In continuous speech, phonemes overlap and run together.

From a larger perspective, understanding speech involves a 'psycholinguistic guessing game' such as is characterized by Goodman (1967), Tovey (1976), Flurkey et al. (2008) and Bever (2009) in the case of reading. It usually has a major social component (What does this refer to? Where did that take place? Why are they saying this? Is there a hidden agenda? and so on). The predictive processing underpinnings of this process are explained by Friston and Frith (2015). These enable the process of 'mind-reading' mentioned earlier: a key social skill leading to a theory of mind (Conte et al., 2019).

6.2 Written Language: Meaning Making in Context

The first and foremost point about written language, parallel to **LAN(o)** above (Section 5.1), is

> **LAN(w):** through written text,[21] printed symbols convey information, meaning and emotion.

This enables complex oral and written communicative transactions in social contexts (Vygotsky, 1978; Rosenblatt, 1982) across distance and time.[22]

Predictive reading. Similarly to when processing spoken language, when reading complex texts, there is never enough information in a sentence to fully convey the intended meaning. Thus in order to read or to listen, we use prediction and then comparison with incoming data, as in the case of all other senses, and in agreement with the predictive processing model of

---

[21] And their extensions to electronic versions. LAN(w) should be interpreted in this way, where 'printed' includes hand written and electronic versions of the same text.

[22] This is beautifully described by Carl Sagan here: https://www.youtube.com/watch?v=MVu4du LOFGY.

the mind (Section 2). Competent readers do not read by assembling phonemes into words and words into phrases as Shaywitz (2003) claims. They read phrases as a whole in a way that makes sense in terms of context and making meaning overall, predicting what text will come next as they do so (Goodman, 1967; Bever, 2009). Not all words need to be read (Figure A.3).

Multiple cueing systems. Readers predict meaning using multiple cueing systems (Figure A.2): semantic, directly involving meaning, grapho-phonic, the look and the sound of the language and syntactic, its grammatical structure (Goodman, 1967; Goodman and Burke, 1973; Clay, 1991; Bergeron and Bradbury-Wolff, 2010). Each is drawn on as required to understand the text, even when using a language which has transparent orthography, such as Spanish or isiX-hosa. This is because these cueing systems work together to support the essence of reading. We strongly suggest that when children are first taught to rely only or mainly on decoding and word level accuracy, this hinders or blocks their developing metacognitive abilities to self-monitor and self-correct for meaning using various cues (Clay, 1991; Juliebö et al., 1998). Moreover, children learning multilingually, who have become habituated to mainly attend to decoding accurately, are likely to struggle when they have to start learning to read in an additional language like English. With the combination of its opaque orthography and their emerging understanding of the language, they have a considerable challenge: attending to different cueing systems within flexible languaging practices (García and Wei, 2014; Makalela, 2014) would make their progress in reading with meaning far more likely.

A basic problem: seeing the written page. In *Reading: The Grand Illusion* (Goodman et al., 2016), the authors comment on how our impression of seeing a whole page of text in front of us when reading is an illusion—a construction of the mind—because, in fact, our eyes see only a small part of the page clearly, and see nothing at all in the blind spot. Gregory and Cavanagh (2011) describe the latter:

> The natural blind spot occurs where axons passing over the front of the retina converge to form the head of the optic nerve, and where retinal blood vessels enter and exit the eyeball, resulting in a hole in the photoreceptor mosaic . . . Each eye has a surprisingly large blind region, about 4° of visual angle, the width across your four fingers held at arm's length . . . . Surprisingly, we are normally unaware of these natural blind spots. They are either filled in perceptually (a remarkable phenomenon) or they are ignored and so not seen.

The predictive processing model strongly supports the first option: the brain fills in the missing text, enabled by saccades: the constant movement of the eye focus across the written pages (Dehaene, 2010: 13–15; Goodman et al., 2016) and visual sampling taking place during the process (Findlay and Gilchrist, 2003). This illusion of seeing a complete page when reading provides strong evidence that the predictive processing model of reading text is correct. A linear model proposing translating incoming signals from the optic nerve linearly into what we 'see' simply cannot explain this process.

6.3 The Two Routes to Reading

The two neural routes allowing for reading as referred to previously (Section 1.2) are a direct one and an indirect one (Coltheart, 2000; Rastle et al., 2001; Taylor et al., 2013; Danelli et al., 2015; Willingham, 2017: 57, 65; Buckingham and Castles, 2019). This is described by Castles et al. (2018: 17) as follows:

> The fact that word reading involves more than just alphabetic decoding is reflected in all major theories of skilled reading . . . . The important point is that all of the models converge in that they represent two key cognitive processes in word reading: one that involves the translation

of a word's spelling into its sound and then to meaning, and one that involves gaining access to meaning directly from the spelling, without the requirement to do so via phonology . . . . This dual-pathway architecture for deriving meaning from printed words is also apparent in the neural implementation of the reading system.

In symbolic form, they are

Dorsal (Decoding) Pathway: {Graphemes} → {Phonemes → {Morphemes},
Ventral (Direct) Pathway: {Graphemes} → {Morphemes}.

Only the second is readily available to people who are deaf.

Note that this is characterized by Castles et al. (2018) as theories of skilled reading. Indeed they state,

> One interesting proposal that is consistent with the characterization of reading acquisition that we have put forward is that reliance gradually shifts with increasing reading skill from the dorsal to the ventral pathway. (Pugh et al., 2000; Shaywitz et al., 2002)

We claim rather that the direct path is also possible for young learners from the start, and indeed is a powerful 'natural' way that they begin and can continue learning to read (Gray, 2013) under favourable conditions. Indeed the fact that it is possible is shown by the quote from Dehaene we give in Section 1.2, emergent literacy research evidence (Bissex, 1980; Goodman, 1992; Harste et al., 1984; Gunn et al., 1995; Dooley and Matthews, 2009), the Visual Word Form Area (VWFA) is used 'as a word letterbox' (Dehaene, 2010) but is also used for other purposes (Vogel et al., 2012, 2014; Moore et al., 2014; Martin et al., 2019; Vidal et al., 2021) so it is not uniquely associated with reading.

6.4 Memory Issues

Memory limitations are claimed to justify the need for an essential initial skills focus, to reach automaticity and fluency with letter-sound combinations (most recently, for the South African context, see Ardington et al., 2020). For example, Abadzi's statement about memory are that

> [in terms of] working memory capacity, we are constantly performing in a very narrow time-frame of about 12 seconds. We must recognize letters and other items within a few milliseconds, otherwise we cannot hold the messages they convey in our minds long enough to interpret them or make decisions; by the end of a sentence, we forget the beginning . . . . Higher-order skills emerge only after the very basic skills are tied to the point of automatic and fluent performance. (Abadzi, 2006: 585)
> Novice readers who make conscious decisions about letters can only read small amounts of text and may have to read a message repeatedly to understand its meaning. (Abadzi, 2006: 586)

The problem here arises due to focusing a learner's attention on the imperative to attend to combining and memorizing the small details, which appear meaningless. Of course this will overburden working memory. Attending to meaning using various cueing systems described above orients learners toward reading words, phrases and sentences holistically.

These are stored in working memory as chunks, solving the problem of memory overload. Attending to combining letters into sounds should only be done when necessary in service of this process:

a language user engages in the process of seeking meaning through the grammatical structures. He [*sic*] uses the surface structure, the sequences of sounds and letters, only as signals or means of getting at, or inducing or recreating, the deep structure. (Goodman, 1982: 55)

Abadzi's assumption of working memory overload (also see Adams, 2001) which is claimed to restrict young learners' initial focus (and which is why she claims they have to focus on the letter sounds first) is also challenged by Merlin Donald. He states that the laboratory studies that this assumption is based on look only at the lower limits of conscious experience (Donald, 2001: 47). Working memory in real life is much larger than this and supports the remarkable capacity we know toddlers and young children have for grasping and memorizing new vocabulary and sayings while involved in going about their daily life.

6.5 The Autonomous, Context-Free Linear Model

The reading model proposed inter alia by Abadzi (2006) and Castles et al. (2018) is skills based, 'bottom up' and linear. Castles et al. (2018) discuss it as follows:

> What does the product of successful orthographic learning look like? First, according to Perfetti (1992), it involves having developed fully specified, rather than partially specified, internal representations. By full specification, Perfetti means that the input code is sufficient to uniquely identify the word to be read, without the necessity for discriminating between several competing partially activated candidates . . . in these circumstances, the correct word is specified completely by the input code, context does not need to be used to assist in the identification of the word . . . skilled 'lexical' retrieval is effectively modular, and is only very minimally influenced by factors other than the input code.

This says that reading does not proceed along the nonlinear predictive lines that all perception uses, as we have explained above. They confirm this view by stating,

> Consider once again the example of the word 'face'. Successful discrimination of this word from the many other words in English that differ from it by only one letter (e.g., fact, lace, fame) requires the reader to develop a very precise recognition mechanism, one that attends to all of the letters in the word and their order. Otherwise, identification accuracy and access to meaning will be compromised.

There is no recognition here that a competent reader does indeed recognize a word by its context, even if the word is jumbled (Section 2.4). One can deduce the word is 'face' not 'lace' or 'fact' or 'fame' if it is in a meaningful sentence as is illustrated in Figure A.3. One does not have to read all the letters as they claim.

This requirement of strict precision contrasts sharply with an understanding where the status of reading as a form of perception is recognized, following the same principles as all other forms of perception: missing data is filled in according to context by a predictive model (Section 2). It also contrasts strongly with what Castles et al. (2018) themselves state later: 'Inferences need to be made beyond what is overtly stated to establish meaning within and between sentences, and need to draw on background knowledge.' Just so.

This contextual process assists in word and letter discrimination (Willingham, 2017: 60–63), The nonlinear hierarchical predictive model shown there is in complete contrast to this linear model. It is enabled by predictive generative processes dependent on context. This is simply not a bottom-up linear reading process. Consequently Friston et al. (2017a) state,

> The key thing to take from these results is that the agent can have precise beliefs about letters without ever seeing them . . . it is not necessary to sample all the constituent letters to identify

a word. Conversely, there can be uncertainty about particular letters, even though the subject is confident about the word.

This crucial point to note is that the core of the reading process is one which does not require getting all the details right first. This is not needed for the communication task that is the central purpose of reading (Friston et al., 2020). Furthermore, perception of words and letters depends on context (Rumelhart, 1977).

The Simple View of Reading. The SVR (Section 1.2) is based on a context-free linear model (see Equation (1)). However, first, explicit decoding is only necessary for one of the two reading pathways (Section 5.3). The Ventral (Direct) Pathway functions without such an explicit process. Even though deciphering structural features is happening, it is not a letter-by-letter decoding process. Second, an ability to comprehend early in reading development can be constrained by decoding if reading is taught by methods orienting the learner's attention on decoding, rather than in ways based in meaning (Section 6.3). That is a limitation resulting from a particular teaching method. Third, it is not clear that capacity to read jumbled words can, in fact, be accounted for by the SVR because of its strict reliance on decoding. But we do indeed have that ability (Section 2.4). An interactive model is far more plausible (Rumelhart, 1977).

Decoding first. The SVR is closely associated with the dominant view that decoding must take place first, as stated for example by Patael et al. (2018): 'The ultimate goal of reading is to understand written text. To accomplish this, children must first master decoding, the ability to translate printed words into sounds.' But they then carry on,

> Although decoding and reading comprehension are highly interdependent, some children struggle to decode but comprehend well, whereas others with good decoding skills fail to comprehend. The neural basis underlying individual differences in this discrepancy between decoding and comprehension abilities is virtually unknown.

Indeed their very careful study shows that such a discrepancy is real. We suggest the resolution that the premise is false: when reading takes place by the ventral pathway, such a discrepancy can be expected. The brain then acts in a predictive way, as discussed above.

6.6   Neuroscience Evidence and Reading: Reductionist Research Methods

When considering the neuroscience evidence supporting either of these views, one should be very aware of the strengths and limitations of the evidence provided. Because evidence for skills-based reading models is based on a reductionist view of brain function, it necessarily incorporates the limitations of that view.

More specifically, books like Dehaene (2010) have major limitations in terms or providing evidence regarding the reading process. They study parts of what is involved in reading, but not the integral process of meaningful reading. Thus they can only provide evidence about isolated aspects of reading, not how they are integrated to enable the process as a whole.

Even then the studies are really limited: Castles et al. (2018) state, 'most of the work on spelling-sound relationships has been conducted with monosyllables; researchers are only just beginning to consider spelling-sound relations in letter strings with more than one syllable.' This is hardly sufficient to determine how meaningful language works. Related to this, there is a lot of data on reading nonsense words and phonemes. This gives no data on the integral process of reading meaningful text. That aspect is missed by all brain imaging studies which look only at how phonemes or pseudowords are processed.

An example of such limitations is a study by Cattinelli et al. (2013), who performed a new meta-analysis based on an optimized hierarchical clustering algorithm which automatically groups activation peaks into clusters. They focussed exclusively on experiments based on single words or pseudowords from the following four classes of tasks: reading, lexical decision, phonological decision and semantic tasks. But you can't do a real semantic task based on single words or pseudowords. This kind of study can only be useful to determine isolated parts of the reading process. It should not be taken to give information on the actual reading process. It simply does not have the necessary data and should not be treated as if it does.

6.7 Neuroscience Evidence and Reading: Holistic Research Methods

Extensive work has been done to put the study of real reading on a scientific basis, as summarized in Flurkey and Xu (2003) and Flurkey et al. (2008). The latter state,

> The emerging concepts from [current] research clearly indicate that the higher cortical structures control the transmission of information from the deeper structures. This interpretation is contrary to the classical teaching, in which deeper sensory relay stations determine what will eventually reach the cortex. The emerging view has profound implications for psychological models of mental life. Whereas the classical neuroanatomic view is most consistent with a bottom-up, information processing model, the emerging view supports an interactive, constructivist model. The cortex either promotes or inhibits the very input being transmitted to it from the eyes, ears, and other sensory receptors. . . . the cortex selects evidence to confirm or disconfirm its predictions. It anticipates what will be seen and heard using knowledge stored in memory. Both this new neuroanatomical view and its psychological reflection are consistent with a transactional sociopsycholinguistic model of reading.

This is precisely the predictive processing view discussed above. It is supported by evidence as follows:

First, **eye tracking studies**: evidence comes from eye movement analysis of fixations, omissions and backtracking. Since the most conspicuous motor behaviour in silent reading is eye movement, studying it allows us to 'see' the silent reading process (Flurkey et al., 2008; Seidenberg, 2017: 62–70). We do not, in fact, read every word (Goodman et al., 2016). Not all words are read: some are skipped. Visual sampling takes place during text reading (Findlay and Gilchrist, 2003).

Second, **miscue analysis**: when combined with miscue analysis from oral reading, it is clear that cortical instructions tell the eyes where to look for cues from the signal, lexicogrammatical and semantic levels of language—the three cueing systems (Flurkey et al., 2008; Goodman et al., 2016).

Third, **garbled words and phrases**: the way that we can read sentences when words are misspelled or missing, or when letters are rearranged within a word (Section 2.4) or grammar is wrong is strong evidence of how reading works in a contextual way.

Fourth, **letters**: how letters are sometimes identified in a top-down way is based on what the probable word is (Willingham, 2017: 60–63; and see Figure A.3).

Fifth, **inferring meaning and pronunciation**: we often have to infer in a top-down way what part of speech a word is and what it means through context (e.g. 'plane', 'flies'). Sometimes the way a word sounds may depend on context (e.g. 'wound' has multiple meanings and pronunciations). This is a common feature of many languages, irrespective of orthographical features.

Sixth, **brain imaging studies**: Flurkey et al. (2008) comment that the subjects in the various brain imaging studies of reading at the time they wrote had not been given phonological processing tasks embedded in a context that requires meaning construction, nor have they even considered imaging studies to illuminate the effect of home reading programmes on neural development. Such studies have recently been initiated by J. S. Hutton and co-workers, who have applied MRI studies to better understand the influence on structural and functional brain networks of young children in home reading environments supporting emergent literacy. They are obtaining information on neural processes related to actual reading processes,[23] and the accompanying skills and attitudes which develop; for example fluent reading, were found to be supported by executive function areas (see Horowitz-Kraus and Hutton, 2015; Horowitz-Kraus et al., 2017; Hutton et al., 2015, 2017, 2020).

All this emerging data provides strong evidence for the meaning-construction view of reading. The transactional sociopsycholinguistic character of reading is an instantiation of the nonlinear, integrative memory-prediction model of brain function discussed above (Section 2). Following on Sherman and Guillery (2006), Flurky et al. (2008) emphasize the role in these processes of thalamocortical circuitry, in agreement with Alitto and Usrey (2003).

7. Oral And Written Language Learning

What about the nature of learning to understand and use oral and written language? The similarities between the processes involved in oral speech and written communication suggest that there should be important similarities in the conditions babies require to learn to listen and speak, and young children require as they learn to read and write (Holdaway, 1979; Cambourne, 1995). Without role models who interact with them and surround them with demonstrations of language being used for various purposes, babies would not have the social context that supports and shapes oral language development (Hoff, 2006). The same applies to learning to read and write. It is this which leads to understanding and supporting the growth of literate environments and reading culture development in providing the conducive conditions for literacy learning.

7.1 Basic Principles of Language Learning

The following can be claimed to be basic principles underlying learning both spoken and written language.

(a) **Constructing and conveying meaning.** In learning to speak, the foremost thing babies have to learn is that 'spoken words convey meaning and emotion and information and stories'[24] (this is **LAN(o)**, Section 5.1). This empowers the drive to understand and to listen and attempt to speak, as they try to make sense of and predict the world around—as well as the need to communicate with significant others. Similarly, in learning to read and write, the foremost aspect toddlers/children have to learn is that 'written words convey meaning and emotion and information and stories' (this is LAN(w), Section 5.2). This too powers the intrinsic motivation to explore and communicate in ways which include using print. It is closely tied in to the key process of learning to read minds,

---

[23] Friederici (2017), in particular, 121–141, presents such studies in the case of oral language.
[24] We mean here stories in their broadest sense, incorporating the narrative form.

which, as stated in a very useful paper by Heyes and Frith (2014), is like learning to read print.

(b) **Joint social processes.** Learning to speak and understand and learning to read and write are both joint socially based processes involving attempted efforts and feedback, and with a strong affective component. This means each is as 'natural' as the other (Goodman and Goodman, 2013, and Section 4): neither has to take place in a formal educational context (Bissex, 1980; Taylor, 1983; Bloch, 1997). The processes are culturally shaped, with the carer/teacher expectations themselves being shaped by the adult's own prior experiences and understandings (Heath, 1983).

(c) **Successive approximations.** Both these socially based processes of learning involve successive approximations enabled by the specifics of sociocultural and educational contexts the child encounters. She learns phonological and phonetic principles: the relationships between sounds and meaning in the case of spoken language, and graphemic and alphabetic principles of written language when writing is based in letters drawn from an alphabet. In each case learning is a process of observation, experimentation and successive approximation to reach the correct form (Heyes and Frith, 2014), with errors corrected by feedback through repeated demonstrations of conventional speaking and writing.

(d) **Building on existing strengths.** When they learn to speak, read and write, children draw on all of their learning strengths to move from the known to the unknown (Bruce, 2015). This includes their understandings, knowledge and uses of oral language, its vocabulary, metaphors and grammar in one or more languages, as they begin to include written language in their communicative repertoire (Au, 1980). Thus a major predictor of success in learning to read is the presence of an already reasonably well-developed spoken language and vocabulary in the same language.

(e) **Motivated to engage.** High motivation to learn and practice is a central aspect of both oral and written language learning. Making meaning of the great complexities of written language needs high and consistent levels of motivation and engagement with texts, and affects comprehension (Wigfield et al., 2016). A child's self-confidence, beliefs, values and goals, as well as sense of autonomy and interest, all play a significant part; if intrinsic motivation continues to be encouraged beyond the early years, activities related to positive achievement are greater than with extrinsic rewards (Ryan and Deci, 2009).

7.2 Learning Oral Language

How does learning oral language take place? Shaywitz (2003) claims that oral language does not have to be taught because learning to speak is a natural process. This claim is widely accepted now by policy makers, academics and language specialists as being based in undisputable scientific evidence. But why is it natural, given the complexity of the task? We suggest that this is because it takes place through the predictive processing kind of interaction emphasized in this article, which is one of trial and error followed by feedback and correction. It involves an informal and superbly effective teaching process because babies have the kind of conditions they require to learn when family members speak constantly and consistently to and around them. Babies want to understand and be able to express themselves too; caregivers and others have high expectations that babies are capable of learning to listen and speak, and talk to them as if they already understand as they try to meet their needs and moods. Castles et al. (2018) state:

> **LEARN(o):** If a child is exposed to a rich oral-language environment, that child will almost certainly learn to understand and produce spoken language.

Such an environment involves enormous numbers of everyday verbal interactions, initially with carers, who guide the ongoing reciprocal interaction, experimentation, practice and play as babbling emerges. Over time, and with ever better approximations of the accepted speech of the particular community, it becomes the appropriate form of conventional spoken language. This has three dimensions. Firstly, the child must learn the motor control involved in speaking: shaping the tongue and lips, controlling breathing and so on. Secondly, she must learn to apply phonological principles which transform sounds into words and sentences. Thirdly, she must learn how and when to use the grammatical, lexical and cultural and linguistic conventions to convey the meanings of her speech community.

As we have intimated, from our viewpoint, the key issue overlooked by many is that **this IS a teaching environment.** It is an **informal** teaching environment (Lave and Wenger, 1991; Rogoff et al., 2016), involving the necessary conditions which support learning (Cambourne, 1995). In terms of the discussion in the next subsection, this is an apt example of 'natural learning' (Holdaway, 1979), corresponding to the need to create meaningful, holistically oriented teaching environments.

7.3 Learning Written L78anguage

How does learning to use written language happen? It can take place in both informal and formal teaching environments. It can be oriented to be either a skills-based process, emphasizing the parts first and then building them up to create wholes, as summarized in Figure A.1, or a meaning-based process, emphasizing engaging with and composing whole texts while also appreciating and attending to the contributing parts, as summarized in Figure A.2. Reductionist skills-based approaches insist on getting the details right first before moving on to use reading and writing for authentic reasons (hence the widely used phrase, 'learn to read, then read to learn'). Holistic, meaning-centred approaches support learning through successive approximations toward conventional reading and writing.

According to Castles et al. (2018), '[t]he fundamental insight that graphemes represent phonemes in alphabetic writing systems does not typically come naturally to children. It is something that most children must be taught explicitly, and doing so is important for making further progress in reading.'

The key issue here is the phrase 'come naturally to children'. What is understood as natural depends crucially on cultural context (Rogoff, 1990). If you live in a highly literate environment that uses and displays as normal writing in a language you are comfortable using, what comes naturally is quite different than if you do not. And what does 'taught explicitly' mean? If a mother teaches her child to spell her own name on a sheet of paper, is that explicit teaching? We would suggest yes. It is not part of an explicit teaching programme: but it is teaching nonetheless, just as is being taught to say her name in the case of spoken language.

It is just as natural in both cases, given appropriate conditions. In other words, to learn to read children have to read and be read to (Smith, 2012), while to learn to write, they have to write—and read too—as potential authors, guided by teachers and others who write themselves so that as they begin to write, they come to see themselves as writers (Smith, 1983). Infants and young children struggle to begin a 'natural' process of learning if they are in settings with few relevant role models using written language in ways which interest and draw them in as newcomers to a cultural practice (Rogoff et al., 2016). They are enabled to begin this process effectively by observing and joining in voluntarily to personally relevant activities involving writing and print in relevant languages, be these in homes, community settings or school contexts.

This kind of informal learning is illustrated by a Polish colleague who tells of his induction into reading as follows: he had a brother who was 4 years older than him, and at that time, school started when children were 7 years old. He was 3 when his brother started to learn to read, sitting in their common room at a small table in the middle of the room. The older brother would be reading the letters and words aloud, running his finger below the line of print. Our colleague would be kneeling on a chair at the other side of the table following his brother's finger. Within a year (by age 4) he had learnt to read fluently—upside down! Only later did he learn to read with the 'normal' orientation. No formal skills teaching occurred in this self-motivated, socially contextualized process. This is one of many cases that demonstrate the successful nature of informal teaching; it is not essential to have formal teaching in order to learn to read.

While we are not in any way claiming here that teaching reading is not necessary, it is well documented that children can learn the fundamentals themselves under appropriate conditions (Clark, 1976; Buckingham and Castles, 2019). Indeed up to 5% of children are 'precocious readers' who do this (Olson et al., 2006).

In parallel to **LEARN(o)** in Section 6.2, the following is plausible:

LEARN(w): If a child is exposed to a rich, contextually relevant written-language environment, which involves that child in regular, satisfying reading and writing interactions with significant others, including shared attention to the details of the process, and constant positive feedback, that child is highly likely to learn to understand and produce written language.

7.4 The Similarities between Learning Spoken and Written Language

The predictive processing viewpoint, and more generally the way perception functions as discussed in Section 2, can be claimed to support learning both processes in neural terms, based in the statistical pattern recognition properties of neural networks. discussed in Section 5. Consequently, our view is that learning spoken and written language is underpinned by very similar processes, as indicated in the figure below (Bloch, in Ellis, 2016: 448):

Language is listening, speaking as well as reading (including braille), writing and signing

Baby learns to speak	**Baby learns to read-write**
• Hears, sees/ experiences people who speak (role models)	• Hears, sees/ experiences people who read – write (role models)
• Expresses and communicates as she learns	• Expresses and communicates as he learns
• Learns why she listens and talks at same time as she learns how	• Learns why he reads- writes at same time as he learns how
• Has shared interactions	• Has shared interactions
• Is included, heard, encouraged, praised - connects emotionally	• Is included, heard, encouraged, praised - connects emotionally
• makes' mistakes' - speaks immaturely(babbles) and plays with sounds.	• makes 'mistakes'- reads/writes immaturely (pretends to read, does emergent writing).

7.5 Implications of an Integrative View for Progress in Meaningful Reading

Considering the integrative body of neuroscience discussed above, what could detract from and what could support children learning to read and write with meaning? Whilst in this paper we don't detail early literacy teaching methods, and acknowledge the huge body of existing expertise in this regard, we make the following general points.

A major issue for learning effectively exists in multilingual print-scarce settings, like South Africa, with de facto language policies that move to teach from African to ex-colonial languages after only three years' schooling (Mkhize and Balfour, 2017; Bua Lit, 2018). Here the potential for compromised understanding already exists to such an extent that it can feel normal. This makes it easy to accept that teaching and assessing reading doesn't involve comprehension until later. The drive to search for meaning can thus be minimized, deflected or hidden when the broad initial orientation is toward separate skills teaching, where phonics automaticity and fluency must be mastered as an initial imperative.

This is particularly so if access to compelling fiction and nonfiction material in preferred languages is absent or positioned as supplementary, and there are few or no reading and writing role models to interact with. Limited vocabulary books, which have been 'levelled' are used far more[25] than materials which stimulate curiosity, challenge imaginations and encourage inference and problem solving. Such materials don't necessarily hinder the progress of children who engage elsewhere with emotionally satisfying texts which build vocabulary and language knowledge as they conjure awe and excitement. But children who have to rely on school for such motivation and enrichment may wait for so long that they give up and never get what they need.

Though learning letter-sound combinations and relationships is integral to learning to read and to write in alphabetic languages, we contest the validity of teaching it in prescriptive ways, dissociated from the wider fields of meaning and personal relevance and agency. Phonics-based methods are acknowledged to possibly delay the relation to meaning until automaticity and fluency have been attained (Seidenberg et al., 2020). The interim learning is often low level and mind-numbing; it is highly questionable whether this can contribute to the much desired recipe for success. Telling children that this will change once they have learned to read does not necessarily help: the experience of meaninglessness is real.

Apart from not fully assessing elements which indicate reading progress, assessments using non-words (Castles et al., 2018: 19; Bua Lit, 2018) and meaningless phonemes, such as the widely used Early Grade Reading Assessments, reinforce the message that reading is not related to anything personally useful or interesting. Again this can be highly demotivating.

In contrast, an orientation which provides a relevant base for meaningful learning emphasizes the value of children's languages, emotional and personal knowledge and connections from home and community. From this place of respect and belonging, stories which can be fictional, factual or historical stimulate imaginative engagement. Teachers can learn to teach phonics and other skills as and when needed by children as they read and write (Figure A.2). Regular, interactive experiences with worthwhile[26] texts, involving plenty of teacher read

---

[25] A wonderful diatribe against such books is given in the section on education in *Let us Now Praise Famous Men* by James Agee (Agee, 1988).

[26] We use the term 'worthwhile' to reiterate the benefits of teachers and teacher educators engaging in an ongoing investigation of books, with discussion about what 'worthwhile' means in diverse cultural contexts. It points to the extraordinarily important role adults have in curating the texts children encounter, and also to their observing and consequently learning from and about the children who explore the books.

alouds and conversations with children to motivate and stimulate imaginative thinking and use of language, should begin early and continue to be supported and overtly valued.

Horowitz-Kraus and Hutton (2015) confirm this by stating[27]

> Children utilising imagery during stories listening will have greater success in reading later in life, which is consistent with findings suggesting that better utilisation of imagery during stories listening improves comprehension. Studies citing quotes of children's experience when listening to stories confirm that imagery supports this process, even more intensely for stories without pictures, perhaps via more intense activation of the visual association cortex.

Castles et al. (2018) states,

> The single most effective pathway to fluent word reading is print experience: Children need to see as many words as possible, as frequently as possible . . . statistics point to the huge value of fostering a love of reading in children and a motivation to read independently.

We agree and suggest that an assumption in this statement needs to be overt: a love of reading is made possible when teachers orient themselves to appreciate the importance, legitimacy and power of becoming well-informed, interactive role models who read aloud well and frequently to children, encouraging curiosity, imaginative and critical thinking and real conversations about what's being read. This ought to be normalized as the essential orientation for all early literacy teachers. Even in the highly print-saturated settings of the UK, reading for pleasure has declined (National Literacy Trust, 2019), and fresh evidence is emerging as to the rich literacy teaching benefits of ensuring that teachers themselves read for pleasure and indeed are readers in their own right (Cremin et al., 2008; Cremin, 2020). This deceptively simple notion helps teachers to awaken the desire to read in children by harnessing the pleasure, enriched language and other opportunities literature holds for learning (Krashen, 1989; Arizpe and Styles, 2016; McQuillan, 2019; Wissman, 2019; Bloch, 2015). It is also the springboard from which to support teachers to encourage children to apply strategies which include multiple cueing systems, a focus on authentic composing and writing and to consider related assessments which address multiple dimensions of literacy.

7.6 Conclusion

Far too many young children's literacy learning opportunities are being compromised daily by the increasingly wide acceptance of a restricted, reductionist body of neuroscience evidence as being the true and unquestionable basis for teaching reading. The following statement referring to teaching in South African schools summarizes how this view is interpreted for teachers:

> Unlike learning to speak, decoding does not come naturally; it is a method that must be taught systematically. It is important to emphasize that reading is produced by the product of vocabulary and decoding: If one has a perfect vocabulary but has not been taught the method of decoding one will not be able to read at all. Letter recognition and phonemic awareness are mastered through systematic teaching and consistent practice. This leads to the next stage of reading acquisition: word recognition. Through practice and appropriate progression from simpler sounds and words to more complex ones, word recognition becomes established leading to the next phase of reading acquisition: fluency. It is only once decoding and word

---

[27] In stark contrast to Dehaene (2010).

recognition have become fluent, even to the point where it becomes automatic and unconscious, that it is possible to reach the ultimate goal of reading comprehension. (Taylor et al., 2019: 20)

What allows children to achieve this perceived initial mastery? They continue (Taylor et al., 2019: 21):

> In order to learn the basics of decoding, a child requires a teacher who is present, capable and motivated to deliver systematic reading instruction. In order for decoding to become fluent a child requires suitable graded materials and the discipline (perhaps imposed) to practice a lot.

This rigid and foreboding vision of what it could mean for teachers in overcrowded and under-resourced classrooms to (perhaps impose) discipline on young children to practice their graded materials (if they even have these) is a depressingly common consequence of relying on this reductive model.

We have contested this vision with the body of integrative neuroscience which supports the view that all understanding is contextual. Learning starts at birth: young children's brains are capable of handling complexity and learning meaningfully from the outset, outside of exceptional cases. This is confirmed by the body of early literacy evidence detailing young children's emergent reading and writing prior to formal schooling (Whitmore et al., 2004; Nutbrown, 2018; Carroll et al., 2019; Teale et al., 2020). Observations of young children reveal much time and effort spent with voluntary skills practice when these skills interest children and form part of play or other authentic purposes—and this includes children from poor communities (Sibanda and Kajee, 2019; Bloch and Mbolekwa, 2021).

School literacy teaching should continue to develop such foundations and build on them, in ways which respond sensitively to children's ongoing meaning-making endeavours. Integrative neuroscience offers evidence to support this, implying the value of teacher education programmes which problematize narrow interpretations of the science (Hoffman et al., 2020), renewing attention to and research on teaching approaches and methods currently eschewed or straitjacketed to fit reductive neuroscience understandings. All teachers, especially those in underserved settings, need overt, systemic support to provide children in their first years of formal school with the kind of culturally responsive, rich learning opportunities that are currently afforded in reasonable quality only to children from affluent communities. Among many others, Cambourne (2000, 2017) and Whitmore and Meyer (2020) provide solid foundations for this endeavour.

Acknowledgements

We thank Eva Bonda, Thomas Parr and Tina Bruce for helpful comments, Mark Solms for providing Figure A.3, Mandy Darling for (re)drawing the figures, and Roland Eastman for extremely helpful discussions that shaped Figure A.1, as well as regarding details of the text. We thank two referees for comments that have materially improved the paper.

Disclosure Statement

No potential conflict of interest was reported by the author(s).

References

Abadzi, H. 2006. *Efficient Learning for the Poor: Insights from the Frontier of Cognitive Neuroscience.* Washington, DC: World Bank.

Abadzi, H. 2008. Efficient learning for the poor: new insights into literacy acquisition for children. *International Review of Education* 54: 581–604.

Abadzi, H. 2017. *Turning a Molehill into a Mountain? How Reading Curricula are Failing the Poor Worldwide.* Unesco, IBE Springer.

Abadzi, H. 2006. *Efficient Learning for the Poor: Insights from the Frontier of Cognitive Neuroscience.* Washington, DC: World Bank. © World Bank. https://openknowledge.worldbank.org/handle/ 10986/ 7023 License: CC BY 3.0 IGO

Adams, M. J. 2001. Alphabetic anxiety and explicit systemic phonics instruction: a cognitive science perspective. In Neuman, S. B. & Dickinson D. K. (eds), *Handbook of Early Literacy Research.* New York: Guilford, 66–80.

Agee, James. 1988. Let us Now Praise Famous Men. New York: Houghton Mifflin, 289–307.

Alexander, N. & Bloch, C. 2010. Creating literate communities–the challenge of early literacy. In Krüger-Potratz, M., Neumann, U. & Reich, H. (eds), *Bei Vielfalt Chancengleichheit.* New York, Münster. 197–212.

Alitto, H. J. & Usrey, W. M. 2003. Corticothalamic feedback and sensory processing. *Current Opinion in Neurobiology* 13: 440–445.

Altwerger, B., Jordan, N. & Shelton, N. R. 2007. *Rereading Fluency. Process, Practice and Policy.* Portsmouth, NH: Heinemann.

Amodei, D., et al. 2016. Deep speech 2: End-to-end speech recognition in English and Mandarin. International conference on machine learning. pp. 173–182.

Ardington, C., Wills, G., et al. 2020. *Technical Report: Benchmarking Early Grade Reading Skills in Nguni Languages.* Stellenbosch, ReSEP, Stellenbosch University.

Arizpe, E. & Styles, M. 2016. *Children Reading Picturebooks.* London and New York: Routledge.

Au, K. H. 1980. Participation structures in a reading lesson with Hawaiian children: analysis of a culturally appropriate instructional event. *Anthropology & Education Quarterly* 11: 91–115.

Bar, M., Kassam, K.S., et al. 2006. Top-down facilitation of visual recognition. *Proceedings of the national academy of sciences* 103(2): 449–454.

Barton, D., Hamilton, M. & Ivanic, R. (eds). 2000. *Situated Literacies: Reading and Writing in Context.* London and New York: Routledge.

Bell, N. 1991. Gestalt imagery: a critical factor in language comprehension. *Annals of Dyslexia* 41: 246–260.

Berent, I. 2020. Talk is cheap. *Inference* 5(3). https://inference-review.com/article/talk-is-cheap

Bergeron, B. & Bradbury-Wolff, M. 2010. 'If it's not fixed, the staples are out!' Documenting young children's perceptions of strategic reading processes. *Reading Horizons* 50: 1–22.

Bever, T. G. 2009. All language understanding is a psycholinguistic guessing game: explaining the still small voice. In Anders, P. (ed.), *Issues in the Present and Future of Reading.* London, Routledge. pp. 249–281.

Bever, T. G. 2013. The biolinguistics of language universals: The next years. In Sanz, M., et al. (eds), *Language Down the Garden Path: The Cognitive and Biological Basis for Linguistic Structures.* Oxford: Oxford University Press.

Bever, T. G. 2017. The unity of consciousness and the consciousness of unity. In de Almeida, R.G. & Gleitman, L. (eds), *Minds on Language and Thought.* Oxford: Oxford University Press. pp. 87–112.

Bever, T. G. & Poeppel, D. 2010. Analysis by synthesis: a (re-)emerging program of research for language and vision. *Biolinguistics* 4: 174–200.

Bialystok, E., Luk, G. & Kwan, E. 2005. Bilingualism, biliteracy, and learning to read: interactions among languages and writing systems. *Scientific Studies of Reading* 9: 43–61. doi:10.1207/s1532799xssr0901_4

Biber, D., Conrad, S. & Reppen, R. 1998. *Corpus Linguistics: Investigating Language Structure and Use.* Cambridge, UK: Cambridge University Press.

Bishop, C. M. 1995. *Neural Networks for Pattern Recognition.* Oxford: Oxford University Press.

Bissex, G. L. 1980. *Gnys at Wrk: A Child Learns to Write and Read*. Cambridge, MA: Harvard University Press.

Bloch, C. 1997. *Chloe's Story. First Steps to Literacy*. Cape Town: Juta and Co. https://www.praesa. org.za/training/.

Bloch, C. 1999. Literacy in the early years: teaching and learning in multilingual early childhood classrooms. *International Journal of Early Years Education* 7: 39–59.

Bloch, C. 2000. Don't expect a story: young children's literacy learning in South Africa. *Early Years: An International Journal of Research and Development. TACTYC* 20: 57–67.

Bloch, C. 2002. A case study of Xhosa and English biliteracy in the foundation phase versus English as a medium of destruction. In Heugh, K. (ed.), *Perspectives in Education. Special Issue: Many Languages in Education* 20: 65–78.

Bloch, C. 2015. Nal'ibali and libraries: activating reading together. library.ifla.org/1282/1/076-bloch-en.pdf.https://www.praesa.org.za/articles/.

Bloch, C. 2018. Story by story: nurturing multilingual reading and writing in South Africa. In Daly, N, Limbrick, L & Dix, P. (eds), *Children's Literature in a Multiliterate World*. London: UCL Institute of Education Press. pp. 161–181

Bloch, C. & Alexander, N. 2003. Aluta continua: the relevance of the continua of biliteracy to South African multilingual schools. In Hornberger, N. (ed.), *Continua of Biliteracy: An Ecological Framework for Educational Policy, Research, and Practice in Multilingual Settings*. Clevedon: Multilingual Matters Ltd. pp. 91–121.

Bloch, C. & Mbolekwa, S. 2021. Apprenticeships in meaning: transforming opportunities for oral and written language learning in the early years. In Erling et al. (eds), *Multilingual Learning in Schools in Sub-Saharan Africa: Critical Insights and Practical Applications*. London: Routledge.

Boyd, B. 2018. The evolution of stories: from mimesis to language, from fact to fiction. *WIREs Cogn Sci* 9: e1444. doi:10.1002/wcs.1444

Briggs, F. & Usrey, W. M. 2008. Emerging views of corticothalamic function. *Curr Opin Neurobiol* 18: 403–407.

Bruce, T. 1991. *Time to Play in Early Childhood Education*. London: Hodder & Stoughton.

Bruce, T. 2015. *Early Childhood Education*, 5th ed. London: Hodder Education.

Bryant, P. E., Maclean, M., Bradley, L. L. & Crossland, J. 1990. Rhyme and alliteration, phoneme detection, and learning to read. *Developmental Psychology* 26: 429–438.

Bua-lit Language and Literacy Collective. 2018. How are we failing our children? Reconceptualising language and literacy education, viewed 28 April 2020. https://bua-lit.org.za/wp-content/uploads/2018/11/bua-lit-language-literacy-education.pdf.

Buckingham, J. & Castles, A. 2019. Why do some children learn to read without explicit teaching? *Nomanis* 7: 18–20.

Buckner, C. & Garson, J. 2019. Connectionism. In Zalta, E.N. (ed.), *The Stanford Encyclopedia of Philosophy* (Fall). URL= https://plato.stanford.edu/archives/fall2019/entries/connectionism/.

Bybee, J. & McClelland, J. L. 2005. Alternatives to the combinatorial paradigm of linguistic theory based on domain general principles of human cognition. *The Linguistic Review* 22: 381–410. doi:10.1515/tlir.2005.22.2-4.381

Cambourne, B. 1995. Toward an educationally relevant theory of literacy learning: twenty years of inquiry. *The Reading Teacher* 49 (3): 182–190.

Cambourne, B. 2000. Conditions for literacy learning. *Reading Teacher* 54: 414–417.

Cambourne, B. 2017. Reclaiming or reframing? Getting the right conceptual metaphor for thinking about early literacy learning. In *Reclaiming early childhood literacies: Narratives of hope, power, and vision*, pp. 17–29.

Cambourne, B. & Kilarr, G. 2020. Helping teachers reframe reading as meaning making. In Whitmore, K. F. & Meyer, R. J. (eds), *Reclaiming Literacies as Meaning Making: Manifestations of Values, Identities, Relationships, and Knowledge*. New York: Routledge. pp. 114–125.

Campbell, S. 2020. Teaching phonics without teaching phonics: early childhood teachers' reported beliefs and practices. *Journal of Early Childhood Literacy* 20(4): 783–814.

Carey, E., Hill, F., Devine, A. & Szücs, D. 2017. The modified abbreviated math anxiety scale: a valid and reliable instrument for use with children. *Frontiers in Psychology* 8: 11.

Carpenter, G. A. & Grossberg, S. (eds). 1991. *Pattern Recognition by Self-Organizing Neural Networks.* Cambridge, MA: MIT Press.

Carroll, J. M., Holliman, A. J, Weir, F. & Baroody, A. E. 2019. Literacy interest, home literacy environment and emergent literacy skills in preschoolers. *Journal of Research in Reading* 42: 150–161.

Castles, A., Rastle, K. & Nation, K. 2018. Ending the reading wars: reading acquisition from novice to expert. *Psychological Science in the Public Interest* 19: 5–51.

Cattinelli, I., Borghese, N. A., Gallucci, M. & Paulesu, E. 2013. Reading the reading brain: a new meta-analysis of functional imaging data on reading. *Journal of Neurolinguistics* 26: 214–238.

Cervetti, G. N., Pearson, D., et al. 2020. How the reading for understanding initiative's research complicates the simple view of reading invoked in the science of reading. *Reading Research Quarterly* 55: S161–S172.

Chall, J. S. 1967. *Learning to Read: The Great Debate.* New York: McGraw-Hill.

Chomsky, N. 1965. *Aspects of the Theory of Syntax.* Cambridge, MA: MIT Press.

Chomsky, N. 1975. *Reflections on Language.* New York: Pantheon Books.

Chukovsky, K. 1968. *From Two to Five* (M. Morton, trans.). Berkeley, CA: University of California Press.

Churchland, P. S. & Sejnowski, T. J. 1999. *The Computational Brain.* Cambridge, MA: MIT Press.

Clark, A. 2013. Whatever next? Predictive brains, situated agents, and the future of cognitive science. *Behavioral and Brain Sciences* 36: 181–204.

Clark, A. 2016. *Surfing Uncertainty: Prediction, Action, and the Embodied Mind.* New York: Oxford University Press.

Clark, M. 1976. *Young Fluent Readers.* What Can They Teach Us? London: Heinemann.

Clark, M. 2020. Independent research into the impact of the systematic synthetic phonics government policy on literacy courses at institutions in England delivering initial teacher *education. Education Journal* 411: 18–23.

Clay, M. M. 1991. *Becoming Literate: The Construction of Inner Control.* Portsmouth, NH: Heinemann.

Coltheart, M. 2000. Dual routes from print to speech and dual routes from print to meaning: some theoretical issues. In Kennedy et al. (Eds.) *Reading as a Perceptual Process,* Oxford: Elsevier. pp. 475–490.

Compton-Lilly, C. F., Mitra, A., Guay, M. & Spence, L. K. 2020. A confluence of complexity: intersections among reading theory, neuroscience, and observations of young readers. *Reading Research Quarterly* 55: S185–S195.

Constant, A., Clark, A., Kirchhoff, M. & Friston, K. J. 2019. Extended active inference: constructing predictive cognition beyond skulls. In *Mind and Language.* New York: Wiley. p. 4.

Conte, E., Ornaghi, V., et al. 2019. Emotion knowledge, theory of mind, and language in young children: testing a comprehensive conceptual model. *Frontiers in Psychology* 10: 2144.

Cooper, P. 2009. *The Classrooms All Young Children Need.* Chicago: University of Chicago Press.

Cremin, T. 2020. Teachers as readers and writers. In *Debates in Primary Education.* London: Routledge. pp. 243–253.

Cremin, T., Bearne, E., Mottram, M. & Goodwin, P. 2008. Primary teachers as readers. *English in Education* 42: 8–23.

Damasio, A. 1999. *The Feeling of What Happens: Body, Emotion and the Making of Consciousness.* London: Vintage.

Damasio, A. 2000. *Descarte's Error.* New York: Harper Collins.

Danelli, L., Marelli, M., Berlingeri, M., Tettamanti, M., Sberna, M., Paulesu, E. & Luzzatti, C. 2015. Framing effects reveal discrete lexical-semantic and sublexical procedures in reading: an fMRI study. *Frontiers in Psychology* 6: 1328.

Davis, K. L. & Montag, C. 2019. Selected principles of Pankseppian affective neuroscience. *Frontiers in Neuroscience* 12: 1025.

Deacon, T. W. 1998. *The Symbolic Species: The Co-evolution of Language and the Brain.* London: Penguin Books.

Deacon, T. W. 2003. Universal grammar and semiotic constraints. In Christiansen, M. H. & Kirby, S. (eds), *Language Evolution.* Oxford: Oxford University Press. pp. 111–139.

Dehaene, S. 2010. *Reading in the Brain: The New Science of How We Read*. London: Penguin.

Department of Education (DOE). 2008. *National Reading Strategy*. South Africa, Department of Education. Online. https://www.education.gov.za/Portals/0/DoE%20Branches/GET/GET%20Schools/National_Reading.pdf?ver=2009-09-09-110716-507

Department of Education (DOE). 2011. *Action Plan to 2014: Towards the Realisation of Schooling 2025*. Pretoria, Department of Basic Education.

Department of Education (DOE). 2015. *The South African National Curriculum Framework for children from Birth to Four*. Pretoria: Department of Basic Education.

Ding, N., Melloni, L., Zhang, H., Tian, X. & Poeppel, D. 2016. Cortical tracking of hierarchical linguistic structures in connected speech. *Nature Neuroscience* 19: 158–164.

Donald, M. 2001. *A Mind So Rare: The Evolution of Human Consciousness*. New York: Norton.

Dooley, C. M. & Matthews, M. W. 2009. Emergent comprehension: understanding comprehension development among young literacy learners. *Journal of Early Childhood Literacy* 9(3): 269–294.

Dowd, A. J. & Bartlett, L. 2019. The need for speed: interrogating the dominance of oral reading fluency in international reading efforts. *Comparative Education Review* 63: 189–212.

Dunbar, R. 1998. Theory of mind and the evolution of language. In Hurford, J. R., Hurford, J. R., Studdert-Kennedy, M., & Knight, C. (eds), *Approaches to the evolution of language: Social and Cognitive Bases*. Cambridge, UK: Cambridge University Press. 92–110.

Dunbar, R. I. 1993. Coevolution of neocortical size, group size and language in humans. *Behavioral and brain sciences* 16: 681–694.

Edelman, G. M. 1987. *Neural Darwinism: The Theory of Neuronal Group Selection*. New York: Basic Books.

Egan, K. 1989. *Teaching as Storytelling*. Chicago: University of Chicago Press.

Elkind, D. 2007. *The Power of Play: Learning What Comes Naturally*. Boston: Da Capo Lifelong Books.

Ellis, G. 2016. *How Can Physics Underlie the Mind? Top-Down Causation in the Human Context*. Heidelberg: Springer.

Ellis, G. 2018. Top-down effects in the brain. *Physics of life reviews* 31: 1–30.

Ellis, G. & Solms M. 2017. Beyond Evolutionary Psychology: How and *Why Neuropsychological Modules Arise*. Cambridge, UK: Cambridge University Press.

Ellis, G. & Toronchuk, J. A. 2005. Neural development affective and immune system influences. In Ellis, R. D. & Newton, N. (eds), *Consciousness and Emotion: Agency, Conscious Choice, and Selective Perception*. Amsterdam/Philadelphia, John Benjamins. pp. 81–119.

Ellis, G. & Toronchuk, J. A. 2013. Affective neuronal selection: the nature of the primordial emotion systems. *Frontiers in Psychology* 3: 589.

Esteban-Guitart, M. & Moll, L. C. 2013. Funds of identity: a new concept based on the funds of knowledge approach. *Culture & Psychology* 20: 31–48.

Evans, V. 2014. *The Language Myth: Why Language is Not an Instinct*. Cambridge, UK: Cambridge University Press.

Evans, V. 2015. *The Crucible of Language: How Language and Mind Create Meaning*. Cambridge, UK: Cambridge University Press.

Evans, V. 2020. *The Evidence There is no Language Instinct*. Aeon. https://aeon.co/essays/the-evidence-is-in-there-is-no-languageinstinct.

Fabry, R. E. 2017. Transcending the evidentiary boundary: prediction error minimization, embodied interaction, and explanatory pluralism. *Philosophical Psychology*. https://doi.org/10.1080/09515089.2016.1272674

Farmer, T. A., Brown, M. & Tanenhaus, M. K. 2013. Prediction, .explanation, and the role of generative models in language processing. *Behavioral and Brain Sciences* 36: 211–212.

Fedorenko, E., et al. 2016. Neural correlate of the construction of sentence meaning. Proceedings of the National Academy of Sciences 113: E6256–E6262.

Feldman, J. A. 2008. *From Molecule to Metaphor: A Neural Theory of* Language. Cambridge, MA: MIT Press.

Ferreiro, E. & Teberosky, A. 1982. *Literacy before Schooling*. Portsmouth, NH: Heinemann Educational Books.

Findlay, J. M. & Gilchrist, I. D. 2003. *Active Vision: The Psychology of Looking and Seeing*. Oxford: Oxford University Press.

Fitch, W.T. 2005. The evolution of language: a comparative review. *Biology and Philosophy* 20: 193–203.

Fleisch, B. & Dixon, K. 2019. Identifying mechanisms of change in the early grade reading study in South Africa. *South African Journal of Education* 39 (3): 1–12.

Flurkey, A. D. & Xu, J. 2003. *On the Revolution of Reading: The Selected Writings of Kenneth S. Goodman*. Portsmouth, NH: Heinemann.

Flurkey, A. D., Paulson, E. J. & Goodman, K. S. 2008. *Scientific Realism in Studies of Reading*. New York: Lawrence Erlbaum.

Francis, D. J., Kulesz, P. A. & Benoit, J. S. 2018. Extending the simple view of reading to account for variation within readers and across texts: The complete view of reading. (CVRi). *Remedial and Special Education* 39: 274–288.

Frankel, K. K., Becker, B. L., Rowe, M. W. & Pearson, P. D. 2016. From 'what is reading?' to 'what is literacy'? *Journal of Education* 196: 7–17.

Frankl, V. E. 1985. *Man's Search for Meaning*. New York: Simon and Schuster.

Freeman, W.J. 2004. How and why brains create meaning from sensory information. *International Journal of Bifurcation and Chaos* 14: 515–530.

Friederici, A. D. 2017. *Language in Our Brain: The Origins of a Uniquely Human Capacity*. Cambridge, MA: MIT Press.

Friston, K. 2003. Learning and inference in the brain. *Neural Networks* 16: 1325–1352.

Friston, K. 2010. The free-energy principle: a unified brain theory? *Nature Reviews Neuroscience* 11: 127–138.

Friston, K. 2012. The history of the future of the Bayesian brain. *NeuroImage* 62: 1230–1233.

Friston, K., Fitzgerald, T., Rigoli, F., Schwartenbeck, P. & Pezzulo, G. 2017. Active inference: a process theory. *Neural Computation* 29: 1–49.

Friston, K. & Frith C. 2015. A duet for one. *Consciousness and cognition* 36: 390–405.

Friston, K. J., Parr, T., et al. 2020. Generative models, linguistic com munication and active inference. *Neuroscience & Biobehavioral Reviews* 118: 42–64.

Friston, K. J., Rosch, R., Parr, T., Price, C. & Bowman, H. 2017a. Deep temporal models and active inference. *Neuroscience & Biobehavioral Reviews* 77: 388–402.

Frith, C. 2007. *Making up the Mind: How the Brain Creates Our Mental World*. Malden: Blackwell.

García, O. & Wei, L. 2014. Language, bilingualism and education. In *Translanguaging: Language, Bilingualism and Education*. London: Palgrave Pivot. https://doi.org/10.1057/9781137385765_4

Gombrich, E. H. 1961. *Art and Illusion*. New York: Pantheon Books.

Goodman, K. S. 1967. Reading: a psycholinguistic guessing game. *Journal of the Reading Specialist* 6: 126–135.

Goodman, K. S. 1982. *Language and Literacy*. Boston: Routledge and Kegan Paul.

Goodman, K. S. 2006. *The Truth about Dibels: What it is, What it Does*. Portsmouth, NH, Heinemann.

Goodman, K. S. & Burke, C. 1973. *Theoretically Based Studies of Patterns of Miscues in Oral Reading Performance*. Washington, DC: U.S. Department of Health, Education, and Welfare.

Goodman, K. S., Fries, P., Strauss, S. & Paulson, E. 2016. *Reading: The Grand Illusion. How and Why Readers Make Sense of Print*. New York: Routledge.

Goodman, K. S. & Goodman, Y. M. 2013. Learning to read is natural. *Theory and Practice of Early Reading* 1: 137.

Goodman, Y. M. 1992. The roots of literacy. *Journal for the Study of Education and Development* 15: 29–42, doi:10.1080/02103702.1992.10822331

Goodwin, A. P. & Jiménez, R. T. 2020. Special issue executive summary: the science of reading. *Reading Research Quarterly* 55: 1–7.

Goswami, U. 2008. Learning to read across languages: the role of phonics and synthetic phonics. In Goouch, K. & Lambirth, A. (eds), *Understanding Phonics and the Teaching of Reading: Critical Perspectives*. Maidenhead, Berkshire: Open University Press. pp. 124–143.

Gottschall, J. 2012. *The Story-Telling Animal: How Stories Make us Human*. Boston, New York: Mariner books.

Gough, P. B. & Hillinger, M. L. 1980. Learning to read: an unnatural act. *Bulletin of the Orton Society* 30: 179–196.

Gough, P. B. & Tunmer, W. E. 1986. Decoding, reading and reading disability. *Remedial and Special Education* 7: 6–10. doi:10.1177/074193258600700104

Gray, P. 2011. *Psychology*. New York: Worth Publishers.

Gray, P. 2013. The reading wars: why natural learning fails in class rooms. https://www.psych ologytoday.com/za/blog/freedom-learn/201311/the-reading-wars-why-natural-learning-fails-in -classrooms

Gray, P. 2017. What exactly is play, and why is it such a powerful vehicle for learning? *Topics in Language Disorders* 37: 217–228.

Greenspan, S. & Shanker, S. 2004. *The First Idea: How Symbols, Language, and Intelligence Evolved from Our Primate Ancestors to Modern Humans*. Cambridge, MA: Da Capo Press.

Gregory, R. & Cavanagh, P. 2011. The blind spot. *Scholarpedia* (accessed 13 October 2020).

Gregory, R. L. 1978. *Eye and Brain: The Psychology of Seeing (World University Library)*. New York: McGraw Hill.

Gunn, B. K., Simmons, D. C. & Kameenui, E. J. 1995. *Emergent Literacy: Synthesis of the Research*. Eugene, OR: National Center to Improve the Tools of Educators, College of Education, University of Oregon.

Gussin Paley, V. 1990. *The Boy Who Would be a Helicopter. The Uses of Storytelling in the Classroom*. Cambridge, MA: Harvard University Press.

Halliday, M. A. K. 1977. Text as semantic choice in social contexts. In van Dijk T. A. & Petöfi, J. (eds), *Grammars and Descriptions*. Berlin: de Gruyter. pp. 176–225.

Halliday, M. A. K. 1993. Towards a language-based theory of learning. *Linguistics and Education* 5: 93–116.

Halliday, M. A. K. 2003. On the 'architecture' of human language. In Webster, J. J. (ed.), *On Language and Linguistics*, vol. 3. New York: Continuum. 1–29.

Harari, Y. 2011. *Sapiens: A Brief History of Humankind*. London: Vintage.

Harrison, G. D. 2020. A snapshot of early childhood care and education in South Africa: institutional offerings, challenges and recommendations. *South African Journal of Childhood Education* 10: a797.

Harste, J. C., Woodward, V. A. & Burke, C. L. 1984. *Language Stories and Literacy Lessons*. Portsmouth, NH: Heinemann.

Hawkins, J. 2005. *On Intelligence*. New York: Times Books, Henry Holt and Co.

Heath, S. B. 1983. *Ways with Words: Language, Life and Work in Communities and Classrooms*. New York: Cambridge University Press.

Heyes, C. M. & Frith, C. D. 2014. The cultural evolution of mind reading. *Science* 344: 1243091.

Hickman, R. 2018. Creating a nation of readers. https://www.shineliteracy.org.za/wp-content/ uploads/2018/11/Creating-a-nationof-readersa-survey-of-evidence-ePDF.pdf.

Hoey, M. 2005. *Lexical Priming: A New Theory of Words and Language*. Abingdon, UK: Routledge.

Hoff, E. (2006). How social contexts support and shape language development. *Developmental Review* 26, 55–88.

Hoffman, J. V., Hikida, M. & Sailors, M. 2020. Contesting science that silences: amplifying equity, agency, and design research in literacy teacher preparation. *Reading Research Quarterly* 55: S255–S266.

Hoffmann, J. 2012. Why EGRA – a clone of DIBELS – will fail to improve literacy in Africa. *Research in the Teaching of English* 46: 340–357.

Hohwy, J. 2013. *The Predictive Mind*. Oxford, Oxford University Press.

Holdaway, D. 1979. *The Foundations of Literacy*. Sydney, Ashton Scholastic.

Horowitz-Kraus, T. & Hutton, J. S. 2015. From emergent literacy to reading: how learning to read changes a child's brain. *Acta Paediatrica* 104: 648–656.

Horowitz-Kraus, T., Schmitz, R., Hutton, J. S. & Schumacher, J. 2017. How to create a successful reader? Milestones in reading development from birth to adolescence. *Acta Paediatrica: Nurturing the Child, January* 106: 534–544.

Howie, S., Combrinck, C., Roux, K., Tshele, M., Mokoena, G. & Palane, N. M. 2017. PIRLS Literacy 2016: South African Children's Reading Literacy Achievement. *Pretoria, Centre for Evaluation and Assessment.* (CEA), Faculty of Education, University of Pretoria.

Hruby, G. G. & Goswami, U. 2019. Educational neuroscience for reading educators. In Alvermann, D. E., Unrau, N. J. & Sailors, M. (eds), *Theoretical Models and Process of Reading.* New York: Routledge. pp. 558–588.

Huey, E. B. 1908. *The Psychology and Pedagogy of Reading.* New York: Macmillan.

Huron, D. 2008. *Sweet Anticipation: Music and the Psychology of Expectation.* Cambridge, MA: MIT Press.

Hutton, J. S., et al., C-MIND Authorship Consortium. 2015. Home reading environment and brain activation in preschool children listening to stories. *Pediatrics* 136: 466–478.

Hutton, J. S., et al. 2017. Story time turbocharger? Child engagement during shared reading and cerebellar activation and connectivity in preschool-age children listening to stories. *PLOS One,* May, 12(5): e0177398.

Hutton, J. S., Dudley, J., et al. 2020. Associations between home literacy environment, brain white matter integrity and cognitive abilities in preschool-age children. *Acta Paediatrica* 109: 1376–1386.

Huybregts, M. A. C., Berwick, R. C. & Bolhuis, J. J. 2016. The language within. *Science* 352: 1286.

Immordino-Yang, M. H., Darling-hammond, L. & Krone, C. R. 2019. Nurturing nature: how brain development is inherently social and emotional, and what this means for education. *Educational Psychologist* 54: 185–204.

Isaac, A.M. & Ward, D. 2019. Introduction: gestalt phenomenology and embodied cognitive science. *Synthese.* https://doi.org/10.1007/s11229-019-02391-7.

Jukes, M., Pretorius, E., Schaefer, M., Tjasink, K., ROPER, M., Bisgard, J. & Mabhena, N. 2020. *Setting Reading Benchmarks in South Africa.* Khulisa Management Services Ltd: USAID and Department of Basic Education. https://www.khulisa.com/wp-content/uploads/2020/12/PA00X1NZ.pdf

Juliebö, M., Malicky, G. V. & Norman, C. 1998. Metacognition of young readers in an early intervention programme. *Journal of Research in Reading* 21: 24–35.

Kandel, E. R. 2012. *The Age of Insight: The Quest to Understand the Unconscious in Art, Mind, and Brain from Vienna 1900 to the Present.* New York: Penguin Random House.

Kandel, E. R. 2016. *Reductionism in Art and Brain Science.* New York: Columbia University Press.

Kearns, D. M., Hancock, R., Hoeft, F., Pugh, K. R. & Frost, S. J. 2019. The neurobiology of dyslexia. *Teaching Exceptional Children* 51: 175–188.

Keitel, A., Gross, J. & Kayser, C. 2018. Perceptually relevant speech tracking in auditory and motor cortex reflects distinct linguistic features. *PLoS biology* 16: e2004473.

Kim, J. S. 2008. Research and the reading wars. *Phi Delta Kappan* 89: 372–375.

Klaas, A. & Trudell, B. 2011. Effective literacy programmes and inde pendent reading in African contexts. *Language Matters* 42: 22–38, doi:10.1080/10228195.2011.569739

Krashen, S. 1989. We acquire vocabulary and spelling by reading: additional evidence for the input hypothesis. *Modern Language Journal* 73: 440–464.

Krashen, S. 2017. The case for comprehensible input. Language Magazine. https://www.sdkrashen. com/content/articles/case_for_comprehensible_input.pdf.

Krashen, S. & Mason, B. 2020. The optimal input hypothesis: not all comprehensible input is of equal value. CATESOL Newsletter, May, 1–2. http://beniko-mason.net/content/articles/2020-the-optimalinput-hypothesis.pdf.

Kuhl, P. 2010. Brain mechanisms in early language acquisition. *Neuron* 67: 713–727.

Kveraga, K., Ghuman, A. & Bar, M. 2007. Top-down predictions in the cognitive brain. *Brain Cogn* 65: 145–168.

Lakoff, G. & Johnson, M. 1980. The metaphorical structure of the human conceptual system. *Cognitive Science* 4: 195–208.

Laszlo, S. & Plaut, D. C. 2012. A neurally plausible parallel distributed processing model of event-related potential word reading data. *Brain and Language* 120: 271–281.

Lave, J. & Wenger, E. 1991. *Situated Learning: Legitimate Peripheral Participation.* Cambridge, UK: Cambridge University Press.

Lewis, D. & Elman, J. L. 2001. Learnability and the statistical structure of language: Poverty of stimulus arguments revisited. *Proc 26th Annual Boston University conference on Language Development.* Vol. 1. Boston: Cascadilla Press.

Longres, J. F. 1990. *Human Behaviours in the Social Environment.* Itasca, IL: F E Peacock.

Makalela, L. 2014. Teaching indigenous African languages to speakers of other African languages: The effects of translanguaging for multilingual development. In Hibbert, L. & Van der Walt, C. W. (eds), *Multilingual Universities in South Africa: Reflecting Society in Higher* Education. Buffalo, NY: Multilingual Matters. pp. 88–106.

Marcus, G. F. 2019. *The Algebraic Mind: Integrating Connectionism and Cognitive Science.* Cambridge, MA: MIT Press.

Martin, L., Durisko, C., et al. 2019. The VWFA is the home of orthographic learning when houses are used as letters. *Eneuro* 6(1): e042517.2019 1–13.

McQuillan, J. 2019. We don't need no stinkin'exercises: the impact of extended instruction and storybook reading on vocabulary acquisition. *Language and Language Teaching* 8: 25–37.

Meek, M. 1988. *How Texts Teach What Readers Learn.* Stroud: Thimble Press.

Meyer, D. K. & Turner, J. C. 2006. Re-conceptualizing emotion and motivation to learn in classroom contexts. *Educational Psychology Review* 18: 377–390.

Miłkowski, M. et al. 2018. From wide cognition to mechanisms: A silent revolution. *Frontiers in Psychology* 9: 2393.

Miller, M. & Clark, A. 2018. Happily entangled: prediction, emotion, and the embodied mind. *Synthese* 195: 2559–2575.

Miller, S. B. 2020. Showdown on the kansas plains: the reading wars continue. *Kansas English* 101(1): 19–23.

Mkhize, D. & Balfour, R. 2017. Language rights in education in *South Africa. South African Journal of Higher Education* 31 (6): 133–150. https://doi.org/10.28535/31-6-1633.

Moll, L. C., Amanti, C., Neff, D. & Gonzalez, N. 1992. Funds of knowledge for teaching: using a qualitative approach to connect homes and classrooms. *Theory Into Practice* 31: 132–141.

Monsalve, I. F., Pérez, A. & Molinaro, N. 2014. Item parameters dissociate between expectation formats: a regression analysis of time-frequency decomposed EEG data. *Frontiers in Psychology* 5:847.

Moore, M. V., Durisko, C., et al. 2014. Learning to read an alphabet of human faces produces left-lateralized training effects in the fusiform gyrus. *Journal of Cognitive Neuroscience* 26: 896–913.

Mullins, D. A., Whitehouse, H. & Atkinson, Q. D. 2013. The role of writing and recordkeeping in the cultural evolution of human cooperation. *Journal of Economic Behavior & Organization* 90: S141–S151.

National Literacy Trust. 2019. *Read on. Get on.* London, National Literacy Trust.

National Reading Panel. 2000. *Teaching Children to Read: An Evidence-based Assessment of the Scientific Research Literature on Reading and Its Implications for Reading Instruction: Reports of the Subgroups.*

NEEDU. 2013. NEEDU National Report 2012: The state of literacy teaching and learning in the foundation phase. Pretoria, National education and evaluation development unit. Department of Basic Education.

Nicholls, J. G., Martin, A. R., Wallace, B. G. & Fuchs, P. A. 2001. *From Neuron to Brain.* Sunderland, MA: Sinauer Associates.

Nicolopoulou, A., Cortina K. S., Ilgaz, H., Brockmeyer Cates, C., De Sáa, B. 2015. Using a narrative- and play-based activity to promote low-income preschoolers' oral language, emergent literacy, and social competence. *Early Childhood Research Quarterly* 31: 147–162,

Nutbrown, C. 2018. *Early Childhood Educational Research.* London: SAGE Publications Ltd https://www.doi.org/10.4135/9781526451811

Olson, L. A., Evans, J. R. & Keckler, W. T. 2006. Precocious readers: past, present, and future. *Journal for the Education of the Gifted* 30: 205–235.

Orbán, G., Fiser, J., Aslin, R. N. & Lengyel, M. 2008. Bayesian learning of visual chunks by human observers. *Proceedings of the National Academy of Sciences of the USA* 105: 2745–2750.

Otten, M., Seth, A. K. & Pinto, Y. 2017. A social Bayesian brain: how social knowledge can shape visual perception. *Brain and Cognition* 112: 69–77.

Panichello, M. F., Cheung, O. S. & Bar, M. 2013. Predictive feedback and conscious visual experience. *Frontiers in Psychology* 3: 620.

Panksepp, J. 1998. *Affective Neuroscience: The Foundations of Human and Animal Emotions.* London: Oxford University Press.

Panksepp, J. & Biven, L. 2012. *The Archaeology of Mind: Neuroevolutionary Origins of Human Emotion.* New York: W. W. Norton and Company.

Parr, T., Corcoran, A. W., Friston K. J. & Hohwy, J. 2019. Perceptual awareness and active *inference. Neuroscience of Consciousness* 2019(1): niz012.

Parr, T., Rees, G. & Friston, K. J. 2018. 'Computational neuropsychology and Bayesian inference' *Frontiers in Human Neuroscience* 12: 61.

Patael, S. Z., Farris E. A., et al. 2018. Brain basis of cognitive resilience: prefrontal cortex predicts better reading comprehension in relation to decoding. *PloS one* 13: e0198791.

Pearson, D. 2004. The reading wars: The politics of reading research and policy. *Educational Policy* 18: 216.

Perfetti, C. A. 1992. The representation problem in reading acquisition. In Gough, P. B., Ehri, L. C. & Treiman, R. (eds), *Reading Acquisition.* Hillsdale, NJ: Lawrence Erlbaum. pp. 145–174.

Pinker, S. 2003. *The Language Instinct: How the Mind Creates Language.* London: Penguin.

Piper, B., Schroeder, L. & Trudell, B. 2016. Oral reading fluency and comprehension in Kenya: reading acquisition in a multilingual environment. *UKLA Journal of Research in Reading* 39: 133–152.

Plaut, D. C. 2005. Connectionist approaches to reading. In Snowling, M. J. & Hulme, C. (eds), *The Science of Reading: A Handbook.* London: Blackwell. pp. 24–38.

Pugh, K. R., Mencl, W. E., et al. 2000. Functional neuroimaging studies of reading and reading disability (developmental dyslexia). *Mental Retardation and Developmental Disabilities Research Reviews* 6: 207–213.

Pullum, G. K. & Scholz, B. C. 2002. Empirical assessment of stimulus poverty arguments. *The linguistic Review* 19: 9–50.

Purves, D. 2010. *Brains: How They Seem to Work.* Upper Saddle River, NJ: Ft Press.

Purves, D., Cabeza R., et al. 2008. *Cognitive Neuroscience.* Sunderland: Sinauer Associates.

Railton, P. 2017. At the core of our capacity to act for a reason: the affective system and evaluative model-based learning and control. *Emotion Review* 9: 335–342.

Rastle, K., Perry, C., Langdon, R. & Ziegler, J. 2001. Drc: a dual route cascaded model of visual word recognition and reading aloud. *Psychological Review* 108: 204.

Rauss, K. & Pourtois, G. 2013. What is bottom-up and what is top-down in predictive coding? *Frontiers in Psychology* 4: 276.

Rawlinson, G. E. 1976. *The Significance of Letter Position in Word Recognition.* Unpublished PhD thesis, Psychology Department, University of Nottingham, Nottingham UK. Summarised here: https://www.mrccbu.cam.ac.uk/personal/matt.davis/Cmabrigde/rawlinson.html.

Rayner, K., White, S. J. & Liversedge, S. P. 2006. Raeding wrods with jubmled lettres: there is a cost. *APS Psychological Science* 17: 192–193. https://doi.org/10.1111%2Fj.1467-9280.2006.01684.

Redhead, G. & Dunbar, R. I. 2013. The functions of language: An exper imental study. *Evolutionary Psychology* 11(4): 147470491301100409.

Reeves, C. 2017. The early grade reading study (EGRS). In *Depth Case Studies of Home Language Literacy Practices in Four Grade 2 Classrooms in Treatment 1 and 2 Schools.* Johannesburg, South Africa: EGRS.

Rogoff, B. 1990. *Apprenticeship in Thinking.* New York, Oxford: Oxford University Press.

Rogoff, B. 2003. *The Cultural Nature of Human Development.* New York, Oxford: Oxford University Press.

Rogoff, B., Gutiérrez, K. & Erickson, F. 2016. The organization of informal learning. *Review of Research in Education* 40: 356–401.

Rolls, E. T. 2016. *Cerebral Cortex: Principles of Operation.* Oxford: Oxford University Press.

Rosenblatt, L. M. 1982. The literary transaction: evocation and response. *Theory into Practice, Children's Literature* 21(4): 268–277.

Roskos, K. A. & Christie, J.F. 2011. Mindbrain and play–literacy connections. *Journal of Early Childhood Literacy* 11: 73–94.

Roskos, K. A., Christie, J. F. & Richgels, D. 2003. The essential of early literacy instruction. *Young Children* 58: 62–60.

Rumelhart, D. E. 1977. Toward an interactive model of reading. In Dornic, S. (ed.), *Attention and Performance VI*. Hillsdale, NJ: Lawrence Erlbaum. pp. 573–606.

Ryan, R. M. & Deci, E. L. 2009. Promoting self-determined school engagement: motivation, learning, and well-being. In Wentzel, K. R. & Wigfield, A. (eds), *Handbook of Motivation in School*. New York: Taylor Francis. pp. 171–196.

Schnelle, H. 2010. *Language in the Brain*. Cambridge, UK: Cambridge University Press.

Seidenberg, M. 2017. *Language at the Speed of Sight: How We READ, Why so Many Cannot, and What Can be Done about it*. New York: Basic Books.

Seidenberg, M. S. 2005. Connectionist models of word reading. *Current Directions in Psychological Science* 14: 238–242. doi:10.1111/j.0963-7214.2005.00372.x

Seidenberg, M. S. 2013. The science of reading and its educational implications. *Language learning and development* 9: 331–360.

Seidenberg, M. S., Cooper Borkenhagen, M. & Kearns, D. M. 2020. Lost in translation? Challenges in connecting reading science and educational practice. *Reading Research Quarterly* 55: S119–S130.

Seidenberg, M. S. & McClelland, L. 1989. A distributed, developmental model of word recognition and naming. *Psychological Review* 96(4): 523–568. doi:10.1037/0033-295x.96.4.523

Seth, A. K. 2013. Interoceptive inference, emotion, and the embodied self. *Trends in cognitive sciences* 17: 565–573.

Seth, A. K. 2014. *The Cybernetic Bayesian Brain*. *Open MIND*. Frankfurt am Main: MIND Group.

Shanahan, T. 2020. What constitutes a science of reading instruction? *Reading Research Quarterly* 55: S235–S247.

Shaywitz, B. A., Shaywitz, S. E., et al. 2002. Disruption of posterior brain systems for reading in children with developmental dyslexia. *Biological Psychiatry* 52: 101–110.

Shaywitz, S. 2003. *Overcoming Dyslexia*. New York: Alfred Knopf.

Shaywitz, S. & Shaywitz, B. A. 2004. Reading disability and the brain: what research says about reading. *Educational Leadership* 61(6): 6–11.

Sherman, S. M. & Guillery, R. W. 2006. Exploring the thalamus and its role. In *Cortical Function*. Cambridge, MA: MIT Press.

Sibanda, R. & Kajee, L. 2019. Home as a primary space: exploring outof-school literacy practices in early childhood education in a township in South Africa. *South African Journal of Childhood Education* 9(1): a686. https://doi.org/10.4102/sajce.v9i1.686

Smith, F. 1983. Reading like a writer. *Language Arts* 60: 558–567.

Smith, F. 2012. *Understanding Reading: A Psycholinguistic Analysis of Reading and Learning to Read*. New York: Routledge.

Snow, C. E. 2018. Simple and not-so-simple views of reading. *Remedial and Special Education* 39(5): 313–316. https://doi.org/10.1177/0741932518770288

Sohoglu, E., Peelle, J. E., Carlyon, R. P. & Davis, M. H. 2012. Predictive top-down integration of prior knowledge during speech perception. *Journal of Neuroscience* 32: 8443–8453.

Spaull, N. & Pretorius, E. 2016. Exploring relationships between oral reading fluency and reading comprehension amongst English second language readers in South Africa. *Reading and Writing*. doi:10.1007/s11145-016-9645-9

Spaull, N. & Pretorius, E. 2019. Still falling at the first hurdle: examining early grade reading in South Africa. In Spaull, N. & Jansen, J. D. (eds), *South African Schooling: The Enigma of Inequality*. Cham:, Springer. pp. 147–168.

Spaull, N., Pretorius, E. & Mohohlwane, N. 2020. Investigating the comprehension iceberg: developing empirical benchmarks for early-grade reading in agglutinating African languages. *South African Journal of Childhood Education* 10(1): a773. https://doi.org/10.4102/sajce.v10i1.773

Spear-Swerling, L. 2019. Structured literacy and typical literacy practices: understanding differences to create instructional opportunities. *Teaching Exceptional Children* 51: 201–211. https://doi.org/10.1177/0040059917-750160

Stanley, S. 2012. *Why Think? Philosophical Play from 3-11*. London: Continuum.

Stevens, A. & Price, J. 2015. *Evolutionary Psychiatry: A New Beginning*. New York: Routledge.

Stone, S. J. & Burriss, K. G. 2016. A case for symbolic play: An important foundation for literacy development. *International Journal of Holistic Early Learning and Development* 3: 59–72.

Strauss, S. L. 2004. *The Linguistics, Neurology, and Politics of Phonics: Silent 'E' Speaks Out*. New York: Routledge.

Strauss, S. L. & Altwerger, B. 2007. The logographic nature of English alphabetics and the fallacy of direct intensive phonics instruction. *Journal of Early Childhood Literacy* 7: 299–319.

Strauss, S. L., Goodman, K. S. & Paulson, E. J. 2009. Brain research and reading: How emerging concepts in neuroscience support a meaning construction view of the reading process. *Educational Research and Review* 4: 021–033.

Street, B. 1984. *Literacy in Theory and Practice*. Cambridge, UK: Cambridge University Press.

Street, B. 2006. Autonomous and ideological models of literacy: approaches from new literacy studies. *Media Anthropology Network* 17: 1–15.

Sugiyama, M. S. 2017. Oral storytelling as evidence of pedagogy in forager societies. *Frontiers of Psychology* 8: 471.

Taylor, D. 1983. *Family Literacy: Young Children Learning to Read and Write*. Portsmouth, NH: Heinemann Educational Books.

Taylor, J. S. H., Rastle, K. & Davis, M. H. 2013. Can cognitive models explain brain activation during word and pseudoword reading? A meta-analysis of 36 neuroimaging studies. *Psychological Bulletin* 139: 766.

Taylor, N. 1989. Falling at the First Hurdle: Initial Encounters with the *Formal System of African Education in South Africa*. Johannesburg, University of the Witwatersrand, Education Policy Unit.

Taylor, S., Cilliers, J., Prinsloo, C., Fleisch, B. & Reddy, V. 2019. *Improving Early Grade Reading in South Africa, 3ie Grantee Final Report*. New Delhi: International Initiative for Impact Evaluation. (3ie).

Teale, W. H., Whittingham, C. E. & Hoffman, E. B. 2020. Early literacy research, 2006–2015: A decade of measured progress. *Journal of Early Childhood Literacy* 20(2): 169–222.

Tickle, C., Arias, A. M., Placzek, M. & Wolpert, L. 2002. *Wolpert's Principles of Development*. Oxford:, Oxford University Press.

Tomasello, M. 2000. *The Cultural Origin of Human Cognition*. Cambridge, MA: Harvard University Press.

Tomasello, M. 2003. *Constructing a Language: A Usage-Based Theory of Language Acquisition*. Cambridge, MA: Harvard University Press.

Tovey, D. R. 1976. The psycholinguistic guessing game. *Language Arts* 53: 319–322.

Trettenbrein, P. C., Papitto, G., Friederici, A. D. & Zaccarella, E. 2021. Functional neuroanatomy of language without speech: An ALE meta-analysis of sign language. *Human Brain Mapping* 42: 699–712.

Van Der Berg, S., Spaull, N., Wills, G., Gustafsson, M. & Kotzé, J. 2016. *Identifying the Binding Constraints in Education*. Report commissioned by the South African Presidency Pro-Poor Policy Development. (PSPPD). Initiative.

Van Rooy, B. & Pretorius, E.J. 2013. Is reading in an agglutinating language different from an analytic language? An analysis of isiZulu and English reading based on eye movements. *SA Linguistics & Applied Language Studies* 31: 281.

Vidal, Y., Viviani, E., Zoccolan, D. & Crepaldi, D. 2021. A general-purpose mechanism of visual feature association in visual word identification and beyond. *Current Biology* 31(6): 1261–1267.

Vogel, A. C., Petersen, S. E. & Schlaggar, B. L. 2012. The left occipitotemporal cortex does not show preferential activity for words. *Cereb. Cortex* 22: 2715–2732. doi:10.1093/cercor/bhr295

Vogel, A. C., Petersen, S. E. & Schlaggar, B. L. 2014. The VWFA: it's not just for words anymore. *Frontiers in Human Neuroscience* 8: 88. doi:10.3389/fnhum.2014.00088

Vygotsky, L. S. 1978. *Mind in Society: The Development of Higher Psychological Processes*. Cambridge, MA: Harvard University Press.

Wagemans, J., Elder, J. H., et al. 2012. A century of gestalt psychology in visual perception I. Perceptual grouping and figure-ground organization. *Psychological Bulletin* 138: 1172–1217.

Wagemans, J., Elder, J. H., et al. 2012a. A century of gestalt psychology in visual perception II. Conceptual and theoretical foundations. *Psychological Bulletin* 138: 1218–1252.

Wandell, B. A., Rauschecker, A. M. & Yeatman, J. D. 2012. Learning to see words. *Annual Review of Psychology* 63: 31–53.

Whitmore, K. F., Martens, P., Goodman, Y. M. & Owocki, G. (2004). Critical lessons from the transactional perspective on early literacy research. *Journal of Early Childhood Literacy* 6(2): 291–325.

Whitmore, K. F. & Meyer, R. J. (Eds.). 2020. *Reclaiming Literacies as Meaning Making: Manifestations of Values, Identities, Relationships and Knowledge*. New York: Routledge.

Wigfield, A., Gladstone, J. & Turci, L. 2016. Beyond cognition: reading motivation and reading comprehension child development perspectives. *Society for Research in Child Development.* doi:10.1111/cdep. 12184

Willingham, D. T. 2017. *The Reading Mind: A Cognitive Approach to Understanding How the Mind Reads*. San Francisco: John Wiley.

Willis, J. 2006. *Research-Based Strategies to Ignite Student Learning*. Alexandria, VA: ASCD.

Willis, J. 2008. *Teaching the Brain to Read: Strategies for Improving Fluency, Vocabulary and Comprehension*. Alexandria, Virginia: ASCD.

Willson, A. M. & Falcon, L. A. 2018. Seeking equilibrium: in what ways are teachers implementing a balanced literacy approach amidst the push for accountability? *Texas Journal of Literacy Education* 6(2): 73–93.

Wissman, K. K. 2019. Reading radiantly: embracing the power of picturebooks to cultivate the social imagination. *Bookbird* 57: 14–25. IBBY.org.

Wolf, M. 2008. *Proust and the Squid: The Story and Science of the Reading Brain*. Thriplow, Cambridge, UK: Icon Books.

Wolf, M. 2018. The science and poetry in learning (and teaching) to read. *Phi Delta Kappan* 100: 13–17.

Yon, D. 2019.Now you see it. Aeon 4. July 2019.

Zarcone, A., Van Schijndel, M., Vogels, J. & Demberg, V. 2016. Salience and attention in surprisal-based accounts of language processing. *Frontiers in Psychology* 7: 844.

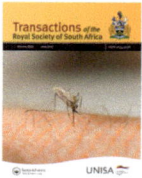

Transactions of the Royal Society of South Africa

ISSN: (Print) (Online) Journal homepage: www.tandfonline.com/journals/ttrs20

Neuroscience and literacy: an integrative view

George Ellis & Carole Bloch

To cite this article: George Ellis & Carole Bloch (2021) Neuroscience and literacy: an integrative view, Transactions of the Royal Society of South Africa, 76:2, 157-188, DOI: 10.1080/0035919X.2021.1912848

To link to this article: https://doi.org/10.1080/0035919X.2021.1912848

Published online: 14 May 2021.

Submit your article to this journal ☑

Article views: 688

View related articles ☑

View Crossmark data ☑

Citing articles: 3 View citing articles ☑

South African Yearly Meeting of the Quakers

Support for a Universal Basic Income (UBI)

Whither South Africa?

We are in a crisis in South Africa—an emergency that needs an urgent response. We cannot carry on as normal or there will be dire consequences.

It is common cause that South Africa experiences high unemployment, inequality and extensive poverty. The threat this poses to the stability of South African society and the economy is seen every day in high levels of crime, gender-based violence, protest action, the wholesale theft of infrastructure, increases in mental health problems, reduced levels of trust and large numbers of skilled people of all races wishing to raise their children elsewhere. The poverty-stricken masses have nothing going for them, see no hope for their future and have nothing to lose. South Africa is a tinderbox that will catch fire if a match is thrown into it, as happened in the Durban riots in July 2021.

South Africa may be considered a real-world example of where the world is going. Globally, quality employment or indeed, any employment, is becoming a scarcer and scarcer commodity, and inequality is hugely deepening. Economic growth, job creation and wage work are becoming a less and less likely prospect of providing a solution to unemployment, inequality and poverty. In South Africa we have demonstrated an inability to reverse this trend—so the crisis deepens.

What should we do?

As Quakers in Southern Africa, we embrace the principles of Ubuntu. We seek solutions that recognize the interdependence of all our citizens, that each person is worthy of dignity, being recognized, being heard, and that we have a collective responsibility to ensure that all members of our communities have the means of subsistence; and we must do so as soon as possible.

In the spirit of Ubuntu, we believe that to make a rapid difference to our current crisis levels of large-scale poverty, we need to implement a universal basic income (UBI) for all people legally in the country, regardless of age, gender or income, as soon as is practicable. This should be done in the context of a strong public sector delivering quality basic services, infrastructure and institutions, providing programmes targeted at supporting micro, small and medium enterprises, and introducing active labour market policies including training and development, life-long learning and access to labour markets, so as not to obviate the state's responsibility to provide an enabling environment for inclusive economic activity.

The Possible Benefits

A UBI would be a huge signal for the poor and dispossessed that their plight has been heard, and action has been taken. At a local level, both urban and rural, it will immediately give hope and dignity to each person. It will support better nutrition, improve access to education, allow people to set up small businesses and participate more fully in social and economic life.

In pilot trials elsewhere, the UBI has been shown to increase happiness levels and improve both physical and mental health. It would also contribute to decreasing levels of crime and violence. There is evidence that it strengthens demand in local and national economies, which would have a further positive impact if the increased demand was met by locally produced and more labour-intensive commodities. Increased economic activity and greater social stability would engender confidence in the country and stimulate both local and foreign investment. Evidence internationally has shown it enables families to undertake economic activity that was otherwise not possible—it is an enabler of entrepreneurial activity that transforms people's lives.

The Basic Income Grant (BIG) Pilot Campaign in Namibia introduced in January 2008 transformed the community of Otjivero, enabling small businesses (brickmaking, dressmaking, shoe repairing, bread baking, small spaza shops) to thrive. It led to the establishment of a local market by increasing households' buying power. The BIG resulted in a huge reduction of child malnutrition; school attendance and pass rates improved significantly; school drop-out rates fell from almost 40% to 5% in seven months. The general health of the community improved as residents could pay the small amount for clinic visits. Crime rates fell by 42%, stock theft by 43% and other theft by 20%. Women were no longer so dependent on men, and were relieved of the burden of engaging in transactional sex. Most tellingly, the social fabric of the community was mobilized by an elected committee who advised residents on how to spend the BIG money wisely. It was overall of great benefit to the community.

Why a Universal Basic Income and Not Social Grants?

Firstly, a UBI is less intrusive than a social grant, as it allows each person to maintain their dignity, and avoids the humiliation involved in defining and proving poverty or any other criteria, and having to prove it to officials via access to your bank account. Secondly, it eliminates the need for the vast, inefficient bureaucracy necessary to implement such criteria on potential recipients. This hugely reduces administration costs, and also removes the possibility of corruption related to the officials controlling these payments.

A simple pay-out system could be via a smart ID card given to everyone who is a citizen or legal immigrant. If a person pays income tax, the money paid out in UBI would revert to the state via SARS.

There is a difficulty: many see the UBI as basically worthwhile but may rather propose an unemployment grant, which is conditional. **But in our view as Quakers, unconditional universality is pivotal.** Firstly, this emphasizes the equality and worth of all. Secondly, it avoids introducing a massive bureaucracy and associated procedures, often carried out by uncaring officials and involving lengthy waits in queues in the hot sun or freezing cold. Furthermore, as it includes all in the family, including babies (registered at the hospital or clinic as soon as they are born), even a relatively small grant makes a real difference to a family as a whole, as it goes to all in the family. And it makes a profound difference to the communities where they live right across the country, because it goes to all families in the community. There is no stigma or envy associated with being included or left out.

Therefore the ideal is a universal payment to each citizen and legal immigrant irrespective of age, gender or income. We do not want a means test which means your grant may be taken away. For a microeconomy to function well, it is really important to have a reliable income.

Affordability

A UBI is desirable, but is it affordable? The Quaker answer is that we cannot afford not to do it. However, we accept that the UBI might require incremental steps in implementation.

The question then becomes, how much? We propose that the UBI would be at least equivalent to the upper-bound poverty line. But the upper-bound poverty line is acceptable only if the South African public sector provides well-functioning health care, education with school feeding, water, electricity and refuse removal at an affordable rate, as well as infrastructure, safe transportation, security and public communication systems. Otherwise a higher level is necessary.

The Institute for Economic Justice has produced detailed models and calculations, and it proposes a number of possible financing sources; namely, taxing or raising taxes on resource rents, luxury goods, carbon emissions, wealth, dividends, estate duty, currency transactions, financial transactions; abolishing medical tax credits and retirement fund contributions for high-income earners (over R1m per annum); cancelling employment tax incentives; reducing irregular and wasteful expenditure in the public sector; and reducing profit shifting by multi-nationals. Some countries and economists also propose a social wealth fund which is amassed through corporate dividends paid in recognition that wealth is socially produced and reliant on the expropriation of the commons.

A key aspect of affordability is that the money raised and paid does not vanish from the economy. It is distributed across the country to every town, township and village and thereby stimulates the economy everywhere. Every small village will experience an economic revival due to the total monies received, as the Namibian experiment evidenced; and that leads to many social and health benefits.

Support for UBI

Support for UBI is widespread across many organizations. Among these is the trade union movement, the Department of Social Development, the Institute for Economic Justice, the Black Sash, the Studies in Poverty and Inequality Institute (SPII), the Climate Justice Charter Movement and the South African Food Sovereignty Campaign. Crucially, the Minister of Social Development Lindiwe Zulu strongly supports it. While all these groups support some form of a UBI, they may differ in approach. As Quakers in Southern Africa, we offer our support to all these voices, while holding strongly to our position of a UBI, of at least the upper-bound poverty line, unconditionally to be given to every citizen.

We recognize that a UBI is no panacea for South Africa's woes. Much work needs to be done to create the Ubuntu vision of a peaceful, harmonious and prosperous society. The UBI must be supported by changes across our regulatory framework that encourage entrepreneurship, and training and support in such activities should be made available.

Overall, as the Southern African Quaker Community, we urge the South African Government, the private sector and the people of South Africa to be moved by the spirit of compassion and to act now to introduce a UBI and thus create an inclusive society.

For more information on the UBI, and links to where it has been tried elsewhere, see the following:

- The Rise and Fall of the Basic Income Campaign: Lessons from Namibia https://www.researchgate.net/publication/283039237_The_Rise_and_Fall_of_the_Basic_Income_Grant_Campaign_Lessons_from_Namibia

- NANGOF, *Making the Difference! The BIG in Namibia*: https://archive.ids.ac.uk/eldis/document/A50498.html
- SPII has many links: https://iej.org.za/universal-basic-income-guarantee/
- A link showing how it can boost the economy: https://www.businesslive.co.za/bd/opinion/2023-04-03-basic-income-grant-modelling-shows-how-to-do-it-as-a-boost-to-the-economy/
- Institute for Economic Justice: https://iej.org.za/resources/
- Hein Marais, *In the Balance* (Johannesburg: Wits University Press, 2022) is a very complete study
- Professor Alex van den Heever (Wits) has written two reports showing UBI's financial viability. See https://www.dsd.gov.za/index.php/documents/category/58-basic-income-support

Gregory Mthembu Salter and Benonia Nyakuwanikwa: SAYM Co-Clerks

About the Author

George F. R. Ellis, FRS, is professor emeritus of applied mathematics at the University of Cape Town. He was a lecturer in the Department of Applied Mathematics and Theoretical Physics (DAMTP) at Cambridge University, where he wrote *The Large Scale Structure of Spacetime* with Stephen Hawking. He has been a visiting professor at the University of Texas, University of Chicago, Hamburg University, Boston University, University of Alberta, Queen Mary (London University) and Oxford University, and at the International School of Advanced Studies in Trieste. In 1999 he was awarded the Star of South Africa Medal, presented by President Nelson Mandela, and in 2006 the Order of Mapungubwe, presented by President Thabo Mbeki.

About the Illustrator

Mauro Carfora is a professor emeritus of mathematical physics at the University of Pavia. He obtained his PhD in Physics at the University of Texas at Dallas (W. Rindler, 1981). He was a lecturer in the Department of Mathematics at the University of Rome La Sapienza and in the Department of Physics at the University of Pavia, and later an associate professor of mathematical physics at the International School of Advanced Studies in Trieste, Italy. The 2016 recipient of the Tullio Levi-Civita Prize for the mathematical and mechanical sciences (with Tudor Ratiu), his books include *The Geometry of Dynamical Triangulations*, written with A. Marzuoli and J. Ambjorn; *Quantum Triangulations, Non-linear Sigma Models and Ricci Flow*, and the monograph *Einstein Constraints and Ricci Flow: A Geometrical Averaging of Cosmological Initial Data Sets*, both with A.Marzuoli. He often doubles as an illustrator and is the author of the Italian stamp commemorating the World Year of Physics 2005. He collaborated with George F. R. Ellis and Ruth Williams to illustrate their book, *Flat and Curved Spacetimes*.

Index